Growth and Characterisation of Semiconductors

Growth and Characterisation of Semiconductors

Papers contributing to a short course organised by the Departments of Materials and Physics with The Continuing Education Centre at Imperial College, in association with the London University Interdisciplinary Research Centre for Semiconductor Materials

Edited by

R A Stradling

and

P C Klipstein

Imperial College of Science, Technology and Medicine
University of London

Adam Hilger, Bristol and New York

British Library Cataloguing in Publication Data

Growth and characterisation of semiconductors.
1. Semiconductor devices. Manufacture
I. Stradling, R. A. II. Klipstein, P. C.
621.38152

ISBN 0-85274-131-6

Library of Congress Cataloging-in-Publication Data are available

Published under the Adam Hilger imprint by IOP Publishing Ltd
Techno House, Redcliffe Way, Bristol BS1 6NX, England
335 East 45th Street, New York, NY 10017-3483, USA

Printed in Great Britain by J W Arrowsmith Ltd, Bristol

Contents

Introduction

The radical changes brought about by the applications of microelectronics in Information Technology have been based on the control of the electronic properties of semiconductor materials. The last decade has seen the emergence of optoelectronics as a companion technology to microelectronics, coupled with dramatic advances in semiconductor growth techniques brought about by developments in the fields of high vacuum technology and organo-metallic chemistry. The new growth techniques are now capable of growing semiconductor structures with constituent layers only a few atoms thick. These 'artificial' materials exhibit novel properties of interest not only for the fundamental science such as produced the Quantum Hall Effect, but also in device applications that now include High Mobility Transistors and Quantum Well Lasers.

The new growth techniques such as Molecular Beam Epitaxy and Metallo-Organic Chemical Vapour Deposition have given rise to the need for new techniques to characterise the structures grown and to optimise the electronic and optical properties that arise from quantum confinement in such small scale structures. As these techniques have developed, a demand has been created for new skills in graduates and need for retraining and updating in the associated sciences and technologies.

The current Course on the Growth and Characterisation of Semiconductors collects the papers presented at a Short Course of the same title given in the week starting 24th April 1989. This short course is one of several self-contained courses which also act as specialist options for the Masters Course on Semiconductor Science & Technology offered annually by Imperial College of Science, Technology and Medicine, which is a constituent college of London University.

Imperial College is a leading European Centre for advanced research and training in the field of semiconductor materials. The Departments of Physics, Materials and Electrical Engineering cooperate both in relevant research and in teaching this interdisciplinary MSc Course which prepares postgraduates engaged in or seeking careers in industries dependent on semiconductors and their applications.

In 1988 the Science and Engineering Research Council decided that the UK's National Interdisciplinary Research Centre for Semiconductor Growth and Characterisation should be established within the University of London with its base on the Imperial College campus. This Centre is acting as a forum for the organisation of a wide-ranging programme of research involving MBE, MOCVD and MOMBE, and for the convening of high-level meetings, symposia and courses.

Staff from the Interdisciplinary Research Centre and other national experts have combined with academics from the Imperial College Teaching departments to produce this volume. In 1989 the Short Course attracted 25 external participants, many from abroad, to join a similar number of MSc and PhD students who attended the course. It is expected that the Course will continue in a similar form for a considerable number of years. These Proceedings will provide a base for development of the Course.

A book of this size cannot cover all parts of the subject in great detail but we aimed to select what we feel to be some of the key areas in both the Growth and Characterisation fields. However, we are aware of two omissions: X-ray and Raman techniques for characterisation. The text for the first topic could not be prepared in time for the publication deadline for the book. It was not felt necessary to include the latter because of the existence of several excellent reviews. For a review of X-ray techniques related to multilayer structures, see the article by D B McWhan in *Synthetic Modulated Structures* (ed. L L Chang and B C Giessen, Academic Press, pp 43–74 (1985)). For texts on Raman techniques see *Light Scattering in Solids* vols 1–5 (ed. Cardona and Guntherodt, Berlin: Springer-Verlag) and also articles by M V Klein and by G Abstreiter, R Merlin and A Pinzuk, *IEEE Journal of Quantum Electronics* **QE–22** 1760–1908 (1986).

We should like to express our appreciation to the contributors to this volume and to others who assisted in making the Course a success, particularly Mr P Combey, Director of Continuing Education at Imperial College, and his staff.

R A Stradling
P C Klipstein

October 1989

List of Contributors

Professor D W Pashley
Department of Materials, Imperial College of Science, Technology and Medicine, London SW7 2AZ

Professor J O Williams
Department of Chemistry, University of Manchester Institute of Science and Technology, Manchester M60 1QD

Professor B A Joyce
Interdisciplinary Research Centre for Growth and Characterisation of Semiconductor Materials, Department of Physics, Imperial College of Science, Technology and Medicine, London SW7 2BZ

Dr C T Foxon
Philips Research Laboratories, Cross Oak Lane, Redhill, Surrey RH1 5HA

Dr D B Holt
Department of Materials, Imperial College of Science, Technology and Medicine, London SW7 2AZ

Dr J B Clegg
Philips Research Laboratories, Cross Oak Lane, Redhill, Surrey RH1 5HA

Professor R C Newman
Interdisciplinary Research Centre for Growth and Characterisation of Semiconductor Materials, Department of Physics, Imperial College of Science, Technology and Medicine, London SW7 2BZ

Professor E C Lightowlers
Department of Physics, Kings College, The Strand, London WC2

Professor R A Stradling
Department of Physics, Imperial College of Science, Technology and Medicine, London SW7 2BZ

Dr D W Palmer
Department of Physics, University of Sussex, Falmer, Brighton, Sussex BN1 9QH

Dr J L Hutchison
Department of Metallurgy and Science of Materials, University of Oxford, Parks Road, Oxford OX1 3PH

The Basics of Epitaxy

D W PASHLEY

1. INTRODUCTION

Epitaxy is now widely used for preparing thin single crystal layers of semiconductors, both as part of the manufacturing process for semiconductor devices and for providing samples for basic measurements of electronic properties. But the science of epitaxy began over 150 years ago. Epitaxy refers to the ordered growth of one crystal upon another crystal, such that the orientations of the two crystals bear some well defined relationship to each other. There are numerous examples of such orientation relationships in minerals, so that all of the earlier studies were carried out by mineralogists. It was difficult for any clear understanding of epitaxy to be obtained until the discovery of X–ray diffraction started to provide detailed knowledge of the arrangement of atoms within crystals. However, some remarkable progress was made by Barker (1906,1907,1908), who found, by growing alkali halides on each other, that epitaxy is more likely if the molecular volumes of the two alkali halides are nearly equal.

The first major advance in the subject was made by Royer, who carried out extensive studies of the growth of one crystal on another, mainly from aqueous solution. He put forward several rules for the occurrence of epitaxy, Royer (1928), the most important of which can be expressed as:

(1) There must be a matching of symmetry between the contacting crystal planes of the substrate and overgrowth

(2) The misfit between parallel lattice rows at the interface must be less than a limiting value of about 15 per cent.

Royer was responsible for introducing the term epitaxy, which he constructed from the Greek, to mean "arrangement on".

The scope for studies of epitaxy increased considerably when electron diffraction was discovered in 1927, thus providing a means of making structural studies on very thin surface layers, by the technique now known as RHEED. As a result, it was found that epitaxy can occur with very much larger misfits than 15 percent, and although there are now many examples of this, the concept that a small misfit is required for epitaxy to occur has continued to be believed to this day. Comprehensive tabulations of the known cases of epitaxy have been given by Seifert (1953), Pashley (1956) and Grunbaum (1975).

All of the early work on epitaxy was concerned with growing one substance on a substrate of a second substance. This is now commonly called heteroepitaxy, to distinguish it from growth of a material on itself (e.g. silicon on silicon), which is called homoepitaxy. The importance of homoepitaxy is that it provides a means of growing successive layers of the same substance containing different traces of impurity for doping purposes, so that homoepitaxy is used widely, for example to produce p–n junctions.

The purpose of this article is to present an account of the basic facts about epitaxy, together with a summary of the various approaches which provide understanding of the occurrence of epitaxy. This includes consideration of the modes of growth of thin surface films, and how a particular mode of growth influences epitaxial growth. The occurrence of imperfections in epitaxial layers, especially dislocations, is considered in some detail because of the importance of eliminating dislocations from semiconductor epitaxial layers prepared for use in electronic devices.

2. THE GEOMETRY OF EPITAXY

For the vast majority of the known cases of epitaxy, the crystallographic orientation relationship between the substrate and the deposit is determined by a crystal plane of the deposit being parallel to the crystal plane of the substrate forming its surface. For the common situation of the substrate surface being flat and parallel to a set of crystal planes, the relationship is simple. For more complex substrate surfaces, such as one containing various inclined microfacets, the relationship can be more complex. The complete epitaxial relationship is determined by the way in which the contacting crystal planes of the substrate and overgrowth are aligned with each other, usually by particular crystal directions or zone axes in the two planes being parallel. Thus alkali halides grow on each other in simple parallel orientation, so that growth of KBr on the (001) surface of NaCl is described by

(001) KBr parallel to (001) NaCl
with [100] KBr parallel to [100] NaCl

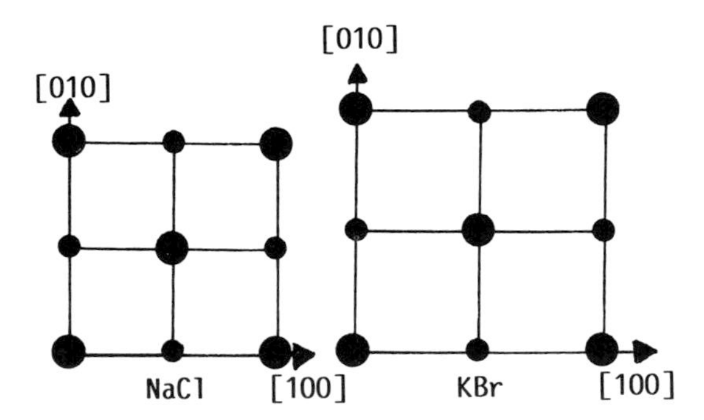

Figure 1. The parallel orientation of KBr on NaCl (001). Large circles represent positive ion sites and small circles represent negative ion sites. The misfit is +17 percent.

This is illustrated in figure 1. It is not necessary for the substrate and deposit to have the same crystal structure, and orientations other than parallel orientations do occur.

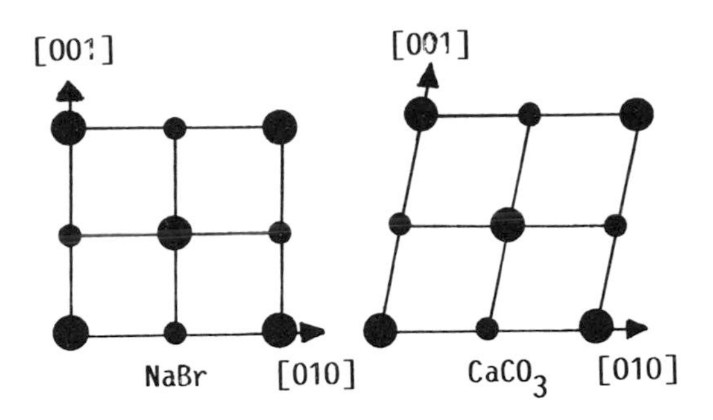

Figure 2. The orientation of NaBr on the cleavage surface of calcite ($CaCO_3$). Large circles represent positive ion sites and small circles represent negative ion sites. The misfit along [010] is −7 percent.

Figure 2 illustrates the growth of NaBr on $CaCO_3$ (calcite). Here a square network is aligned with a rhombic network. There is often a good symmetry match at the interface, which is well illustrated by the growth of various cubic materials on the (001) cleavage face of monoclinic mica. The cleavage face has hexagonal symmetry and cubic substances (e.g. ammonium iodide, see figure 3) grow with their (111) planes parallel to the cleavage plane.

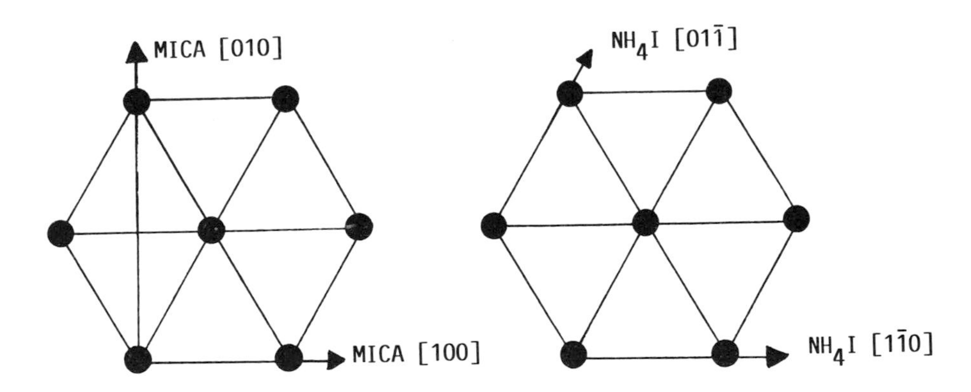

Figure 3. The orientation of NH_4I on the cleavage surface of mica. The circles represent ammonium ion sites in NH_4I and potassium ion sites in mica. The misfit is −1 percent.

It follows automatically from the way in which the interfacial alignment occurs that there will, in some cases, be two or more equivalent ways in which the same matching takes place. The growth of NaBr on $CaCO_3$ (figure 2) is a good example. The diagram shows the NaBr [010] axis parallel to the $CaCO_3$ [010] axis. A rotation of the (100) NaBr plane to cause the NaBr [001] axis to become parallel to the $CaCO_3$ [001] axis would give a fit which is completely equivalent to the one shown. Consequently both of these two orientations occur simultaneously when NaBr is grown on $CaCO_3$, leading to what is termed double positioning. This is a special case of the more general multiple positioning which can occur. The growth of ammonium iodide on mica provides a less obvious example of double positioning. In this case a rotation of the ammonium iodide (111) plane through 180 degrees (or 60 degrees) results in identical matching at the interface but distinct orientations, as can be seen by considering a small pyramidal shaped crystallite with a (111) base and three inclined faces of {100} type (figure 4).

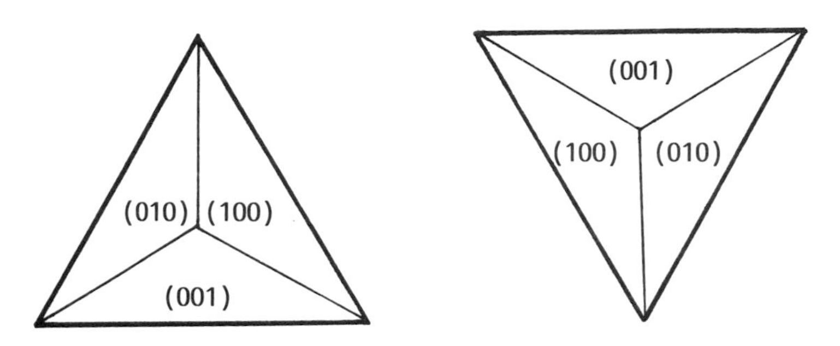

Figure 4. The two equivalent orientations of NH_4I on the cleavage surface of mica. Each orientation is represented by a pyramid with {100} side faces.

Because the epitaxial orientation relationship is determined by the way in which the matching occurs at the interface plane between the substrate and the deposit, it follows that the actual relationship can depend upon the crystal plane of the substrate on which growth occurs. Thus when AgBr is grown on the Ag(001) face, the orientation is

(001) AgBr parallel to (001) Ag
[110] AgBr parallel to [100] Ag

but when growth occurs on the Ag (110) face, the orientation is

(110) AgBr parallel to (110) Ag
$[1\bar{1}0]$ AgBr parallel to [001] Ag

These are distinct orientations.

Perhaps the most important of Royer's rules for epitaxy concerns the misfit between the two crystals. The misfit is normally defined as m where

$$m = \frac{(b-a)100}{a} \quad \text{percent} \qquad (1)$$

where 'b' is the spacing between atoms or ions in a row in the deposit interface plane parallel to the row in the substrate interface plane of inter-atomic spacing 'a'. It follows that, for full symmetry matching at the interface (e.g. figure 1), the misfit is the same in all directions in the interface. In other cases (e.g. figure 2), the misfit varies with direction.

The misfit value(s) do have some important consequences for the growth of epitaxial layers, but there is much evidence that good epitaxy can occur for very high misfit values. This is well illustrated by the result of depositing alkali halides on alkali halides by vacuum evaporation. The compounds have the same NaCl type structure, and each grows on one of the others in parallel orientation (Schulz, 1951, 1952), and the misfits range from −39% for LiF on KBr to +90% for CsI on LiF.

It is also found that the orientation which does occur is not necessarily the one with the best fit (i.e. smallest misfit). The growth of silver on (100) sodium chloride is a good example. The silver grows in parallel orientation with a misfit of −28%, whereas the orientation

(100) Ag parallel to (100) NaCl
with [011] Ag parallel to [010] NaCl

would involve a misfit of only +2%.

In order to provide a better understanding of epitaxy it is necessary to consider the mode of growth involved.

3. GROWTH MECHANISMS OF EPITAXIAL LAYERS

A wide variety of methods can be used for growing a thin layer of one substance on a substrate surface of another substance. These include growth from aqueous solution, electrodeposition or deposition from the vapour phase. Much emphasis, in the case of semiconductor layers, is placed on deposition from the vapour phase by molecular beam epitaxy (MBE) or chemical vapour deposition (CVD) or one of the several variants of these two techniques. These all result in the arrival of mobile deposit atoms on the surface of the substrate, and the mode of growth is classified according to the way in which these atoms arrange themselves and aggregate. Three main modes of growth have been identified:

1) Monolayer growth
2) Nucleated growth
3) Nucleation following monolayer formation

Monolayer growth occurs when the deposit atoms are bound more strongly to the substrate than they are to each other. The atoms aggregate to form monolayer islands of deposit which enlarge as deposition continues until a complete monolayer coverage has taken place. Fresh monolayer islands then form on the first monolayer of the deposit and this results in the formation of a second monolayer. Thus the deposit grows monolayer by monolayer up to a total thickness of many monolayers.

Nucleated growth occurs when the initial deposit atoms aggregate as small three-dimensional islands, which increase in size as further deposition continues until they touch and intergrow to form a continuous film. Such nucleation occurs when the binding of the deposit atoms to each other is stronger than their binding to the substrate. It is commonly known as the Volmer-Weber mode.

The third mode of growth involves the formation of one monolayer, or a small number of monolayers, just as for the first growth mode followed by subsequent nucleation of three-dimensional islands on top of these monolayers, in just the same way as occurs on the substrate for Volmer-Weber nucleation. This is known as the Stranski-Krastanov mode.

3.1 Monolayer growth

Frank and van der Merwe (1949 a,b,c,) considered the consequences of the monolayer mode of growth and introduced the idea of strained, or pseudomorphic monolayers. They concluded that the lowest energy state of a monolayer would be one matching in spacing with the substrate surface plane provided the natural misfit is less than some limiting value. Their theory gave a typical limiting misfit of about 9% for complete matching of periodicities, with misfits up to about 14% giving rise to a metastable equilibrium of perfect matching. For natural misfits greater than these values the monolayer would, according to the theory, grow with their natural spacing. Initially, the theory was presented as providing a criterion for epitaxy, so that only deposits for which the natural spacing corresponds to a misfit less than the limiting value would grow as pseuodomorphic monolayers leading to epitaxy. It was postulated that for greater misfits no pseudomorphic monolayers would form and that epitaxy would not occur, i.e. the pseudomorphic monolayer was regarded as an essential stage of the formation of an epitaxial deposit. It is now known that monolayers, although not pseudomorphic monolayers, can form above the limiting misfit and that epitaxy can occur for misfits above the limiting value, both by monolayer growth and by the nucleation mode of growth. However, it is now well established that strained, or pseudomorphic, monolayers do form at low misfit values and that there is a limiting misfit above which strained monolayers will not be formed. The exact value of the limiting misfit will vary depending upon the particular substrate and deposit being considered, and it is convenient, and realistic, to use an average limiting misfit value of 10–15 percent, as an approximate guide to what happens.

Once a strained monolayer has been formed for a natural misfit value below the limiting misfit, further strained monolayers can form on top. This results in a pseudomorphic three-dimensional structure, in which the atomic planes parallel to the substrate surface will match the substrate planes, but where the spacing of these planes is changed from the normal bulk value in such a way that the atomic volume remains approximately constant. Thus for a positive misfit value the planes are compressed to match the substrate and their inter-planar spacing is increased from its natural value. Such a strained, or pseudomorphic, thin film deposit is only stable up to a limiting value of thickness, and above this limiting thickness the lowest energy state would be a layer with its natural spacings. The value of this limiting thickness has been determined by several workers, following the earlier work of van der Merwe (1963). Olsen and Ettenburg (1975) have derived an approximate version which is helpful to provide a quick estimate of the critical thickness h_c.

$$h_c = \frac{50b}{m} \qquad (2)$$

where b is the edge component of the Burgers vector of the misfit dislocations required and m is the percentage misfit. It gives reasonably good values for m less than about 5.

Before considering what happens when the critical thickness is exceeded, we need to introduce the concept of misfit dislocations. The concept was introduced by Frank and van der Merwe (1949 a,b,c) to describe the situation once a deposit of at least several atomic layers in thickness had reverted to its normal spacings. This results in a mismatch between the substrate and deposit crystal lattices, but it was argued that the

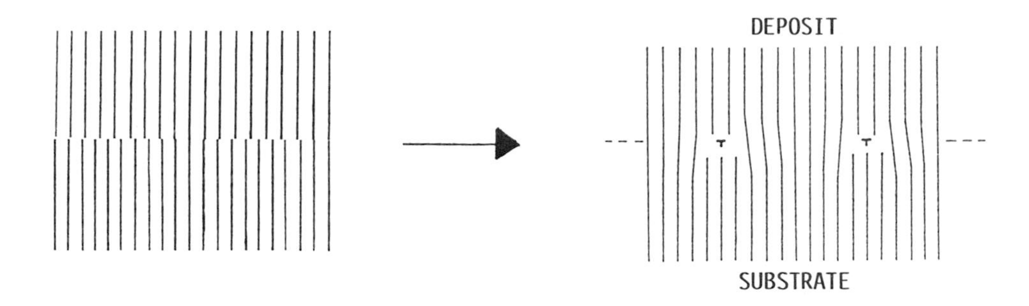

Figure 5. The formation of misfit dislocation by the redistribution of the strain at the interface between a substrate and an epitaxial deposit with a misfit of about +11 percent.

mismatch would not be uniformly distributed in the interface plane, see figure 5, but would be concentrated in regions between which there would be relatively perfect matching. The mismatched regions resemble edge dislocations, so that the misfit is effectively accommodated by an array of edge dislocations. In order to accommodate a two dimensional planar misfit, a two dimensional network of edge dislocations is required. This network of misfit dislocations accommodates the misfit between substrate and deposit right from the earliest stages of growth if the deposit forms initially with its natural spacing. The higher the misfit, the more closely spaced are the misfit dislocations. The spacing S between adjacent dislocations is given by

$$S = \frac{ab}{b-a} = \frac{100b}{m} \qquad (3)$$

where a, b and m have the meanings defined in section 2.

The concept that an elastically strained pseudomorphic layer is the minimum energy configuration up to a limiting thickness, and is therefore unstable above that limiting thickness, does not necessarily mean that the elastic strain disappears once the critical thickness is exceeded. The strain can only be eliminated if misfit dislocations are placed at, or close to, the interface between the substrate and deposit. For an extensive, effectively semi-infinite, thin deposit there has to be a mechanism for generating these dislocations and without such a mechanism the deposit will remain elastically strained as it thickens further. Consequently the critical thickness observed in practice will not necessarily match with the value given by relation (2) or any other equivalent relation based upon minimum energy. The possible mechanisms for generating the required misfit dislocations are given below.

3.2 Nucleated Growth

For many systems, experimental evidence shows that the deposit forms as three dimensional nuclei distributed over the substrate surface, right from the earliest stages of growth. Between these three dimensional nuclei the substrate surface remains more or less free of deposit material. The classical thermodynamic approach to the understanding of this nucleation process, for example in condensation from the vapour phase, is to consider the energy of the nucleus as a sum of its volume energy and its surface energies, including the interfacial energy of its surface in contact with the substrate, and to compare it with the energy of the volume of vapour from which the nucleus would form. The volume energy reduces when condensation occurs, and this reduction is proportional to the size, or volume, of the nucleus. The surface energies increase with the size of the nucleus, and are proportional to the square of the linear dimensions of the nucleus, for a given nucleus shape. For very small nuclei sizes there is a net gain in energy on condensation, hence such small nuclei are not stable. Above some critical size, there is a net loss in energy on condensation, so that nuclei above this critical size are stable. Hence any nuclei above the critical size will grow whilst those below the critical size will re-evaporate. The interfacial surface energy is an important factor in distinguishing between the occurrence of nuclei in diferent orientations. The critical size of a nucleus will increase as the interfacial energy increases, so that orientations with a low interfacial energy are favoured because the nuclei of such orientations are stable, and can grow in preference to nuclei with higher interfacial energy. The equilibrium shape of a nucleus will be determined by the variation of surface energy with the crystallographic indices of the surface, so that the nucleus will become bounded by crystal planes of low energy.

In this way, it is energetically favourable for nuclei of a given shape and orientation with respect to the substrate to form and to increase in number, per unit surface area, until a stage is reached at which it becomes more favourable for newly arriving deposit atoms to condense on the existing nuclei rather than to lead to the formation of any new nuclei. This results in a saturation surface density of nuclei which then increase in size as further deposit atoms arrive.

The growth of these nuclei, without change in shape, continues until the nuclei become sufficiently large to touch and coalesce. The compound nucleus will not then have an equilibrium shape and restoration of the equilibrium shape can occur by two processes:

(1) Newly arriving deposit atoms condense on the nucleus in such a way as to produce an enlarged nucleus or island of equilibrium shape.

(2) There is a sufficient rate of surface or volume diffusion of the existing atoms in the compound nucleus to allow the shape to change, without the arrival of any new deposit atoms.

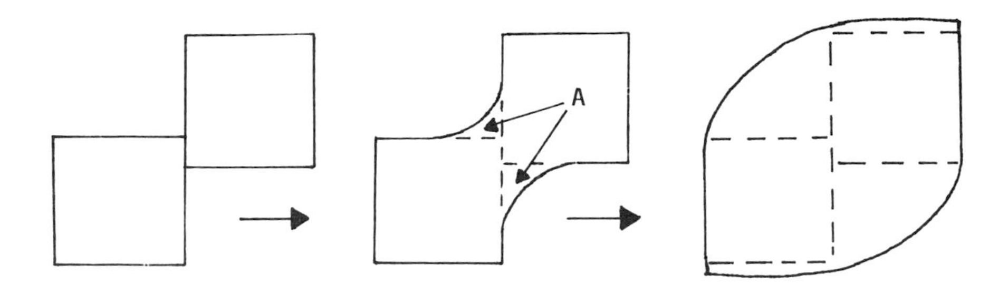

Figure 6. The restoration of the equilibrium shape of a compound island by selective deposition of freshly arriving atoms following the coalescence of two crystallographically shaped islands.

Figure 6 shows the coalescence by process (1) for two nuclei whose equilibrium shape is a square platelet with a particular ratio of height to area. It will be energetically favourable for the earliest arriving atoms to condense in the regions A, since this brings about the most rapid decrease in total surface energy. This is therefore a selective deposition process. As the total required base area of the nucleus is approached, the crystallographic shape will become restored. Whilst this change in base shape is occurring there is a gradual thickening in the compound nucleus until eventually the compound nucleus, or island, has the same shape, including the same

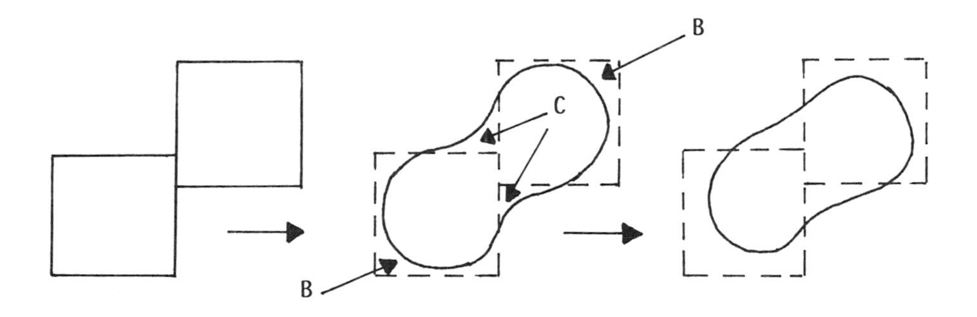

Figure 7. The restoration of the equilibrium shape of a compound island by liquid-like coalescence of two crystallographically shaped islands.

ratio of height to area, as that of the two primary nuclei. Figure 7 shows process (2) for the same nuclei. The reduction in surface energy provides the driving force for a shape change to occur by the migration of atoms from the regions such as B to the regions C. The early stages of the diffusion process result in the complete loss of a crystallographic shape, since this causes the biggest immediate reduction in surface energy, but subsequently the crystallographic shape is restored during the later stages of the process. The process has been termed "liquid-like coalescence" because observation of the process in the electron microscope gives the impression of liquid behaviour (Pashley et al 1964), although the islands remain solid during the shape changes involved. If this process is completed before a significant number of newly depositing atoms condenses on the island, the equilibrium shaped island has a smaller size than the one produced by selective deposition.

In practice, liquid-like coalescence and selective deposition occur simultaneously. Selective deposition will occur more rapidly, the greater is the rate of deposition. The rate of liquid-like coalescence increases rapidly with temperature, so that the relative importance of the two processes depends upon the surface and volume self diffusion coefficients of the deposit material at the substrate temperature. These determine the time t_c required to obtain completion, or near completion, of liquid-like coalescence. The relative contribution of selective deposition will be determined by the additional volume added to the compound nucleus by freshly arriving deposit atoms during the time t_c. Electron microscope observations have demonstrated examples where coalescence occurs mainly by selective deposition or mainly by liquid-like coalescence or by various combinations of the two (Honjo and Yagi 1980).

Once an island has acquired an equilibrium shape, it can continue to grow with this shape as further deposition continues until it touches another nucleus or island when a fresh coalescence takes place. These coalescence processes continue to result in larger and larger equilibrium shaped islands but because the time to attain an equilibrium shape increases as the size of the islands increases, a stage is reached at which restoration of an equilibrium shape is not attained before the island touches another island and a new coalescence process is initiated. The crystallographic form of the boundaries of the islands therefore reduces, and can disappear as growth continues. The joining together of the islands eventually results in a continuous network structure and as a result of further deposition, the holes in this network structure can become filled-in so that a continuous hole-free film is produced. The minimum deposit thickness at which this stage is reached varies appreciably from one substrate deposit combination to another, and also depends upon the deposition parameters such as substrate temperature and rate of deposition.

If the misfit m between substrate and deposit is sufficiently low, the small initial nuclei can be strained elastically just as for pseudomorphic monolayers. But as the nuclei or islands grow and exceed the critical thickness the elastic strain is largely eliminated. This occurs readily since misfit dislocations can easily be moved into the interface from the edge of the islands. In effect, the island is free to expand or contract parallel to the substrate surface, as necessary, to allow it to take up its equilibrium lattice spacings. In practice, the nucleation mode of growth tends to occur when the misfit is large, say above five percent or more, so that nuclei either have close to the bulk lattice parameters right from the beginning or after a relatively small increase in their size (height), so that pseudomorphic nucleated deposits are not commonly observed.

3.3 Nucleation Following Monolayer Formation

The occurrence of this mode of growth appears to be far less common than the other two modes, and it has therefore been much less studied. Examples involving the deposition of metals on metals, especially refractory metals, and metals on semiconductors have been identified (Venables et al 1984). Consequently the mode of growth could be of some importance to the preparation of semiconductor devices. It is to be expected that the main characteristics of the growth mode can be deduced directly from the characteristics already described for the monolayer and nucleation growth modes. No attempt to do so is made in this paper.

4. THE FORMATION OF IMPERFECTIONS IN EPITAXIAL DEPOSITS

Thin films grown by epitaxy commonly have large numbers of lattice defects in them, and special conditions have to be fulfilled to avoid the formation of defects during growth. There are various mechanisms which are known, or which are believed, to introduce imperfections, especially dislocations, in a growing film. These will now be reviewed.

4.1 Copying Dislocations in the Substrate

When a pseudomorphic monolayer or a pseudomorphic island forms over a point on the substrate where a dislocation emerges, it would be expected that the dislocation would extend into the deposit, since the deposit is effectively continuing the structure of the substrate, and the position and spacing of the deposit atoms is determined by the substrate. However, if the deposit, either as a monolayer or in the form of a nucleus, has its own lattice constant which is different from that of the substrate it would be expected that any substrate dislocation would not extend into the deposit, because the relative positions of the deposit atoms are determined primarily by binding to other deposit atoms. There has been no systematic study aimed at putting these conclusions to the test. Since there is much interest in growing semiconductor films by the pseudomorphic monolayer mechanism, it is clearly important to ensure that high quality substrates, with low dislocation densities, are used, in order to avoid dislocations being introduced into the growing film by this mechanism.

4.2 The Formation of Misfit Dislocations

The presence of misfit dislocations at the interface between two aligned crystals with different lattice spacings is inevitable, see figure 5, although the width of the dislocations is not necessarily narrow in relation to the distance between adjacent dislocations. This applies particularly when the misfit is high and the distance between dislocations is small. If the deposit layer forms with its bulk lattice spacings from the earliest stages of growth, the misfit dislocations will be present from the beginning and are likely to be pure edge dislocations, or dislocations with a large edge component. They could well be non-glissile dislocations which would not move as growth proceeds, and they would normally be in the form of a two-dimensional network.

If growth occurs in the form of pseudomorphic islands, the necessary dislocation network can form easily if the deposit reverts to its normal bulk lattice spacings whilst the deposit is still in the form of isolated islands. These dislocations can simply move in from the edge of the islands. If growth occurs by the pseudomorphic monolayer mechanism, the subsequent introduction of misfit dislocations at or near the interface must involve more complex mechanisms. A variety of different mechanisms has been proposed, with at least partial direct experimental evidence that they occur. This evidence always involves transmission electron microscopy.

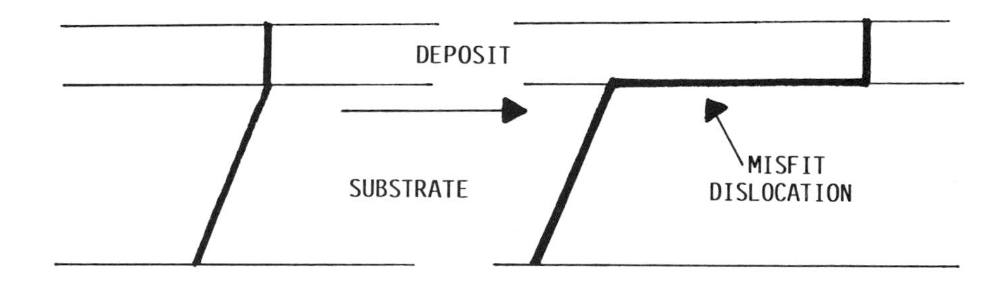

Figure 8. The formation of misfit dislocations at the interface between substrate and deposit by the motion of a threading dislocation formed by extension of a substrate dislocation.

The first mechanism, proposed by Matthews (1975), involves the copying of substrate dislocations. If such a dislocation exists, and that part of it in the deposit layer moves whilst that in the substrate stays fixed, a length of dislocation line will be produced in the interface, as shown in figure 8. For the dislocation to be effective as a misfit dislocation, it must have a large edge component and this requires that the movement occurs perpendicular, or nearly perpendicular, to the direction of its Burgers vector. If a threading dislocation of appropriate Burgers vector for this mechanism exists, then the stress present in the pseudomorphic layer would be in the required direction to cause this movement, and hence this mechanism could be expected to operate, and has in fact been observed. However, it could not provide more than a small proportion of the misfit dislocations required, unless the substrate had an unrealistically high initial dislocation density, or the misfit is very low.

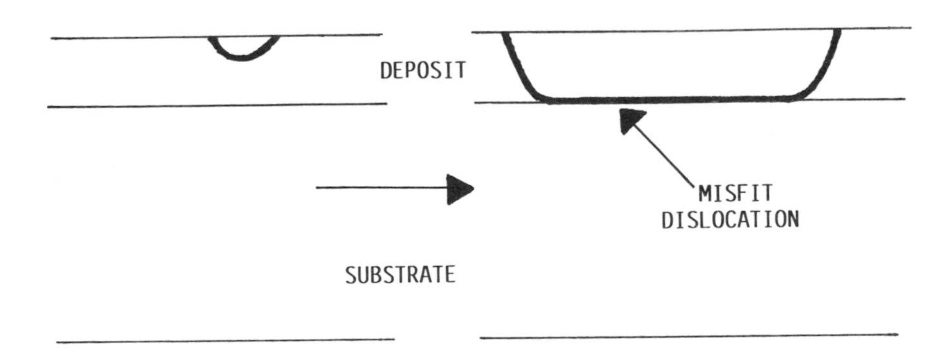

Figure 9. The formation of misfit dislocations at the interface between substrate and deposit by the surface nucleation of a dislocation loop at the growing surface of an epitaxial deposit.

The second mechanism, also proposed by Matthews (1967), involves the nucleation of a half loop of dislocation at the surface of the deposit, and the movement of this dislocation down to the interface with the substrate by glide, as shown in figure 9. More recently, variations on this mechanism have been proposed as a result of observations on films being grown inside the electron microscope so that the formation of misfit dislocations can be observed directly as they occur. Yagi et al (1971) have suggested that the growth of dislocation loops can occur by climb, rather than slip. This would involve diffusion of point defects. Cherns and Stowell (1975a,b, 1976) have interpreted detailed observations of the growth of palladium on gold in terms of more complicated mechanisms, also involving dislocation climb, whereby partial dislocation loops are nucleated and expanded in sequence. The required density of dislocations in the neighbourhood of the interface is produced by these mechanisms, but the dislocation configurations are more complex than this.

It is important to note that the elimination of the elastic strain associated with a pseudomorphic layer requires the addition (in the case of a negative misfit) or subtraction (in the case of a positive misfit) of planes of atoms perpendicular to the interface. If this is brought about by slip processes, which effectively represent plastic extension of contraction parallel to the interface, it will result in a complex array of surface steps on both sides of the deposit. The extent to which such steps could be accommodated at the interface would limit the extent to which full relief of elastic strain could be achieved. The addition of climb processes, i.e. diffusion processes, provides more scope for achieving the requirements without serious disturbance at the interface. Recently, Eaglesham et al (1989) have proposed a pure slip mechanism for generating misfit dislocations, based upon their observations on the growth of GeSi alloys on Si. They suggest that the mechanism will operate only for low misfit systems.

The introduction of misfit dislocations by mechanisms such as those just mentioned also results in other dislocation features being present at or near the interface. If the deposit layer is thickened further, these dislocation features could lead to the formation of some kind of dislocation structure threading through the deposit. There is little evidence to indicate what actually occurs, and no detailed models have been proposed.

4.3 Imperfections Resulting from Island Coalescence

When growth occurs by the nucleation mode, with a non-zero misfit, the nuclei are separated from each other by distances which are not integral multiples of the appropriate lattice distances. Hence as two adjacent nuclei grow towards each other some of their atomic planes are not aligned so that elastic strain, e.g. bending, occurs as coalescence takes place. If three nuclei coalesce together simultaneously, the misalignments can result in the formation of a dislocation, which threads through the thickness of the triple compound island (Pashley, 1965, Jacobs et al, 1966). This process has been observed during growth carried out inside an electron microscope. It means that the presence of a misfit causes the formation of dislocations during the coalescence stages of nucleated growth, and that the presence of dislocations in such epitaxial films is inevitable. In practice the density of such dislocations in films grown by the nucleation mode can be very high, up to $10^{8}mm^{-2}$, although they are not necessarily formed by this mechanism.

In addition to dislocations, coalescence can also result in the formation of stacking faults or thin platelets of microtwins. It has been suggested that these can also form as a result of the misalignment of coalescing islands (Jacobs et al, 1966). Whatever the mechanisms involved, it is found that stacking faults and microtwins are present in many epitaxially grown films, sometimes in large numbers.

5 SEMICONDUCTOR EPITAXY

5.1 Silicon

Much of the early interest was centred on the growth of silicon on silicon (homoepitaxy). Good epitaxy, with or without doping, is readily achieved but care is needed to avoid the presence of defects in the layers. It is necessary to use substrates with very low dislocation densities in order to reduce the density of 'copied' dislocations, and the substrates have to be extremely clean since contamination seems to be effective in causing the nucleation of defects. Without such precautions it is found that the epitaxial deposits contain high densities of defects, especially microtwins.

When silicon is required on an insulating substrate, it is often grown on sapphire (Al_2O_3), which has a rhombohedral structure. Deposition is usually carried out on the sapphire $(1\bar{1}02)$ plane, which results in the orientation

(100) Si parallel to $(1\bar{1}02)$ Al_2O_3
and [010] Si parallel to $[0\bar{1}11]$ Al_2O_3

The aluminium atom sites in the sapphire $(1\bar{1}02)$ plane form a rectangular network which is nearly, but not quite, a square network. Thus there is not a perfect symmetry match at the interface between the silicon and the sapphire, and the misfit varies with direction in the interface. Along the Si[010] the misfit is +4.2%, and along the Si[001] it is 12.5%. Such a large misfit is unlikely to allow pseudomorphic monolayer growth to take place, and the growth actually occurs by the nucleation mode. Partly due to the large misfit, a high density of defects is commonly observed in silicon grown on sapphire (Abrahams et al 1976), with twins and stacking faults prominent. However, whilst this commonly occurs with thin layers, and these defects persist in regions near the interface as the deposit is thickened, the defect density is much lower in regions of thicker deposits further away from the substrate. Layers of about $1\mu m$ thick have quite low defect densities close to their surface.

5.2 Semiconducting Compounds

Currently there is considerable interest in growing III–V or II–VI semiconducting compounds on each other, since these substances have electronic and optoelectronic properties which can be exploited in a range of different devices. All of the III–V compounds have the same sphalerite crystal structure which is cubic and similar to the silicon structure. Some of the II–VI compounds exhibit the sphalerite structure, but also occur with a related hexagonal structure. One of the advantages of a group of compounds, such as the III–V compounds, is that they have a range of electronic properties (e.g. bandgap values) as well as a range of lattice parameters, so that there is flexibility in influencing both properties and growth characteristics by choice of substrate and overgrowth. This flexibility is much increased by using ternary or quarternary alloys such as $Ga_xAl_{1-x}As$ or $Ga_xIn_{1-x}As_yP_{1-y}$.

By suitable choice of alloys it is possible to arrange for the substrate and deposit to have the same lattice parameter so that the misfit is zero and epitaxy can take place without any defects being being formed in the deposit, other than the extension of defects already present in the substrate. There will be a range of compositions which satisfy the zero misfit condition for a given pair of alloys, and within that range the electronic properties will vary, so that some choice of properties is available.

In order to obtain a misfit of exactly zero during growth, to avoid the formation of any misfit dislocations, the substrate and deposit must have identical lattice parameters at the deposition temperature. Since the usual requirement is for a sharp boundary between the substrate and overgrowth, interdiffusion across the boundary must be minimised and therefore the substrate temperature needs to be kept as low as possible consistent with other requirements, such as sufficient lateral surface diffusion of the depositing atoms to allow them to take up their positions of lowest energy.

5.3 Solid State Superlattices

Following the ideas put forward by Esaki and Tsu (1970), much attention has been given to the growth of superlattices consisting of thin alternating layers of two semiconductor compounds. The novel electronic properties of these superlattice structures are exciting great interest, and still remain to be explored in detail. This exploration requires that techniques be developed for producing the superlattice structures, and these inevitably involve epitaxial growth because individual layer thicknesses as low as a few atomic layers are required.

The superlattices are of two main types. In the first type the two alternating layers consist of two different compounds or alloys, such as alternating GaAs and $Al_xGa_{1-x}As$. In the second type the compound or alloy is the same for each of the two alternating layers but the doping alternates (e.g. between Si and Be in GaAs) so that there is an alternation between n and p type layers. The preparation of this second type of superlattice is fairly straightforward from the point of view of epitaxy, because the misfit between adjacent layers is zero since the same semiconducting compound is involved. For the first type, the requirement to have zero misfit, in order to avoid misfit dislocations and other defects, can be achieved by varying the composition of the two alternating layers so that they have identical lattice parameters. Provided it is also possible, within this condition, to choose a pair of alloys with the required electronic properties this technique for growing superlattices can be completely satisfactory. However, the condition on lattice matching does put severe restrictions on choice. Consequently a different approach, involving the growth of strained layer superlattices, provides additional flexibility.

Strained layer superlattices involve growing alternating layers which have a small misfit with each other, and rely on growth occurring as pseudomorphic layers. If the layer thicknesses are kept below the critical thickness in each case, it is possible to arrange for the lattice periodicities in all directions parallel to the interface planes to be the same in the two layers and hence in all layers through the superlattice. The thicker the layers the smaller is the limiting misfit, as determined by the expression for the critical thickness. It is interesting to note that the elastic strain present in the layers causes a change in the band gap for that layer, so that strained layer superlattices also provide the opportunity for some controlled changes to be made to the electronic properties.

The growth of these strained layer superlattices requires a substrate which has a lattice spacing intermediate between that of the two alternating layers. This ensures that the average elastic strain of the superlattice is very low, certainly much less in magnitude than the elastic strains in the individual layers. Thus the required natural lattice parameter of the substrate would depend upon the relative thicknesses of the two component layers. Since it is not always possible to provide a bulk substrate crystal which has both the required lattice parameter and sufficient perfection, a technique has been developed for using a good quality substrate with a slightly different lattice parameter. This is brought about by chosing a suitable compound for the substrate and growing an epitaxial layer of graded composition on top of this until the required lattice parameter is achieved. For example, a strained layer superlattice made up of successive layers of $GaAs_xP_{1-x}$ and $GaAs_yP_{1-y}$ can be grown on a bulk substrate of GaP on which a graded layer of $GaAs_zP_{1-z}$ is first grown. The value of z in this graded layer is increased from zero to a suitable value between x and y, through its thickness, so that a surface of the required composition and lattice parameter is produced. Since the purpose of this graded layer is to produce changes in lattice parameter through its thickness, misfit and other dislocations will form in the layer. However, there is evidence that these dislocations tend to be deflected to one side as the superlattice is grown on the graded layer, with the result that the upper layers of the superlattice are relatively perfect.

6 REFERENCES

Abrahams M S, Buicchi C J, Smith R T, Corboy J F, Blanc J and Cullen G W 1976. *J. Appl. Phys.* **47** 5134

Barker T V 1906 *J. Chem. Soc. Trans.* **89** 1120

Barker T V 1907 *Mineral Mag.* **14** 235

Barker T V 1908 *Z. Kristallogr.* **45** 1

Cherns D and Stowell M J 1975a *Thin Solid Films* **29** 107

Cherns D and Stowell M J 1975b *Thin Solid Films* **29** 127

Cherns D and Stowell M J 1976 *Thin Solid Films* **37** 249

Eaglesham D J, Maher D M, Kvam E P, Bean J C and Humphreys C J 1989 *Phys. Rv. Letters* **62** 187

Esaki L and Tsu R 1970 *IBM J. Res. Develop.* **14** 61

Frank F C and van der Merwe J H 1949a *Proc. Roy. Soc. A* **198** 205

Frank F C and van der Merwe J H 1949b *Proc. Roy. Soc. A* **198** 216

Frank F C and van der Merwe J H 1949c *Proc. Roy. Soc. A* **200** 125

Grunbaum E 1975 *Epitaxial Growth* (J W Matthews Ed., Academic Press, New York) 611

Honjo G and Yagi K 1980 *Current Topics in Materials Science* **6** 195

Jacobs M H, Pashley D W and Stowell M J 1966 *Phil. Mag.* **13** 129

Matthews J W 1967 *Physics of Thin Films* **4** 137

Matthews J W 1975 *Epitaxial Growth* (J W Matthews Ed., Academic Press, New York) 560

van der Merwe J H 1963 *J. Appl. Phys.* **34** 117

Olsen G and Ettenburg M 1975 *Crystal Growth Theory and Techniques* (C H Goodman Ed., Plenum Press)
Pashley D W 1956 *Advanc. Phys.* **5** 173
Pashley D W, Stowell M J, Jacobs M H and Law T J 1964 *Phil. Mag.* **10** 127
Pashley D W 1965 *Advanc. Phys.* **14** 327
Royer L 1928 *Bull. Soc. franc. Mineral* **51** 7
Seifert H 1953 *Structure and Properties of Solid Surfaces* (R Gomer and C R Smith Eds., University Press, Chicago) 218
Schultz L G 1951 *Acta Crystallogr.* **4** 487
Schultz L G 1952 *Acta Crystallogr.* **5** 130
Venables J A, Spiller G D T and Handbucken M 1984 *Rep. Progr. Phys.* **47** 399
Yagi K, Takayanagi K, Kobayashi K and Honjo G 1971 *J. Cryst. Growth* **9** 84

Metal Organic Chemical Vapour Deposition (MOCVD) for the Preparation of Semiconductor Materials and Devices

J O WILLIAMS

INTRODUCTION

This paper reviews the progress that has been made over the past decade in the use of metal organic chemical vapour deposition (MOCVD) for the preparation of high purity, epitaxial layers of a wide range of semiconductor materials:

It is now over a decade since the pioneering work of Manasevit and co-workers [1, 2] gave rise to the remarkable interest subsequently shown in processes that use metal-organic compounds for the production of semiconductor materials. Since 1981 four international conferences have been held on the topic [3], numerous workshops have been organised and nowadays, for certain key III-V materials systems (and to a lesser extent narrow band gap II-VI analogues), MOCVD or metal organic vapour phase epitaxy (MOVPE) is an accepted production technique yielding novel electronic and opto-electronic devices [4-6]. In many of these cases, MOVPE compares favourably with molecular beam epitaxy (MBE); for certain materials particularly those containing phosphorous, MOVPE can produce superior materials. MOVPE is capable of producing thin layered structures - the so-called quantum wells and superlattices - and the technique can be used for atomic layer epitaxy (ALE) [7] and in preparation of layered crystals (LC) [8].

One area where MOCVD has not developed to the same extent as MBE is the in-situ monitoring of the deposition reactions. Despite recent progress [9,10] our understanding of the basic processes occurring in MOCVD is not very advanced and monitoring techniques are only currently being developed unlike the case of, for example, reflection high energy electron diffraction (RHEED) in MBE to monitor and control atomic layer deposition [11]. In addition, since compounds are used as precursors in MOCVD, a great effort has to be expended in order to achieve the required levels of high purity and safety/convenience of the growth process. Many innovative preparation and purification routes have been developed [12,13] for the most common precursors and over the past two years, safer and less toxic chemicals have been studied as alternative starting materials [14,15]. With an improved understanding of the growth mechanisms, we can look forward to a period of considerable innovation in the use of the MOCVD technique for the preparation of III-V semiconductor devices.

Equations

1a $$CH_3\text{--}Ga(CH_3)_2 + H\text{--}As(H)\text{--}H \xrightarrow[ca\ 700\,^{\circ}C]{\Delta} GaAs + 3\,H\text{--}C(H)(H)\text{--}H$$

1b $$TMGa + AsH_3 \longrightarrow GaAs + 3CH_4$$

1c $$C_2H_5\text{--}Ga(C_2H_5)_2 + H\text{--}As(H)\text{--}H \xrightarrow[ca\ 700\,^{\circ}C]{\Delta} GaAs + H\text{-}CH_2\text{-}CH_2\text{-}H + H_2C{=}CH_2$$

1d $$TEGa + AsH_3 \longrightarrow GaAs + C_2H_6 + C_2H_4$$

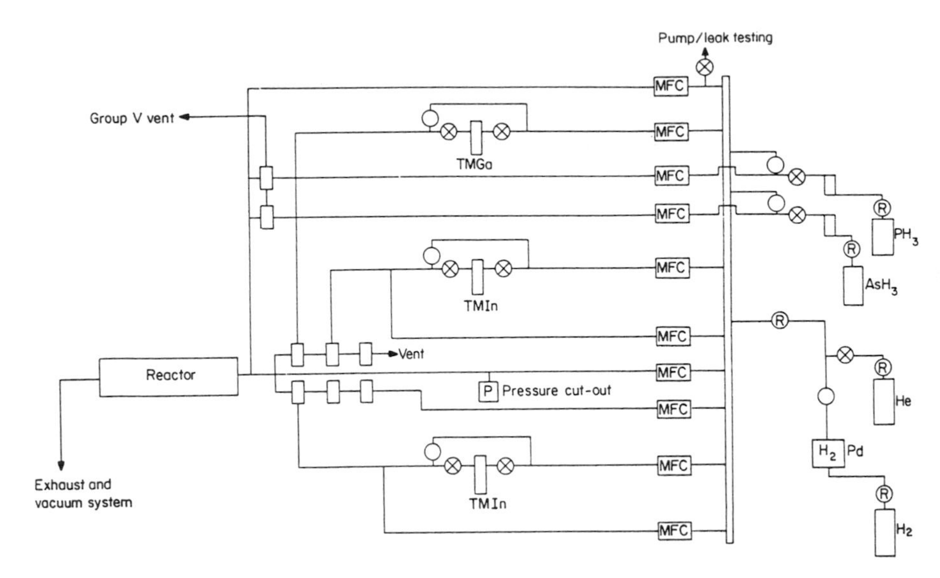

Schematic representation of the MOVPE reactor

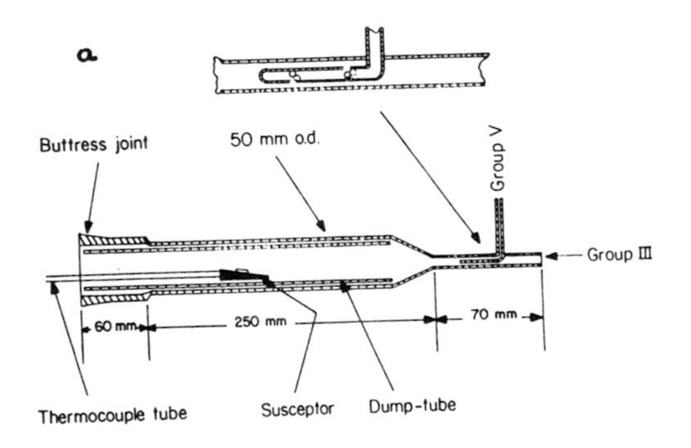

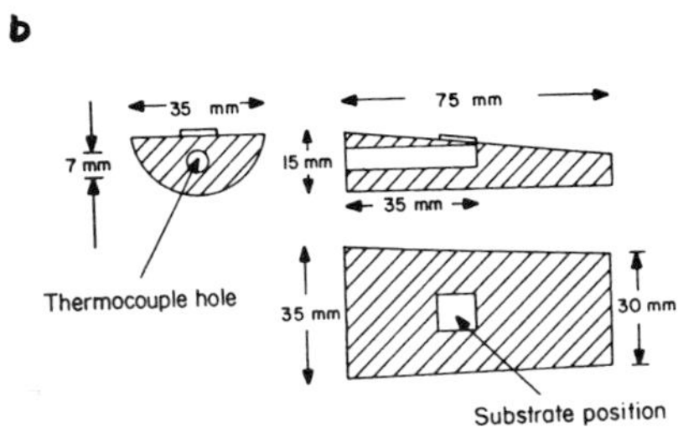

(a) Reactor chamber and (b) susceptor design

Figure 1. **Schematic representations of an MOVPE reactor.**

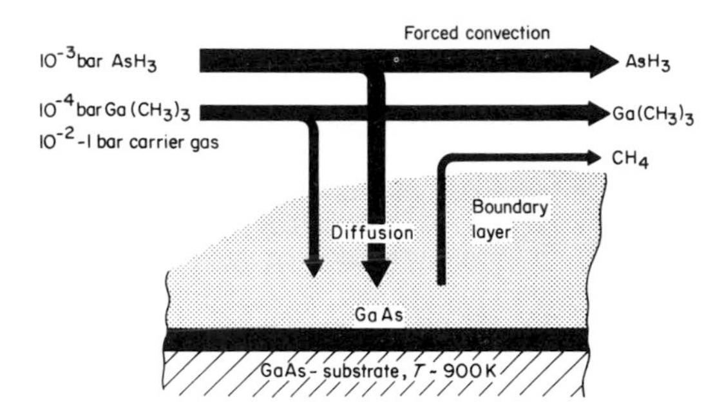

Figure 2. Schematic diagram of the MOCVD process. (Reproduced by kind permission of Butterworths Scientific Ltd; after Ref. 50)

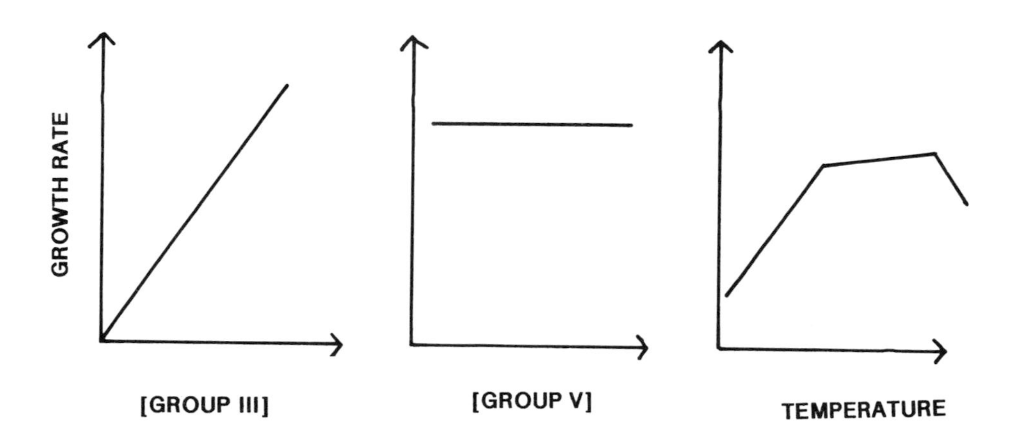

Figure 3. Growth rate dependencies on [Group III], [Group V] and temperature.

THE MOCVD PROCESS

Equations (1) a-d represent, to a first approximation, the reactions that occur during MOCVD growth of GaAs, a typical III-V semiconductor. These reactions are carried out thermally with the flowing gases and the reactor cell maintained at or near atmospheric pressure, ca 1 to 9 x 10^4 Pa. A typical horizontal reactor suitable for the growth of GaAs is shown in Figure 1 and Figure 2 schematically shows the gas phase processes that are believed to control layer deposition. Further details await mechanistic investigations currently underway in several laboratories.

From growth rate studies and variations with [Group III], [Group V] and temperature (see Figure 3) the general picture depicted in Figure 2 can, to a large extent, be justified. The growth rate increases linearly with [Group III], is largely independent of [Group V] and shows three regimes in its temperature dependence - a linear increase at lower temperatures, an essentially temperature-independent, intermediate regime and then a decrease at the higher temperatures. In the intermediate temperature regime where GaAs epitaxial layers exhibit optimum morphology, electrical and optical properties the growth is controlled by diffusion of the Ga-precursor across an intermediate (boundary layer?) region.

Even though present in large excess, the concentration of the Group V precursor does not influence the growth process. Its concentration is, however, critical in eliminating carbon from the Group V lattice sites in the epilayers.

ICP analysis for impurities in TMGa1 and TMGa2*

Elements	TMGa1	TMGa2
Si	0.2 ppm	0.05 ppm ND
Zn	0.6 ppm	0.3 ppm ND
Sn	< 0.5 ppm ND	0.5 ppm ND
Mg	0.05 ppm	0.05 ppm ND
Mn	< 0.05 ppm ND	0.05 ppm ND
Fe	< 0.3 ppm ND	0.3 ppm ND

*ND = not detected

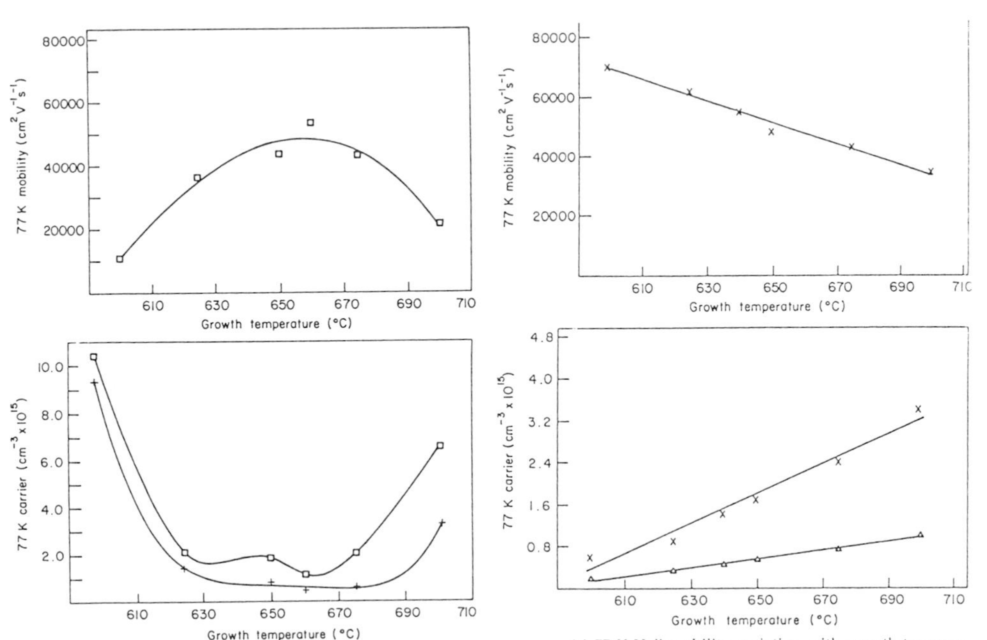

(a) 77 K Hall mobility variation with growth temperature for GaAs grown from TMGa1 at constant V/III ratio of 75:1. (b) 77 K Hall carrier concentration yielding values of (□) N_D and (+) N_A plotted against growth temperature for GaAs grown from

(a) 77 K Hall mobility variation with growth temperature for GaAs grown from TMGa2 at constant V/III ratio of 75:1. (b) 77 K Hall carrier concentration yielding values of (X) N_D and (Δ) N_A plotted against growth temperature for GaAs grown from TMGa2 at constant V/III ratio of 75:1

Figure 4. Growth results for GaAs using two different TMGa sources and the same high purity arsine. (Reproduced by kind permission of Butterworths Scientific Ltd; after Ref. 18).

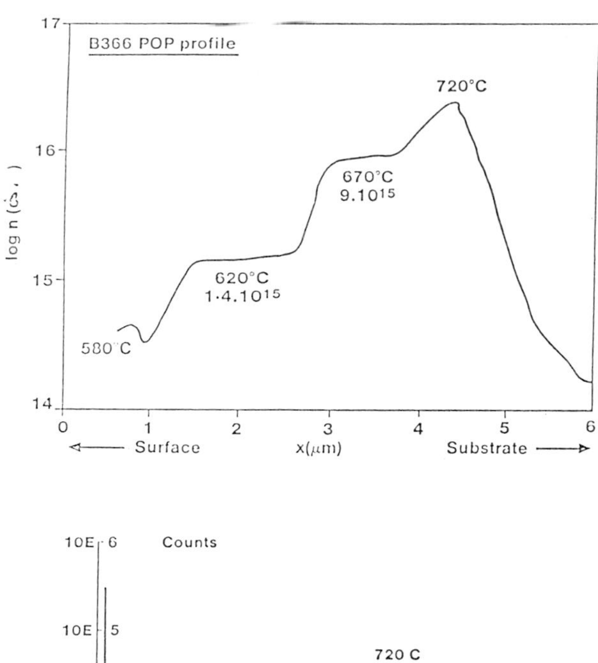

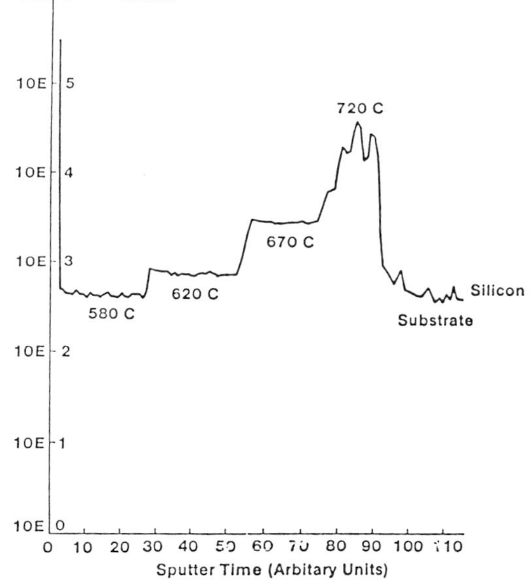

Figure 5. Undoped MOCVD-InP. Carrier concentration versus depth and SIMS depth profile identifying donor as silicon. (Reproduced by kind permission of E J Thrush STC Ltd).

ATTAINMENT OF HIGH PURITY IN SEMICONDUCTOR MATERIALS

Most advances in semiconductor devices rely on the availability and subsequent doping of high purity materials. In recent years there have been significant advances in this field. MOCVD has its own difficulties in this area additional to those of other conventional preparative techniques since the precursors that are used do not lend themselves to ready purification; they are often pyrophoric, hazardous chemicals. It is now possible to analyse directly the metal organic precursors for elemental impurities at the ppb level using variants of inductively coupled plasma (ICP) analysis [16].

Parallel studies to characterise impurities in the grown epitaxial layers by such techniques as secondary ion mass spectrometry (SIMS) and low temperature photoluminescence (PL) allow correlation between impurities in the starting materials and those in the epitaxial layers [17 - 19]. This relationship may be illustrated by reference to some recent work on the growth of GaAs. Figure 4 summarises the results. Two batches of trimethylgallium (TMGa) were used with the same, high purity arsine source. The ICP analysis of the TMGa showed different amounts of common impurities as shown in the table. The level of Si and Zn, recognised n- and p-type impurities, respectively, in GaAs is lower in source TMGa2 than in TMGa1. The electrical characteristics of GaAs grown from TMGa2 and arsine are much improved over those from TMGa1 and the same arsine. The low temperature (77K) electron mobility and concentration are higher and lower, respectively, and for TMGa2 the lowest carrier concentration and highest mobility are obtained at the lower growth temperature of 600 $^{\circ}$C. This low growth temperature can be adventageous for good morphology and crystallinity of the epilayers. The electrical properties are dominated by the Si and the activated incorporation of this impurity is consistent with doping studies using disilane [20].

A similar situation is observed in the case of InP preparation by MOCVD where Si is again found to be the dominant impurity in the highly purified organometallic precursor, trimethylindium (TMIn). The presence of this impurity in the grown layers has been confirmed by depth profiling studies and analysis of low temperature photoluminescence spectra [17]. Figure 5 shows the electron concentration as a function of depth through the InP epilayers grown with increasing temperatures. The corresponding silicon profile in SIMS is also shown. Low temperature (4K) PL spectra confirm the presence of donors and measurements under a magnetic field (where the donor related transitions are resolved) indicate the presence of S and Ge impurities in addition to the Si. Removal of Si yields extremely high quality InP material with theoretical values of electron mobility and concentration reached for epilayers ca 10 μm in thickness. Indeed, one of the highest reported electron mobility values for InP ca 400,000 $cm^2 V^{-1} s^{-1}$ at 60K has been reported for MOCVD material using purified TMIn [21]. The two examples cited are for binary materials. There has also, as a result of this success with GaAs and InP, been significant improvements in the purity of ternary and quaternary materials. This will not be covered in this review, but the interested reader is referred to the recent literature [22].

DOPING OF III-V EPILAYERS DURING MOVPE

Either organometallic Group II sources or Group IV/VI hydrides may be used as p- and n-type dopants in III-V semiconductor growth. As typical examples let us consider the doping of GaAs using Zn and Si precursors. The ideas which have been developed for the binary semiconductors may also, with modifications in specific cases, be applied to ternary and quaternary doping experiments. Figures 6a and b show the carrier concentration versus dopant concentration dependencies for GaAs. Two zinc precursors have been used (dimethyl- and diethyl zinc; DMZ and DEZ) [23] and disilane is the preferred source of silicon [24]. The silicon is well behaved with nett electron concentrations of between 10^{16} and 5×10^{18} cm^{-3} achievable at a growth temperature of 700 $^{\circ}$C and values varying linearly with the disilane concentration.

The Zn shows a complex dependence exhibiting an almost quadratic dependence of carrier concentration on dopant mole fraction at low concentrations with an almost corresponding linear dependence at higher concentrations. The decomposition of these two precursors increases as temperature increases, the sticking coefficient decreases and the nett incorporation rate decreases with increasing temperature. This complex behaviour substantiated by temperature dependence measurements is not ideal for accurate control of p-type GaAs doping. Recently, there have been reports that controlled p-type doping of both GaAs and InP may be possible with carbon emanating from CCl_4 [25]. Diethylberyllium has been used for p-type doping of GaAs and $Al_{1-x}G_{1-x}$ As [26] and tetraethyltin is useful as an n-type dopant in GaAs [27].

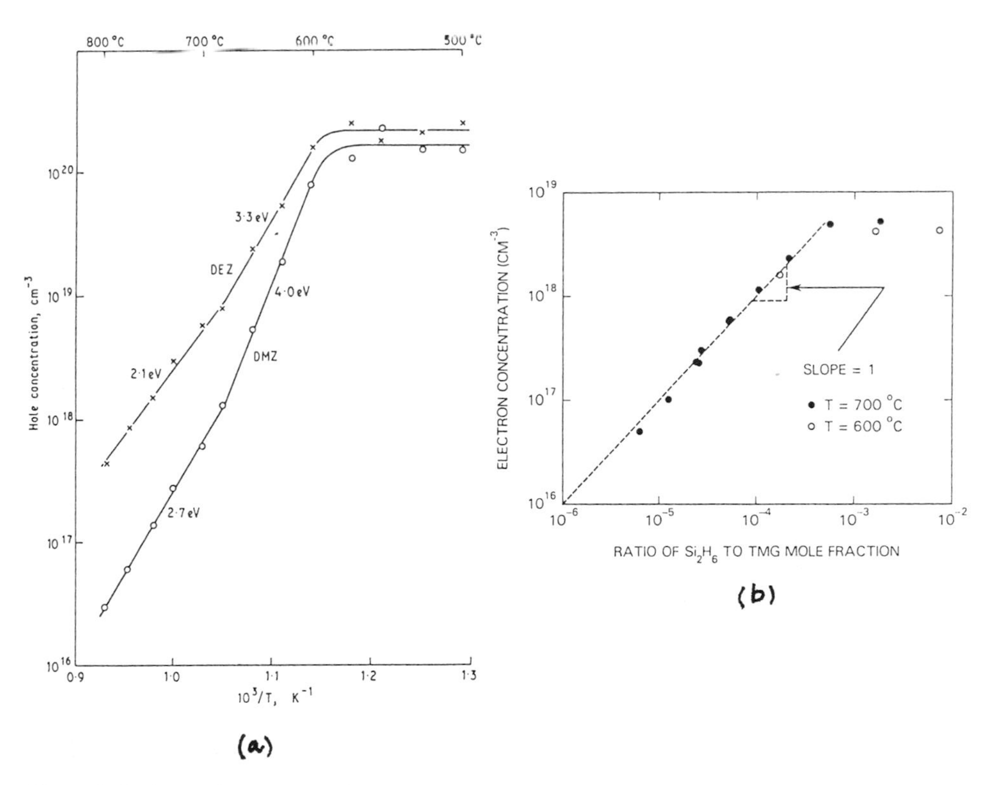

Figure 6. GaAs carrier concentration versus dopant precursor concentration for (a) DMZ and DEZ (b) disilane (Si_2H_6). (Reproduced by kind permission of North Holland Publishing Ltd; after Ref. 23, 24).

HETEROEPITAXY

Heteroepitaxy describes the situation where an epitaxial layer of one material is grown upon another in strict crystallographic registry. In addition to providing considerable fundamental scientific interest such materials combinations have enormous technological significance. Table 1 summarises the properties of a few simple heteroepitaxial systems which are prepared by MOCVD. The list is confined to binary and elemental materials although, as we shall see later in sections 6 and 7, systems including ternary and quaternary materials may also be classified as heteroepitaxial.

It is instructive to consider these pairs of materials in turn. Since Si, dominates the commercial semiconductor market, there has been interest in preparing GaAs on Si. There appear to be significant advantages if this could be done properly including: the relative cheapness of Si substrates as compared to GaAs, the better thermal conductivity of Si with relevance to power devices, the ability to combine high speed GaAs devices with high density Si circuits and the possibility for fabrication of GaAs opto-electronic devices in combination with Si integrated circuits.

However, two major difficulties arise. First the two materials differ in the symmetry of their respective crystal structures and second the unit cell dimensions differ leading to a "lattice mismatch" at the interface between epilayer and substrate. The symmetry difference (sometimes referred to as different colour symmetry) leads to the formation of anti-phase boundaries (APB) when growth proceeds on substrate surfaces of particular orientations with catastrophic consequences for electrical behaviour at the interfaces. Because of these difficulties it is now accepted that any devices made in the GaAs/Si system and relying on the GaAs must utilise material well away from the interface unless the APBs and other defects can be eliminated. Furthermore, these limitations are likely to preclude the development of devices that utilise both Si and GaAs simultaneously.

Simple Heteroepitaxial Systems

Material	Crystal Structures	Unit Cell dimension a_o/pm	Lattice Mismatch $[a/a]_o \times 10^2$	Band Gap /eV
GaAs (e)	sphalerite (fcc)	0.565331	+4.09	1.35
Si (s)	diamond	0.543072		1.14
GaAs (e)	sphalerite (fcc)	0.565331	-3.68	1.35
InP (s)	"	0.586928		1.28
ZnSe (e)	sphalerite (fcc)	0.56670	+0.24	2.67
GaAs (s)	sphalerite (fcc)	0.565331		1.35

(e) epitaxial layer (fcc) face-centred cubic.

(s) substrate

$[a/a]_o$ is the measurement for a stress free epitaxial layer and is defined as $(a_o - a_s)/a_s$ where a_s is the unit cell dimension of the substrate.

Considerable progress has been made recently in the MOCVD growth of GaAs on Si without necessarily employing buffer layers to accommodate the lattice mismatch. Preparation of Si surfaces prior to growth is a prerequisite and a complex series of nucleation and crystal growth steps appear necessary for the preparation of GaAs with acceptable crystallinity [28]. Experiments to reduce interfacial defects have also been carried out employing substrate surfaces deviating from (100) orientation and good properties obtained at 4° off towards (110) [29]. In addition to the problems encountered in the GaAs/Si system the other two pairs of materials referred to in Table 1 suffer the disadvantage of possible interdiffusion of elements with catastrophic consequences if it occurs to an appreciable extent. The situation is worst for ZnSe/GaAs since Zn acts as a p-type dopant in GaAs and Ga an n-type dopant in ZnSe. Excellent quality ZnSe has been grown on (100) GaAs surfaces by reduced and atmospheric pressure MOVPE [30].

Room temperature photoluminescence is observed indicating the low concentrations of shallow and deep centres, but the structural, electrical and optical properties are highly dependent on epilayer thickness and distance away from the interface. The X-ray and PL data for ZnSe epilayers grown on GaAs and varying in thickness from 0.5 μm to 8.0 μm are given in figure 7. In the thinnest layers, deep centre emission appears in the PL and in the X-ray data the ZnSe diffraction maxima are shifted to lower 2Θ values.

SIMS analysis on the ZnSe reveals the presence of Ga diffusion from the substrate (fig 7c) and cross sectional transmission electron microscopy (fig 7d) confirms the defective nature of the interface. It is unlikely that the ZnSe/GaAs interface can be made sufficiently good to support efficient devices relying on optical/electronic properties. However, the ternary $ZnSe_{0.95}S_{0.05}$ grown on GaAs [31] offers a lattice-matched system and studies following MOCVD growth indicate that its properties are superior to those of ZnSe/GaAs and adequate to support devices. The distortion at the interface of ZnSe/GaAs renders the structure tetragonal as opposed to cubic [32] and this has been confirmed recently by electron spin resonance studies of Mn-doped ZnSe using the Mn^{2+} ions to probe the local symmetry [33].

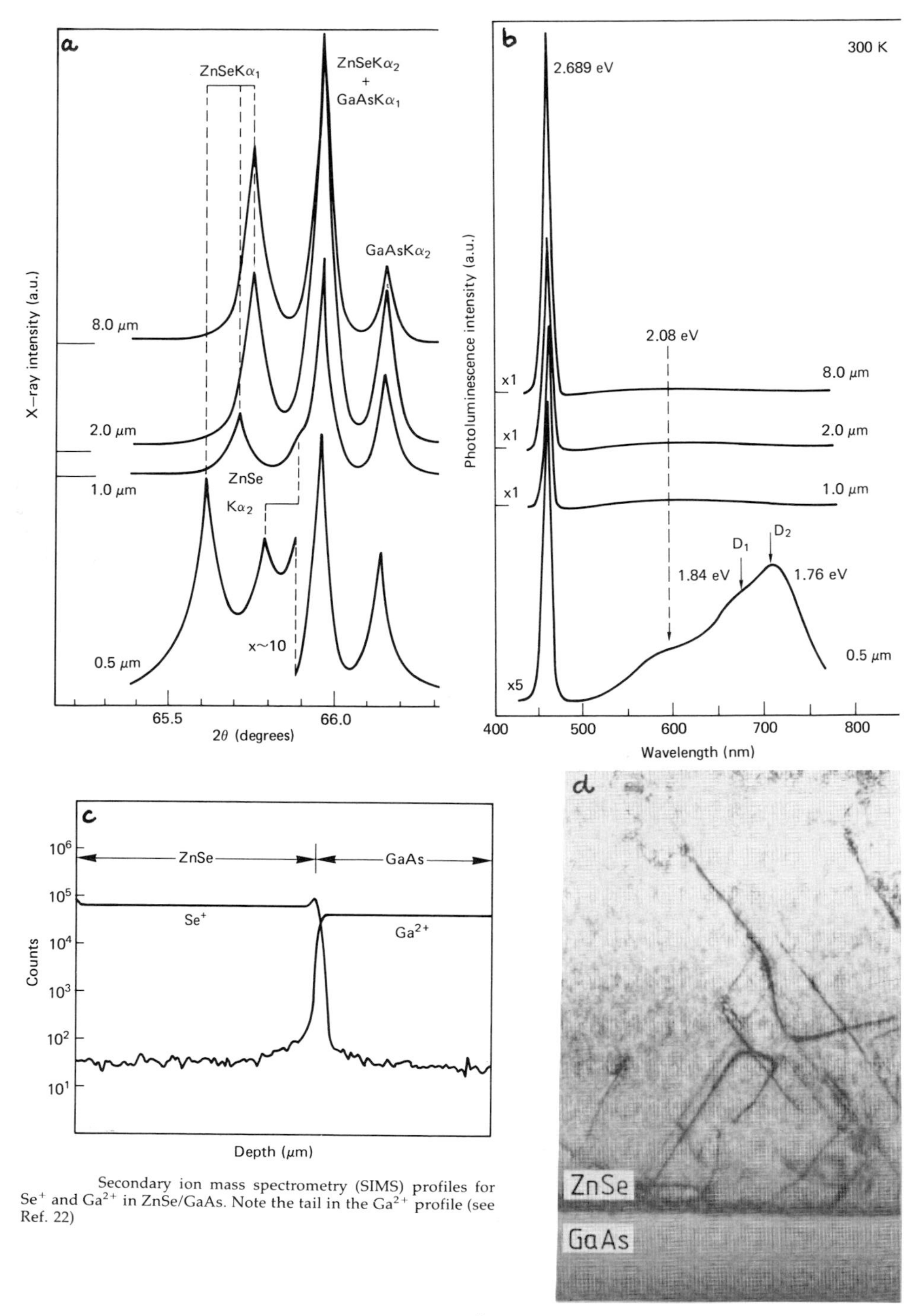

Secondary ion mass spectrometry (SIMS) profiles for Se^{+} and Ga^{2+} in ZnSe/GaAs. Note the tail in the Ga^{2+} profile (see Ref. 22)

Figure 7. ZnSe on GaAs grown by MOCVD at 280 °C (a) X-ray data (b) PL data (c) SIMS profiles (d) cross sectional TEM.

GaAs has been grown successfully on InP substrates [34] and efficient devices such as solar cells and MESFETS have been fabricated. As with the other two heteroepitaxial systems these do not rely on charge carrier transport across the interface or on optical generation of carriers in that region. We can, therefore, conclude that in these three heteroepitaxial systems we have important building blocks for the development of integrated opto-electronic devices that will combine important electronic and optical functions on a common substrate ie Si, GaAs or InP.

TERNARY AND QUATERNARY MATERIALS

Figure 8 shows how the band gap of III-V materials varies with the lattice constant and composition. Such diagrams are often referred to as phase diagrams and they aid the crystal grower in his choice of material system appropriate to a particular electronic/optical requirement. (The term band gap engineering is commonly used to describe this approach). The tie-lines represent ternary compositions eg $Ga_{1-x}In_xAs$ and $InAs_{1-y}P_y$ and areas on the diagram contained within points representing four binary compounds represent quaternary materials eg $Ga_{1-x}In_xAs_{1-y}P_y$. It is apparent that one particular ternary composition will have the same lattice constant as a binary compound and this may be found by drawing vertical lines through the binary point to intercept the tie line for that particular ternary eg $Ga_{0.47}In_{0.53}As$ is lattice-matched to InP (line AB). Similarly, a range of quaternary compositions eg $Ga_{1-x}In_xAs_{1-y}P_y$ with $0 < x < 1$ and $0 < y < 1$ will be matched to InP and will be represented by the same vertical line AB. Almost the entire range of III-V direct band gap, ternary and quaternary materials have been prepared by MOCVD. For many of these the problem of compositional uniformity both within the bulk and across the surface of the grown layers has been solved and lattice matching is achieved with good morphology and crystallinity of the grown layers. One of the most important III-V ternary systems is $Ga_{1-x}Al_xAs$/GaAs and the other is $Ga_{1-x}In_xAs$/InP. The 'AlGaAs' system is unique in that lattice-matching to GaAs is retained for a wide range of x values since the radii of Al (1.43Å) and Ga (1.41Å) are similar. In contrast only one composition of "InGaAs" is lattice-matched to InP. Double crystal X-ray rocking curve analysis confirms that a lattice mismatch a/a_0 of less than 8×10^{-5} may be routinely obtained with a FWHM of < 40 arc/sec for this system. Several device structures including junction field effect transistors (JFETs) and double heterostructure pin diodes have been fabricated with excellent characteristics. Figure 9 gives details of the structure, fabrication, uniformity and characteristics of a typical JFET structure grown by MOCVD [35]. Table 2 summarises some properties of devices that have been fabricated by MOCVD over the past few years.

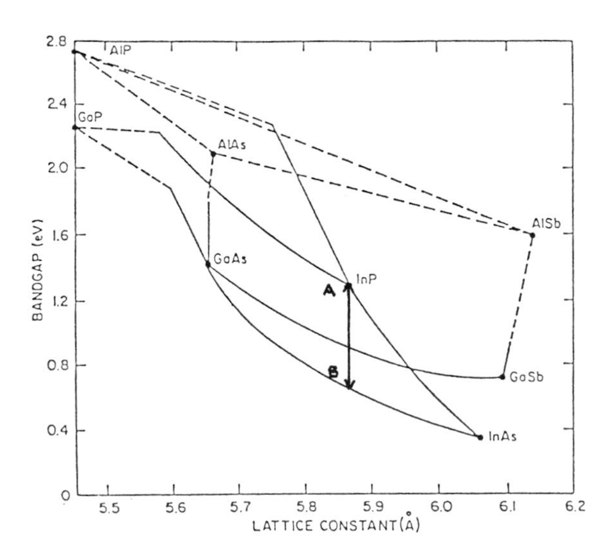

Figure 8. Variation of the band gap as a function of lattice constant for III-V binary and alloy semiconductors.

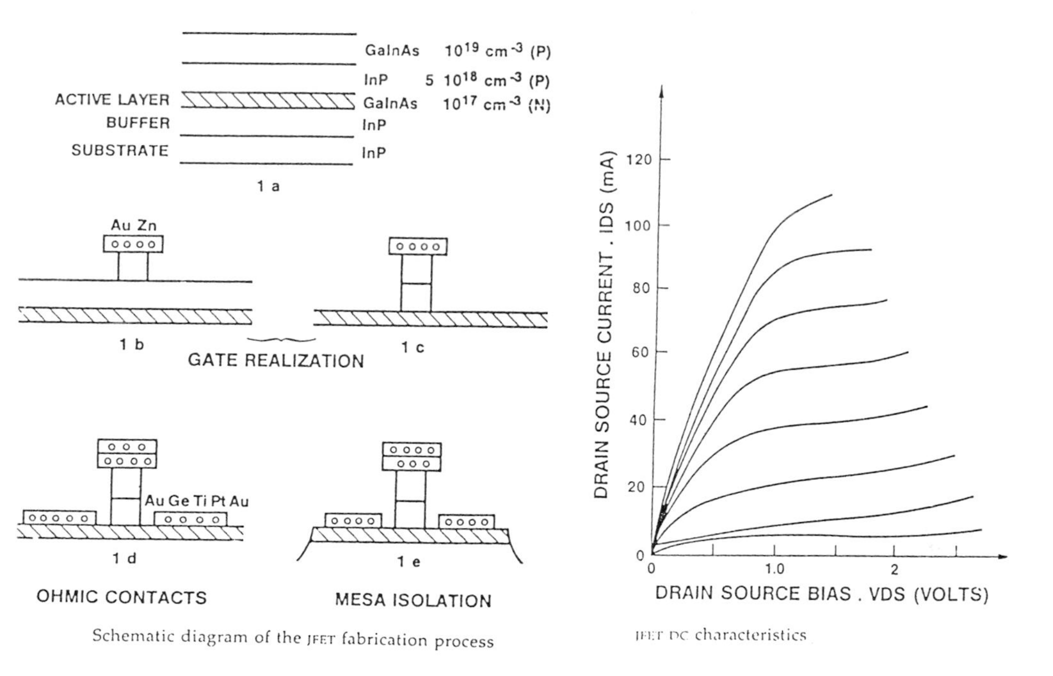

Figure 9. Details of the structure, fabrication and characteristics of a typical JFET structure. (Reproduced by kind permission of Butterworths Scientific Ltd; after Ref. 35).

The quaternary system $Ga_{1-x}In_xAs_{1-y}P_y$ covering the entire composition range has also been prepared by MOCVD [36] and the field has been reviewed recently [6]. There are also important advances presently being made with ternary and quaternary materials containing antimony [37].

Over the past three years, Dupuis and collaborators have made significant advances in the fabrication of low threshold, high efficiency $Al_{1-x}Ga_xAs/GaAs$ double heterostructure injection lasers directly on Si substrates [38]. As mentioned in section 3, this is another example of heteroepitaxy but in this work, the lattice mismatch has been accommodated by the preparation of Si/Ge buffer layers. Four types grown by MBE have been used (i) single Ge layers (1 μm thick) deposited on Si (ii) a single Ge layer grown on Ge/Si strained layer superlattice, (iii) a Ge layer deposited on linearly graded Ge/Si alloy buffer layer, (iv) a combination of (ii) and (iii). It is found that best device performance is found using procedure (i) [39] and this work illustrates the sophistication of layered structures that can be prepared by a combination of MBE and MOCVD.

LOW-DIMENSIONAL STRUCTURES AND DEVICES

Over the past decade there has been considerable progress in the study of low dimensional structures prepared by both MOCVD and MBE. It is not my intention in this review to compare the two techniques and the task of presenting a comprehensive account of 'LDS/LDD' is an impossible one because of the enormity of the subject area. I shall, therefore, be highly selective in my choice of topics and will attempt to address particular materials problems. The interested reader is referred to two recent excellent reviews which cover other aspects [40, 41].

TABLE 2
III-V DEVICE STRUCTURES

1. HEMT — AlGaAs ¦ GaAs. RT transconductance 240 mS mm^{-1} for an 1 x 300 μm gate;charge density of ~ 10^{12} /electrons cm^{-2} or RT mobility of ~ 6500 $cm^2 V^{-1} s^{-1}$.

2. InP Gunn Diodes — N^+ using Si doping.
N^+ - 10^{18} cm^{-3} N - 10^{16} cm^{-3}

F	P(mw)	η/ %
94 GHz	110	3.5

3. InGaAs / InP
   (a) *Junction FET*
   transconductance 260 mS mm^{-1}; microwave gain 10 dB at 18 GHz.

   (b) *Double HET. Pin Diode*
   At 20 V dark current < InA sometimes ~ 1 pA. Chip capacitance 0.2 pF; detection/emission 1000-1600 nm.

4. GaAs / InP
   (a) *Solar Cells*
   Spectral response 500 - 850 nm; efficiency ~ 15%

   (b) *MESFET*
   Metal semiconductor FET.
   GaAs on GaAs 100 $mSmm^{-1}$ }trans
   GaAs on InP 90 " " } conductance
   Cut off freq (input current = output current) = 7.1 GHz.

   (c) *GRINSCH Lasers*

   Graded index separate confinement heterostructure laser; single quantum well of:
   GaAs on GaAs : GaAs on InP. 8 nm GaAs SQW.

   GaAs/InP. Cw operation at RT ~ 15 mA for 10 μm stripe.
   GaAs/GaAs. Cw operation at RT ~ 250 mA cm^{-2}

Initial studies dealt with the "AlGaAs/AlAs" or "AlAs/GaAs" quantum well and multiple quantum well structures. Here GaAs acts as the well material and AlAs or "AlGaAs as the material in the barrier layers. Lasers operational in the red spectral region ca 850 nm are now in commercial use. A particularly interesting structure is that of a graded refractive index separate confinement heterostructure single quantum well laser (GRINSCH SQW laser). The laser consists of a 500 nm GaAs buffer layer, 1 μm $Al_{0.6}Ga_{0.4}As$ cladding, 340 nm graded "AlGaAs" confinement region within 8 nm thick GaAs SQW active layer, 500 nm $Al_{0.6}Ga_{0.4}As$ cladding and 100 nm GaAs contact layer. This type of laser gives CW operation at room temperature with a broad area threshold current density of ca 300 Acm^{-2} a single longitudinal mode operation at 843 nm with a peak width of less than 1 nm [34].

The advent of devices such as high electron mobility transistors (HEMTs) based on a two dimensional electron gas (2 DEG) formed in the "AlGaAs"/GaAs system has further emphasised the dimensionality of these thin layer structures. Figure 10 gives a schematic diagram of a typical heterojunction structure grown by MOCVD. The two dimensional electron gas is formed in the undoped GaAs layer just below the thin ca 15Å undoped "AlGaAs" layer and sustains a high electron mobility in two dimensions. The doping in the upper "AlGaAs" layer is achieved using selenium. A HEMT structure requires good doping uniformity, low carrier concentration in undoped material and sharp interfaces. These requirements can be met by MOCVD. Multiple quantum well structures in InP-based ternary and quaternary materials have also been prepared by MOCVD. Figure 11 shows the actual structure grown and the corresponding photoluminescence spectra for the GaInAsP/InP system. We note that each quaternary well yields a PL emission peak which shifts to lower wavelengths as the well width decreases.

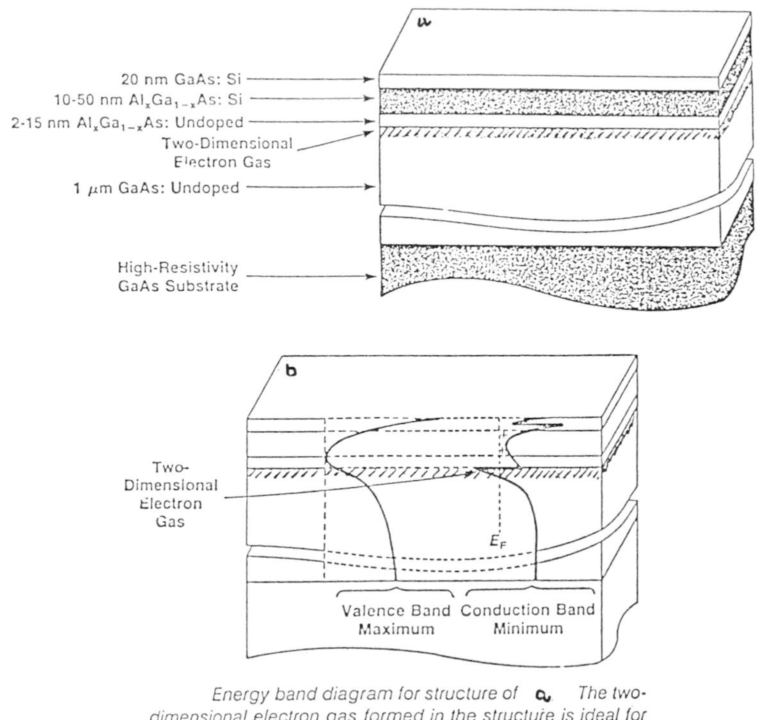

Figure 10. Schematic diagrams of a typical 2DEG heterostructure grown by MOCVD and its energy band diagram.

Examination of Figure 8 allows us to understand the concept of strained layer superlattices (SLS) and pseudomorphic structures. It has been shown that dislocations in ternary materials grown on binary semiconductors can be confined to the interfacial regions when the layers are thin ca ≤ 200Å. Furthermore, the effective lattice-matched composition of the ternary can be achieved by growing successively thin layers of different binary compounds of the required thickness. Thus, the composition $Ga_{0.47}In_{0.53}As$ can be built up of GaAs and InAs layers of appropriate relative thickness. In a similar way, the quaternary of a given composition can be equivalent to two ternaries, neither of which is lattice-matched to the substrate. Typical materials combinations that have been grown successfully by MOCVD are indicated in Table 3. In order to achieve uniform compositions and sharp interfaces, careful control of reactor geometry is required.

NOVEL PRECURSORS FOR III-V GROWTH

Conventional Group III organometallic and Group V hydrides used as precursors for MOCVD suffer severe disadvantages despite the fact that good quality epitaxial layers can be produced in the majority of cases. These disadvantages arise from the toxic and hazardous nature of arsine and phosphine, for example, and the pyrophoric nature of TMGa and TMIn. TMIn suffers further limitations due to its crystalline nature which can severely limit the levels of compositional uniformity that can be obtained in several ternary and quaternary systems.

Another important consideration when preparing complex systems is the relative thermal behaviour of a pair of Group III and/or Group V precursors. Arsine and phosphine, for example,

different rates. Such considerations have prompted several research groups worldwide to seek alternative precursors.

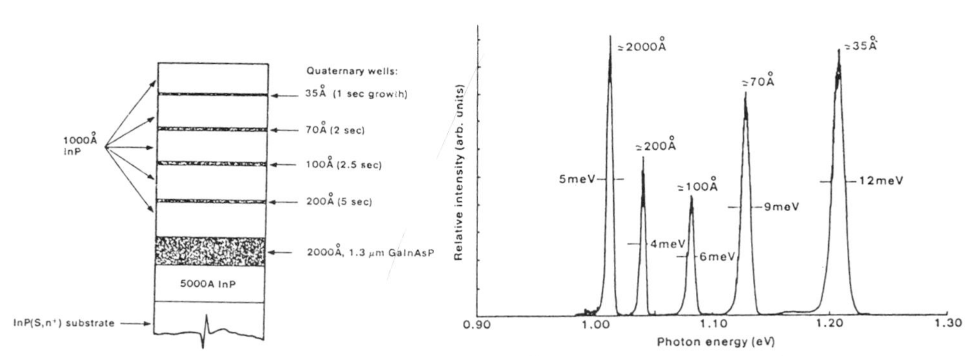

GaInAsP/InP quantum wells grown to assess epitaxy and interface quality

Photoluminescence spectrum of a GaInAsP multiquantum well sample, emission from each of the four different wells and from the bulk epitaxial material can be observed. $T = 4.2$ K, 25 mV, 40 mW, 0.18 W cm^{-2}

Figure 11. An example from the GaInAsP/InP system. (Reproduced by kind permission of Butterworths Scientific Ltd, after Ref. 51).

TABLE 3

Typical Materials Combinations

GaAs	-	$Al_{1-x}Ga_xAs$	InP	-	$In_{1-x}Al_xAs$ - $In_{1-x}Ga_xAs$.
GaAs	-	$In_{1-x}Ga_xAs$.	InP	-	$In_{1-x}Ga_xAs_{1-y}P_y$.
GaAs	-	InAs.	InP	-	InAs - GaAs.
GaAs	-	$GaAs_{1-y}Sb_y$.	GaSb	-	InAs.
GaAs	-	$GaAs_{1-y}P_y$.	GaSb	-	InAs - $AlA_{1-y}Sb_y$.
GaAs	-	$In_{1-x}Ga_xP$ - $In_{1-x}Al_xP$.	GaSb	-	AlSb.
GaSb	-	$In_{1-x}Ga_xAs_{1-y}Sb_y$ - $Al_{1-x}Ga_xAs_{1-y}Sb_y$.			

Table 4 lists alternative precursors for arsine that have been examined in recent years for GaAs growth by MOCVD. With most of these reagents carbon incorporation is a major problem and hitherto only some two precursors, tertiary butyl arsine (TBAs) and phenylarsine (PAS) have shown real promise. Both these are liquids under normal conditions. GaAs with relatively high electron mobilities (ca 40,000 cm^2 V^{-1} s^{-1} at 77K) and low donor concentration for nominally undoped material ($n \leq 1 \times 10^{15}$ cm^{-3}) has been reported [14]. For the case of InP growth, liquid tertiary butyl phosphine (TBP) has also shown promise [15].

TABLE 4
A Selection of Alternative As Precursors used with TMGa to Produce GaAs.

Compound	Vapour Pressure	GaAs n_{77K}/cm^{-3}	Electrical Data $\mu_{77K}/cm^2V^{-1}s^{-1}$
Trimethylarsine (TMAs)	235 Torr at 20 °C	p ~ 10^{17}	-
Triethylarsine (TEAs)	5 Torr at 20 °C	n ~ $5x10^{15}$	13,000
Diethylarsine (DEAs)	0.8 Torr at 18 °C	n ~ $3x10^{14}$	64,000
Ternary butylarsine (TBAs)	96 Torr at -10 °C	n ~ $1x10^{15}$	53,000
Phenylarsine (PAS)	2 Torr at 20 °C	n ~ $1x10^{15}$	40,000

It may be no accident that these liquid precursors with hydrogen atoms linked directly to the Group V element yield epitaxial layers relatively free of carbon impurities in contrast to those precursors that contain no Group V - H bonds, eg trimethyl- and triethyl arsine (TMAs and TEAs respectively). It appears that As-H and/or As-H_2 units are required in order to react with methyl radicals from TMGa to form GaAs without resulting in carbon formation, although final details await mechanistic investigations currently underway in several laboratories. PAS is also being used to grow InAs.

From time to time in the MOCVD field, there has been interest in developing adduct precursors or "magic compounds" where the Group III - Group V bond has already been formed [42]. Thermal dissociation would simply detach side groups leaving behind the III-V binary compounds - a process that would be safe and cheap. However, progress in this area has been disappointing for several reasons. First, the adducts cannot be purified sufficiently; second, they require the presence of other Group III or Group V sources in order to produce stoichiometric material and third, they cannot alone provide ternary and quaternary materials of variable composition.

CRYSTALLOGRAPHY - CONTROLLED GROWTH

The manufacture of the current generation of electronic devices with features on the order of several microns involves complex etching and lithographic processes which are both inefficient and time consuming. Therefore, considerable interest is being shown in local or selected area deposition where the semiconductor material is deposited only where it is required, eg in the active region. There have been several approaches to achieving selected area deposition using MOCVD. One has involved "photo-assisted" deposition using, for example, laser radiation and capitalising on photonic or thermal effects to deposit the desired material. Several review articles have appeared covering these aspects [43, 44]. A much more recent and novel approach is to use precursors selectively to deposit on particular surfaces but not on others. The acronym CODE, compound and orientation dependent epitaxy has been used to describe this technique [48]. Such growth surfaces would vary in their composition and/or structure and crystallography. It is also desirable using MOCVD to overgrow contoured substrates in order to isolate component function and for interfacing. The work of Hersee et al [45], Demeester et al [46], Yoshikawa et al [47], Scott et al [48] and Garrett and Thrush [49] covering both GaAs and InP based systems should be consulted for further details. Much of this work has been concerned with the ability of MOCVD to deposit over mesa type structures and into channels of given crystallography.

Such capabilities would assist in the fabrication of buried heterostructure lasers, waveguides and active/passive waveguide interfaces by single step epitaxy. A common technique to study the behaviour over contoured substrates is to employ a multi-heterostructure growth technique to follow the time evolution of the growing surfaces, ie to grow a superlattice whose individual layer thickness is much less than the total thickness of the grown layer. Although the precise distortion of the growth process introduced by the creation of the superlattice is uncertain, some general preliminary conclusions can be drawn in relation to preference for growth on certain crystallographic planes in the InGaAs/InP and AlGaAs/GaAs systems: (i) InGaAs is reluctant to nucleate on either (111)A or (111)B surfaces, whereas InP grows on all available crystal planes. AlGaAs/GaAs on the other hand nucleate on (111)A surfaces but not on (111)B surfaces. (ii) For both systems nucleation and growth on (100) and (110) surfaces appears to be better than for (111) surfaces. Clearly, the growths are controlled by surface free energies, preferential adsorption, differential sticking coefficients etc and it will be interesting to see the results of controlled experiments involving, for example, different precursors for the Group III elements.

CONCLUSION

In this review, I have attempted to cover certain aspects of MOCVD. What is readily apparent is the tremendous versatility of the MOCVD processes and their applicability to the preparation of a wide range of important materials systems. Thin film epitaxial structures of widely varying but controlled compositions and thicknesses can be prepared with complete certainty and we appear to be confined at present by the limited range and purity for available precursors. We are also beginning to understand the fundamentals of the growth processes that involve complex combination of gas phase and surface phenomena. Progress in this area will remove the empiricism currently dominating our quest for better and novel starting materials. The entire field appears to have an exciting future and should contribute to the development of advanced materials and devices for many years to come.

ACKNOWLEDGEMENTS

I would like to thank members of the UMIST Solid State Chemistry Group and the UK JOERS Consortium on InP-based materials for numerous stimulating discussions. The financial support of the SERC, US Air Force, Thorn EMI CRL, Air Products Ltd and Johnson Matthey Ltd is also gratefully acknowledged.

REFERENCES

1. H.M. Manasveit, Appl. Phys. Letters, 12 [1968] 136.
2. H.M. Manasveit and W.I. Simpson, J. Electrochem Soc., 116, [1969] 1725.
3. a) ICMOVPE - I Ajaccio, France 1981, see J. Cryst. Growth, 55, [1981].
   b) ICMOVPE - II Sheffield, England 1984, see J. Cryst. Growth, 68, [1984].
   c) ICMOVPE - III California, USA 1986, see J. Cryst. Growth, 77, [1986].
   d) ICMOVPE - IV Hakone, Japan 1988, see J. Cryst. Growth, 93, [1988].
4. P.D. Dapkus, J. Crystal Growth, 68, [1984], 345.
5. P.H. Manuel, M. Defour, C. Grattepain, F. Omnes, O. Archer, G. Timms and M. Razeghi, Chemtronics, 4, [1989], 40.
6. The MOCVD Challenge, Vol. 1, M. Razeghi and Adam Hilger, Bristol, 1989.
7. H. Sasaki, M. Tanaka and J. Yoshino, Jap. J. Appl. Phys., 24, [1985], 417.
8. T. Fukui and H. Saito, Jap. J. Appl. Phys., 24, [1985], L774.
9. R. Luckerath, P. Balk, M. Fischer, D. Grundmann, A. Hertling and W. Richter, Chemtronics, 2, [1987], 199.
10. J.O. Williams, R.D. Hoare and M.J. Parrott, Phil. Trans. Roy. Soc. (Lond). 1989, in press.
11. B.A. Joyce in Advance Crystal Growth, ed. P.M. Dryburgh, B. Cockayne and K.G. Barraclough, Prentice Hall, [1987], 337.
12. A.H. Moore, M.D. Scott, J.I. Davies, D.C. Bradley, M.M. Faktor and H. Chudzynska, J. Cryst. Growth, 47, [1989], 15.
13. A.C. Jones, Chemtronics, 4, [1989], 15.
14. G.T. Muhr, D.A. Bohling, T.R. Omsbead, S. Brandon and K.F. Jensen, Chemtronics, 4, [1989], 26.
15. G.B. Stringfellow, J. Electronic Materials, 17(4), [1988], 327.
16. A.C. Jones, G. Wales, P.J. Wright and P.E. Oliver, Chemtronics, 2, [1987], 83.
17. N.D. Gerrard, D.J. Nicholas, J.O. Williams and A.C. Jones, Chemtronics, 3, [1988], 17.
18. N. Hunt and J.O. Williams, Chemtronics, 2, [1987], 145.
19. A.T.T. Briggs and B.R. Butler, J. Cryst. Growth, 85, [1987], 535.
20. T.F. Kuech, E. Veuhoff and B.S. Meyerson, J. Cryst. Growth, 68, [1984], 48.
21. J.M. Boud, M.A. Fisher, D. Lancefield, A.R. Adams, E.J. Thrush and C.G. Cureton, InP Conf. Series No 91, [1987], 801.

22. See e.g. R.J.M. Griffiths in Abstracts of 3rd European Workshop on MOVPE Montpellier, France, 5-7 June 1989, p.16.
23. R.W. Glew, J. Cryst. Growth, 68, [1984], 44.
24. T.F. Kuech, B.S. Meyerson and E. Venhoff, Appl. Phys. Letters, 44 [1984], 986.
25. B.T. Cunningham, M.A. Haase, M.J. Collum, J.E. Baker and G.E. Stillman, Appl. Phys. Letters, 54, [1989], 1905.
26. N. Bottka, R.S. Sillmon and W.F. Tseng, J. Cryst. Growth, 68, [1984], 54.
27. J.D. Parsons and F.G. Krajenbrink, J. Cryst. Growth, 68, [1984], 60.
28. N.R. Dennington and J.O. Williams, to be published.
29. R.W. Kaliski, C.R. Ho, D.G. McIntyre, M. Feng, K.B. Kim, R. Bean, K.Zznio and K.C. Hsieh, J. Appl. Phys., 64, [1988], 1196.
30. J.O. Williams, Chemtronics, 2, [1987], 43.
31. H.M. Yates and J.O. Williams, Appl. Phys. Letters, 51, [1987], 809.
32. J.E. Potts, H. Cheng, S. Mohapatra and T.L. Smith, J. Appl. Phys., 63, [1987], 333.
33. A.H. Reddoch, D.J. Northcott, J.M. Park and J.O. Williams, unpublished results.
34. P. Demeester, A. Ackalrt, M. Van Ackere, F. DePestel, C. Eckhout, Y. Grigase, D. Lootens, I. Moerman, G. Vanden Bossche, R. Baets, M. Bottle, P. Van Daele and P. Lagasse, Chemtronics, 4, [1989], 44.
35. M.A. diForte-Poisson and C. Brylinski, Chemtronics, 4, [1989], 3.
36. P.L. Maurel, M. DeFour, D. Grattepain, F. Omnes, O. Acher, G. Timms, M. Razeghi and J.C. Portal, Chemtronics, 4, [1989], 40.
37. i) S.K. Haywood, private communication.
    ii) S.K. Haywood, A.B. Henriques, D.F. Howell, N.J. Mason, R.J. Nicholas and P.J. Walker, Inst. Phys. Conf. Ser., 91, [1988], 271.
38. R.D. Dupuis, J.P. Van der Ziel, R.A. Logan, J.M. Brown and C. J. Pinzone, Appl. Phys. Letters, 50, [1987], 407.
39. R.D. Dupuis, J.C. Bean, J.M. Brown, A.T. Macrander, R.C. Miller and L.C. Hopkins, J. Electron, Materials, 16, [1987], 69.
40. J.W. Orton, Chemtronics, 3, [1988], 129.
41. M.J. Kelly in "The Physics and Fabrication of Microstructures and Microdevices" ed. M.J. Kelly and C. Weisbuch, Springer Verlag Berlin, 1986, p.174.
42. A.H. Cowley, B.L. Benac, J.G. Ekerdt, R.A. Jones, K.B. Kidd, J.Y. Lee and J.E. Miller, J. Ann. Chem. Soc., 110, [1988], 628.
43. W.E. Johnson and L.A. Schlic, Appl. Phys. Letters, 40(a), [1982], 798.
44. D.J. Ehrlich, J.Y. Tsao and C.O. Bozler, J. Vac. Sci. Technol, 83(1), [1985], 1.
45. S.D. Hersee, E. Barbier and R. Blondeau, J. Cryst. Growth, 77, [1986], 310.
46. P. Demeester, P. Van Daele and R. Baets, J. Appl. Phys., 63, [1988], 2284.
47. A. Yoshikawa, A. Yamamoto, M. Hirose, T. Sugino, G. Kana, I. Teramoto, IEEE. J. Quantum Electron., QE-23(6), [1987], 725.
48. M.D. Scott, J.R. Riffat, I. Griffith, J.I. Davies and A.C. Marshall, J. Cryst. Growth, 93, [1988], 820.
49. B. Garrett and E.J. Thrush, J. Crystal Growth, [1989] in press.
50. W. Richter and L Hunermann, Chemtronics 2, [1987], 175.
51. B.R. Butler, A.T.R. Briggs, E.J. Thrush, B. Garrett and J.P. Stagg, Chemtronics 3, [1988], 31.

Growth of Thin Films and Heterostructures of III–V Compounds by Molecular Beam Epitaxy

C.T. FOXON AND B.A. JOYCE

1. INTRODUCTION

The technique which has become known as Molecular Beam Epitaxy (MBE) is, at its simplest, a refined form of vacuum evaporation. The molecular beams are produced by evaporation or sublimation from heated liquids or solids contained in crucibles. The flux produced is thus determined by the vapour pressure of the element or compound in the MBE source. At the pressures used in MBE equipment, collision-free beams from the various sources interact chemically on the substrate to form an epitaxially related film. Ultra-high vacuum (UHV) techniques are used to reduce the pressure of gases from the ambient background and thus improve the purity of the layers. More recently gas sources mounted outside the equipment have been employed in what are known respectively as gas source MBE (GS-MBE) - metals plus group V hydrides, metalorganic MBE (MOMBE) - metalorganics plus conventional group Vs and chemical beam epitaxy (CBE) - all gas sources.

MBE began as a basic study of the chemical reactions occurring on surfaces during the growth of III-V compounds but quickly evolved into a practical method for the growth of high purity materials. The ability to start and stop a molecular beam in less than the time taken to grow a single atomic or molecular layer has led to the ability to produce complex multilayer structures. The use of UHV technology has enabled the physical and chemical properties of the films to be measured in-situ using reflection high energy electron diffraction (RHEED) and Auger Electron Spectroscopy (AES). Modulated beam mass spectrometry (MBMS) was developed to study the chemical processes involved and the dynamics of film growth have been investigated using the so-called RHEED oscillation technique which can measure the growth rate in-situ, a unique feature of MBE.

In this series of lectures we will review the fundamental aspects of MBE using mostly examples from the best understood system (Al,Ga)As. We will also discuss the techniques required to grow high purity samples such as high mobility two-dimensional electron gas structures (2DEGs), multi-quantum-well (MQW) structures and superlattices (SL). Finally we will examine the problems associated with dopant incorporation.

2. BASIC TECHNOLOGY OF MBE

Growth rates in MBE are typically about 1 monolayer (ML) per second which is equivalent to a pressure of about 10^{-6} Torr arriving at the sample. For any reasonable purity material to be grown by MBE it is essential to maintain the pressure of unwanted impurities as low as possible (this point will be discussed below in relation to the growth of high mobility structures). MBE requires, therefore, conventional UHV techniques and the apparatus is always constructed to the highest standards of cleanliness. An additional requirement of MBE is that the pressure in the system be low enough to ensure that no gas phase collisions occur. Thus homogeneous reactions, which can occur in MOCVD processes, are entirely avoided and the process is determined entirely by heterogeneous reactions on the substrate surface.

The pumping techniques depend to some extent on the materials being used. In conventional MBE where the group III and V elements are supplied from solid or liquid sources in the MBE equipment, the usual combination employed consists of ion, titanium and closed cycle helium pumps. Following bakeout at about 180°C (limited by the vapour pressure of the group V elements) pressures in the 10^{-11} Torr range are normally obtained. For GSMBE, MOMBE or CBE where the group III or V elements are supplied from gaseous sources an appropriately trapped diffusion or turbomolecular pump may be more suitable to handle the increased gas load on the system. Base pressures in the same range can still be achieved but the partial pressure of hydrocarbons may be somewhat higher.

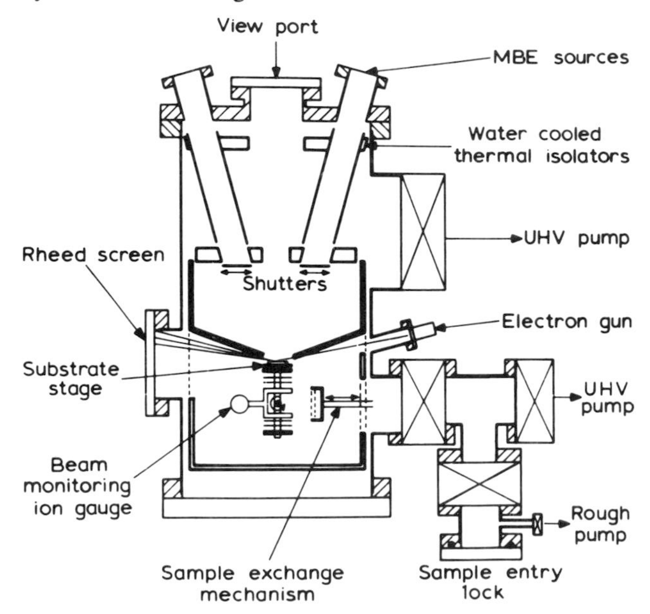

Figure 1: Schematic diagram of an MBE system

The general arrangement of an MBE system is shown schematically in Figure 1. A critical feature is the extensive cryopanelling surrounding both the substrate station and the evaporation sources. The low temperature (usually 77K) reduces the arrival rate of unwanted species and provides heat dissipation for both the evaporation sources and the substrate heater.

The evaporation sources, sometimes referred to as Knudsen cells, have a crucible to contain the material. This is now generally made from pyrolytic Boron Nitride (BN), also a III-V compound, but occasionally graphite is also used. Figure 2 shows a true Knudsen source. In this case the crucible has only a small recessed orifice from which the material effuses. This is necessary to ensure that equilibrium vapour pressure is established in the crucible which can be related to the temperature of the source. The cell temperature can be accurately determined by recessing the thermocouple into the crucible taking care to ensure that the heat loss by conduction via the thermocouple leads is negligible compared to the heat received. Spot welding the thermocouple to a large radiation collector can assist in fulfilling this condition. Surrounding the crucible is a heater usually made of Ta wires with BN or Al_2O_3 insulators. Outside the heater are multiple Ta radiation shields. Typically such a source can be heated to 1200°C using quite modest power (about 150 W).

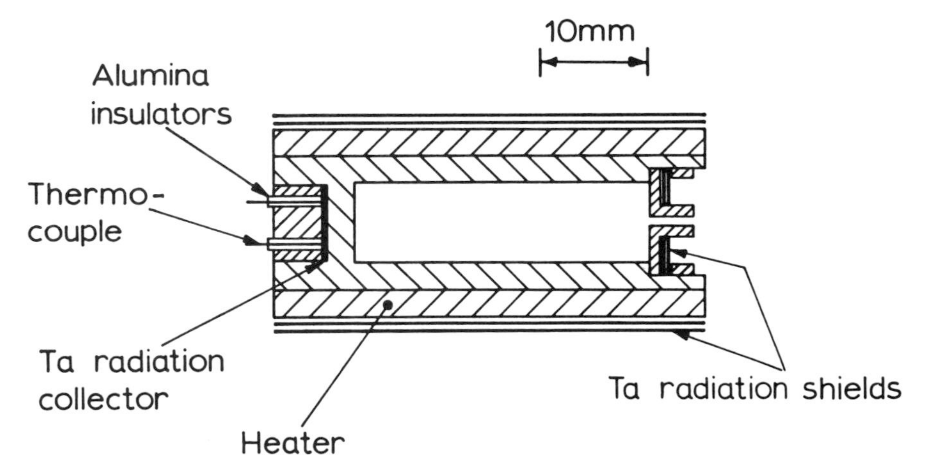

Figure 2: Schematic diagram of a true Knudsen source

From a true Knudsen source with a simple orifice the flux of molecules reaching the sample F, is given by

$$F = 1.118 \times 10^{22} P \times A / [L^2 (M \times T)^{0.5}] \qquad (1)$$

where P is the pressure in the source in Torr, A is the area of the orifice in cm^2, L is the distance from the orifice to the substrate in cm, M is the molecular weight and T the absolute temperature. This relation is obeyed until the point where the mean free path becomes comparable to or less than the diameter of the orifice. Above this pressure (temperature) there is a transition from molecular flow to viscous flow. In conventional MBE this limit is never exceeded but in GSMBE and CBE it is closely approached for the group V sources.

In a practical MBE system a true Knudsen source is seldom used, since by increasing the size of the orifice it is possible to reduce the melt temperature required to give a particular flux at the substrate. This in turn will reduce the power consumption and minimise the number of unwanted impurities associated with outgassing of the cell and its surroundings. In such a source there is no unique relation between temperature T and flux F but from RHEED measurements we can directly determine the fluxes of the group III and V elements (see below). The prime requirement is for a stable reproducible flux and typically over an 8 hour day less than 1% drift is observed with day to day variations of 2 to 5%. Increasing the orifice diameter reduces the pressure within the source and therefore increases the mean free path for the molecules, this ensures molecular flow is maintained.

To start and stop the molecular beams a simple mechanical shutter can be used, typically this can operate in a time short (0.1 to 0.3s) compared to the time taken to deposit a ML (1s). Two practical problems arise; the first comes from the need to operate the shutter many thousands of times for superlattice samples leading to occasional vacuum (bellows) failures; the second point is that the shutter is a heat shield for the source and flux transients associated with the opening of the shutter are often observed. Both lead to problems in growing short period MQW or SL samples.

To obtain crystalline material of adequate quality it is usually essential to heat the substrate to a temperature of between 400 and 700°C. This is usually achieved using a heater which remains within the vacuum system. In some MBE systems the sample is attached to a platen using liquid In and the platen is heated by radiation, but with more recent designs the sample is heated radiatively. III-V compounds are, however, transparent to radiation below the bandgap and this therefore leads to poor coupling.

Since the fluxes from the various MBE sources come from different directions their distributions across the substrate will differ and this can lead to films of non uniform thickness and composition. In practice this can be overcome by rotating the substrate during growth, ideally once per monolayer of material deposited. This is simple in principle but difficult to achieve in practice because no conventional lubricant can be used on the various bearings. The bellows motion usually used to transmit the rotation has also to be capable of many million rotations. One of the major advantages of MOMBE and CBE may be that a uniform flux can be obtained over a large area without the need to rotate the sample.

Sample introduction is via a two or three chamber system. In the first chamber the pressure is reduced to about 10^{-4} Torr using sorption pumps. These are then isolated and an ion pump or small cryopump is used to reach $<10^{-6}$ Torr. The sample is then transferred to the second part of the interlock in which water vapour is removed by heating the sample to about 400C, the limit is set by the temperature at which the III-V compound begins to decompose. After about one hour the pressure in this second chamber is usually below 2×10^{-10} Torr. At this stage the sample can be safely introduced into the growth chamber without degrading the vacuum quality.

In practical MBE systems the diagnostics used are relatively simple. A quadrupole mass spectrometer (QUAD) is needed to ensure that the correct UHV conditions have been achieved. Leak checking before and after the system bakeout is essential and if a QUAD is mounted in the preparation chamber it can be used to ensure that all the water vapour is removed from the sample before transfer to the growth chamber.

A RHEED facility in the growth chamber is also used to determine that the native oxides have been removed before growth and to determine the growth rates for the various binary compounds (see below). This is usually carried out on a small monitor slice kept in the preparation chamber. The RHEED system consists of a 10 to 15 keV electron gun on one side of the chamber with a phosphor screen on a window on the other side. The detection system will be discussed below.

The final item found in most MBE systems is a beam monitoring ion gauge which can be placed in the sample position to measure the intensity of the molecular beams. The absolute sensitivity of the gauge depends to some extent on the materials being deposited on the collector and so cannot easily be used in a quantitative way, but the relative sensitivity to different species is usually maintained and it is a useful tool where RHEED measurements cannot easily be employed.

3. FUNDAMENTAL ASPECTS OF MBE

Several experimental and theoretical investigations have contributed to our present understanding of the processes controlling the growth of films and dopant incorporation in MBE: surface chemical processes have been studied using MBMS methods (Foxon et al 1974), the dynamics of film growth have been examined using the RHEED technique (Joyce et al 1988) and Monte-Carlo (M-C) simulations of growth have added to our knowledge of the factors influencing growth and interface roughness (Madhukar and Ghaisas 1988). In addition thermodynamic calculations have shown fundamental limitations involved in using certain dopants and the factors governing the incorporation of unwanted impurities (Heckingbottom 1985).

MBMS studies

The principle of MBMS is illustrated in Figure 3. By mechanically modulating the molecular beams arriving at, or leaving, the sample surface we are able to distinguish "real" signals in the mass spectrometer from "false" ones arising due to background gases in the vacuum system in the vicinity of the ioniser. By modulating desorbing fluxes (case (a) shown in Figure 3) we are able to determine desorption rates, sticking coefficients, thermal accommodation coefficients of surface species and the order of chemical reactions. If instead we modulate the incident beam (case (b) in Figure 3) we can also determine surface lifetimes and from the temperature dependence deduce binding energies of adsorbates. From such data we can build up a fairly complete picture of the chemistry of the growth process. It is important to note that any species which exist only on the surface will not be detected by this technique, which could be a problem in applying these methods to the study of MOMBE or CBE.

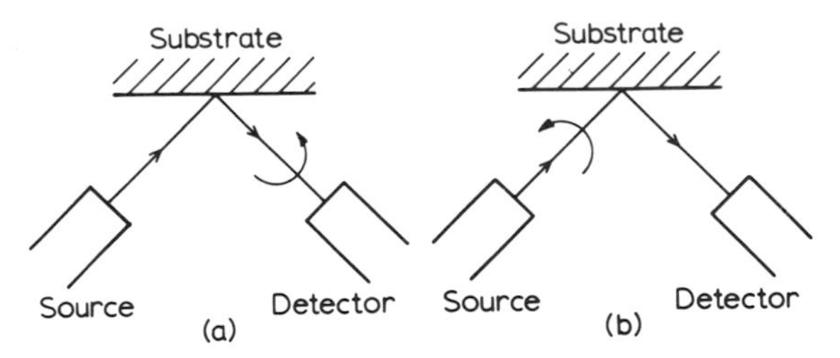

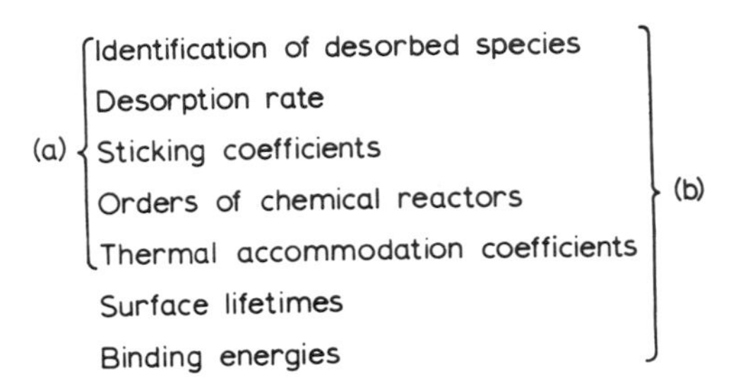

Figure 3: MBMS principle

The type of experimental arrangement used is illustrated schematically in Figure 4. It is a conventional UHV MBE equipment with additional facilities for mechanically modulating the incident and desorbing beams (1 to 100Hz); the detector (Quadrupole Mass Spectrometer - QUAD) is arranged as a single pass density detector, i.e. any molecule not being ionised is condensed on the 77K panel. RHEED and AES facilities may also be included.

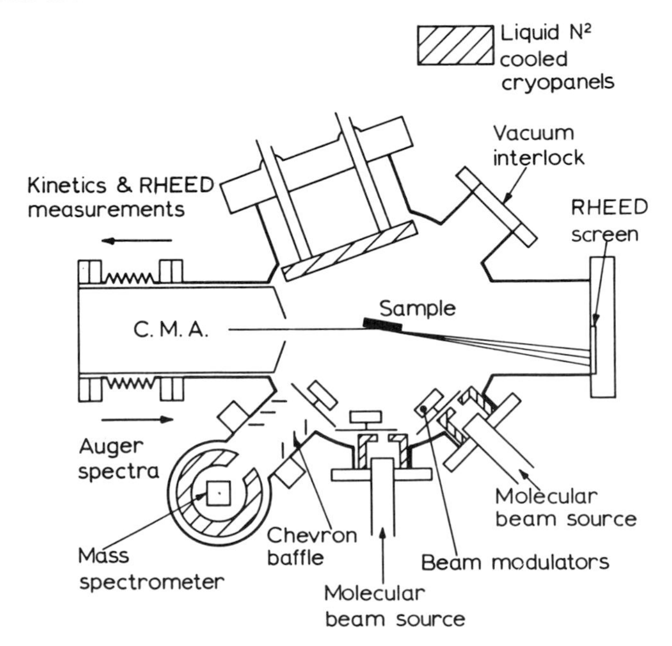

Figure 4: MBMS system

We will consider as a typical example the reaction features involved in the growth of GaAs films on GaAs (001) substrates, for incident beams of Ga with either As_2 or As_4 (Foxon and Joyce 1975, 1977).

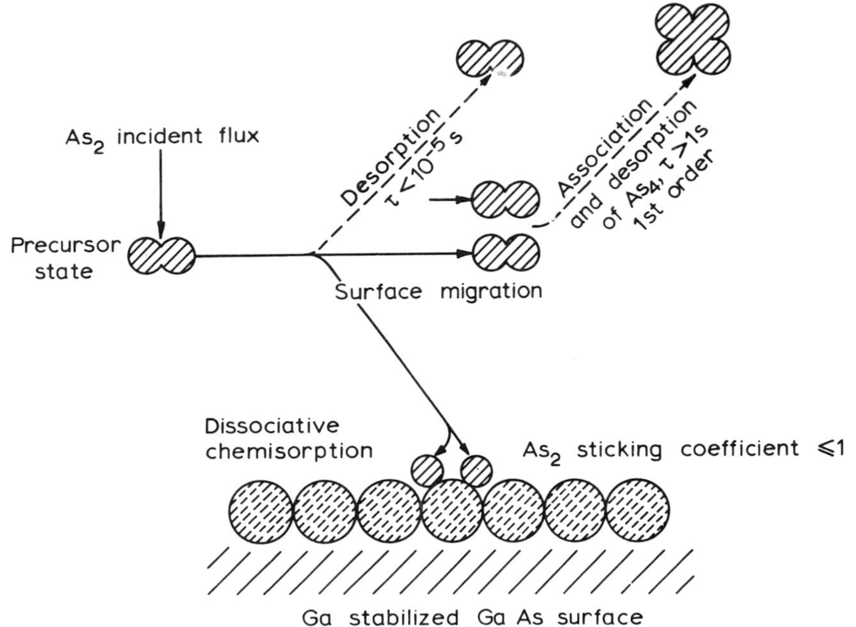

Figure 5: Model of the growth chemistry of GaAs from molecular beams of Ga and As_2

The interactions of an As_2 flux and gallium on a GaAs substrate are summarised in the growth model shown in Figure 5. The As_2 molecules are first adsorbed into a mobile, weakly bound precursor state. The surface residence time of molecules in this state is less than 10^{-5}s. The basic process for As incorporation during thin film growth is a simple first-order dissociative chemisorption on surface Ga atoms. The maximum value of the As_2 sticking coefficient is unity, which is obtained when the surface is covered with a complete monolayer of Ga atoms (as in Figure 5). At relatively low substrate temperatures (< 600K) there is an additional association reaction to form As_4 molecules. These desorb very slowly by a first order reaction. At substrate temperatures above 600K some dissociation of GaAs may occur.

By contrast, the reaction mechanism with an incident As_4 flux is significantly more complex. The model for this is shown in Figure 6. Here the As_4 molecules are first absorbed into a mobile precursor state. The migration of the adsorbed As_4 molecules has an activation energy of about 0.25 eV. The surface residence time is temperature dependent, from which a desorption energy of about 0.4eV may be determined. A crucial finding is that the sticking coefficient of arsenic never exceeds 0.5, even when the Ga flux is much higher than the As_4 flux or when the surface is completely covered with a monolayer of Ga atoms (as in Figure 6). This is explained by assuming a pairwise dissociation of As_4 molecules chemisorbed on adjacent Ga atoms. This second-order reaction is the key feature of the thin-film growth of GaAs with an As_4

flux. From any two As_4 molecules four As atoms are incorporated in the GaAs lattice, while the other four desorb as an As_4 molecule.

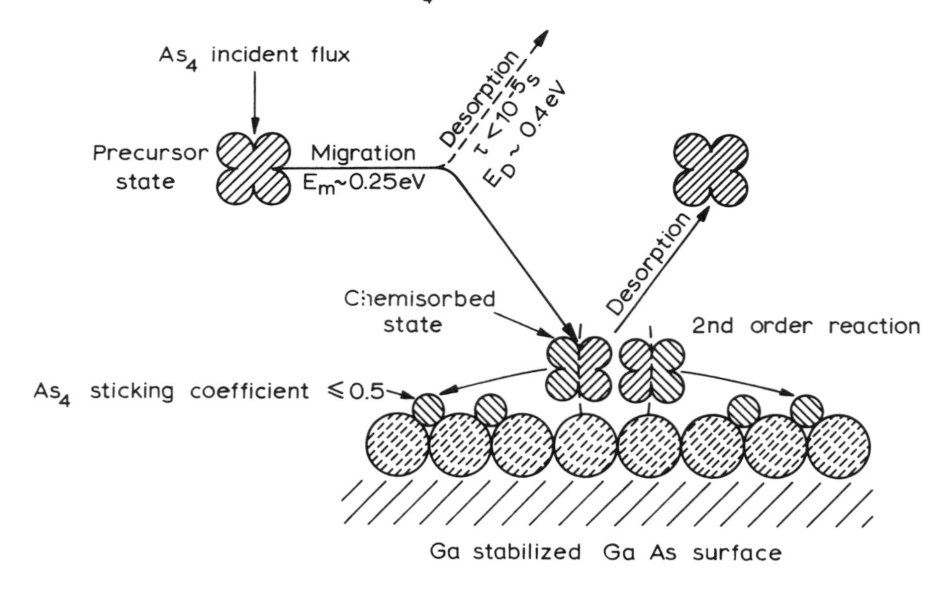

Figure 6: Model of the growth chemistry of GaAs from molecular beams of Ga and As_4

The different chemistry is reflected in the properties of GaAs films grown from As_2 and As_4. In general, for equivalent growth conditions As_2-grown films have lower deep level concentrations and better optical (photoluminescence and interface recombination velocity) properties than those grown from As_4.

Although well over ten years old, these models are still the accepted versions and have in fact received some recent experimental support. Tsao et al (1988) measured changes in sticking coefficient of As_4 (not the As incorporation rate, as they claimed) by a non-modulated beam desorption method which relied on extensive cryopumping to separate beam from background signals in the mass spectrometer. (The As incorporation rate involves the formation of Ga-As bonds in the proper lattice site and this cannot be measured mass spectrometrically). They found no temporal oscillatory behaviour in the signal, which is consistent with the measured chemical process being fast compared with the incorporation rate of As into the lattice via layer-by-layer growth (We will return to this point for a more detailed discussion when we consider growth dynamics). They found, however, in agreement with Foxon and Joyce (1975), a maximum As_4 sticking coefficient of 0.5 and were also able to establish that the surface reconstruction changed from (001)-2x4 to (001)-4x2 just when the sticking coefficient reaches this maximum value, i.e. the surface stoichiometry becomes more Ga-rich.

Harbison and Farrel (1988) have proposed totally speculative configurational models of chemisorption sites and structures at various stages in the growth of a monolayer of

GaAs-(001) from elemental beams. The models are based on consideration of the electron occupancy of surface bonds, but since they have no experimental justification and in fact conflict with existing data obtained by RHEED and angle resolved photoemission spectroscopy (Larsen et al 1982) we prefer not to include them here.

Kinetic effects in alloy growth

For alloys with mixed Group III elements (i.e. III_aIII_bV), reactions with the Group V element are identical to those observed in binary compound growth, so we are only concerned with those factors which influence the Group III ratio. We should note, however, that the thermal stability of the alloy, once produced, is determined by the less stable of the two binary compound end members from which it is formed. In (Al,Ga)As, for example, the thermal stability of GaAs is the limiting factor. This is beautifully demonstrated by the work of Kawai et al (1986), who showed that the Langmuir (free) evaporation of (AlGa)As at 750°C proceeds by the selective evaporation of GaAs until four monolayers of AlAs have accumulated at the surface, when sublimation ceases.

At low growth temperatures, the alloy composition is determined only by the flux ratio, since the sticking coefficients are effectively unity. Where growth is carried out at higher temperatures, which is often necessary to obtain optimum electrical or optical properties, this is no longer the case. The temperature dependence of their sticking

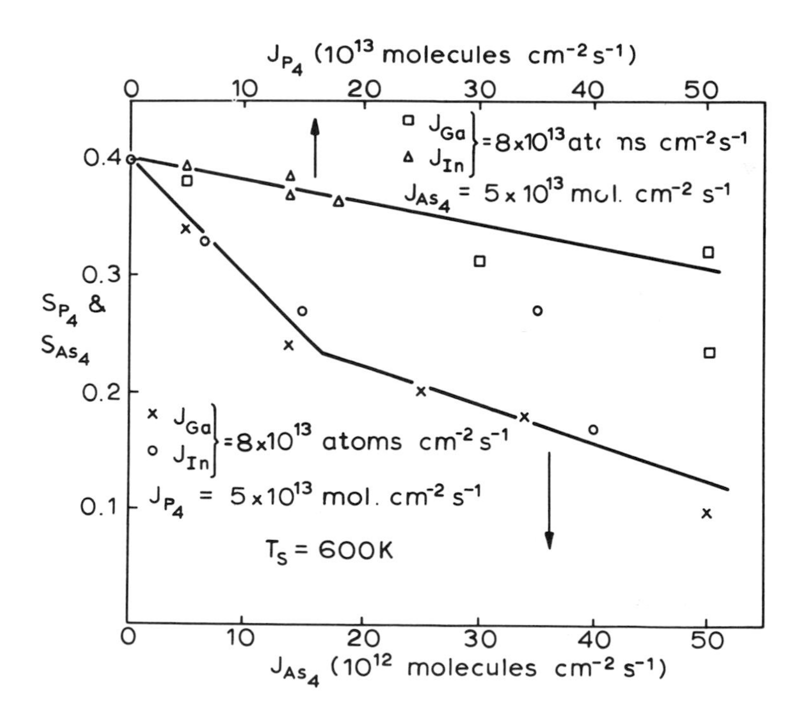

Figure 7: Sticking coefficient of $As_4(P_4)$ as a function of the $P_4(As_4)$ flux at constant fluxes of Ga or In at T_s=600k

coefficients relates closely to that of their vapour pressure, which can therefore be used as an adequate guide to composition control under conditions which correspond to the Group III-rich liquidus. Contrary to some literature reports, the presence of one element does not significantly influence the sticking coefficient of the other during growth.

For mixed Group V element alloys (i.e. III V_x V_y) the situation is much more complex. There is no simple relationship between flux (either of dimers or tetramers) and film composition, because one element is always adsorbed much more readily than the other, the order being Sb>As>P (Chang et al 1977, Foxon et al 1980). This is illustrated in Figure 7 for As_4 and P_4 and is shown as the influence of one molecule on the sticking coefficient of the other, but no serious attempt has been made to understand the mechanism. To control film composition the strongly adsorbed component is supplied in the required ratio to the Group III element and the weakly adsorbed component is provided in excess. The method can break down at high temperature, however, where in, say, Ga(As,P), As_2 is lost preferentially by thermal dissociation but tends to be replaced by phosphorus, which is present in excess (Woodbridge et al 1982).

Growth from gaseous compound sources - MOMBE and CBE

One of the more significant changes in technology is the use of gas sources for the preparation of III-V compounds, referred to as chemical beam epitaxy (CBE) or metal-organic molecular beam epitaxy (MOMBE) or gas source molecular beam epitaxy (GSMBE). Advantages include non-depleting sources which can be changed or recharged from outside the UHV system; no heat source in the growth chamber other than the substrate; uniform layer thickness and doping levels without substrate rotation and reduced surface defect density.

At present, the sources used are Group V hydrides, i.e. PH_3 and AsH_3, with Group III alkyls, principally trimethyls and triethyls. The major disadvantage of these materials is their high toxicity and pyrophoricity, so considerable attention is being given to the preparation and purification of alternative precursors. These include compounds which contain both the Group III and Group V atoms in the same molecule and in the correct ratio to produce stoichiometric films, while also being volatile and non-toxic.

Although very high quality material has been prepared from alkyl and hydride sources, with good control of layer thickness and doping precision, little or no work has been carried out on the chemistry of the growth process, which is clearly much more complex than that occurring with elemental sources. Experimental approaches have all relied simply on measurement of the growth rate as a function of substrate temperature and incident fluxes. Such information is useful for empiric process control, but provides little insight into reaction mechanisms and pathways.

Several underlying concepts can be emphasised however, the most important of which is that since growth occurs in the molecular flow regime, there are no gas phase collisions; the beams impinge directly on to the substrate. This is quite different from the metal-organic chemical vapour deposition (MOCVD) method, where gas phase reactions have been claimed to play an important part, even in the so-called low pressure regime, since flow is still viscous. Some very recent results by Aspnes et al (1988) cast serious doubts on the extent of gas phase decomposition during atmospheric pressure MOCVD. They used in situ reflectance difference spectroscopy to follow the surface coverage of TMG and found the kinetic limits to growth to be determined by surface site availability and subsequent decomposition of TMG. Their experimental approach precluded the possibility of a gas phase component of the decomposition reaction, but the results were nevertheless in excellent agreement with literature values where no attempts were made to restrict possible homogeneous reactions. It is therefore not clear to what extent an understanding of one process implies any understanding of the other.

A second, related factor is that the hydrides do not appear to have adequate dissociation rates on the substrate surface and it is necessary to predissociate them so that they impinge predominantly in the form of elemental dimers. The alkyls, however, require no predissociation and are allowed to impinge directly; nor do they require the presence of Group V species to enable them to dissociate, but the rate is temperature dependent. As determined only from film growth rate measurements, the dissociation rate increases to a maximum with increasing temperature and then decreases rather rapidly as the temperature is further increased. This appears to be generally true, but the exact shape of the curve is dependant on the particular compounds involved.

A third pointer to possible mechanisms is the dependence of carbon incorporation into the growing film on the particular alkyl used. In the case of GaAs, it is well-established that very much higher concentrations of carbon are incorporated from trimethyl gallium (TMG) than from triethyl (TEG). It is assumed that the use of TEG allows the possibility of a reaction pathway for surface decomposition of the metal triethyl via ß-hydrogen elimination to form ethene, which of course cannot occur with TMG.

Robertson et al (1988) have attempted to model the flux and temperature dependence of the growth rate of GaAs from beams of TEG and As_2 (derived from AsH_3). In the absence of any experimental information on surface or desorbing species, surface reaction kinetics or mechanisms, they assume the absorbed molecules to be mono-, di- and triethyl gallium, together with ethyl radicals and the rate limiting step is supposed to be cleavage of the second ethyl-gallium bond. ß-hydrogen elimination from adsorbed ethyl groups to form ethene is also assumed, as is Arrhenius behaviour of rate constants for the elementary reaction steps. The observed non-linear dependence of growth rate on flux is then associated with a second order recombination of adsorbed diethyl gallium with an ethyl radical, followed by desorption of triethyl gallium. The high temperature decrease in growth rate is ascribed to desorption of diethyl gallium. Such speculative approaches area clearly of limited value; kinetic and spectroscopic measurements are essential if any real progress is to be made in understanding the complex surface processes involved in MOMBE (CBE).

4. SURFACE STUDIES

MBE is a unique thin film growth process in that it is UHV based and therefore compatible with a wide range of surface chemical and physical techniques. It also allows in-situ growth of a number of structures not otherwise accessible.

Crystallography and Composition

In general semiconductor surfaces either relax by bond rotation or reconstruct, i.e. display a lower symmetry than that produced by a simple termination of the bulk lattice. In III-V compounds the form of the reconstruction is related to surface stoichiometry. We will treat GaAs(001) as a typical example. It is a polar orientation and the surface is ideally terminated by a complete layer of Ga or As atoms (i.e. it is composed of alternating layers of covalently bonded Ga and As atoms, Figure 8, but a wide range of stoichiometry and structure can occur. The different structures in order of decreasing Ga surface concentration are C(8x2), (otherwise 4x2), 3x1 and 2x4 (otherwise C(2x8)), with high temperatures and high Ga fluxes favouring a high Ga surface population. A particular reconstruction can exist over a moderately wide surface composition range ($\approx$ 10%), but there is an additional structure, the C(4x4), formed by chemisorption of excess arsenic, which has a much wider range, with As present at well over monolayer coverage.

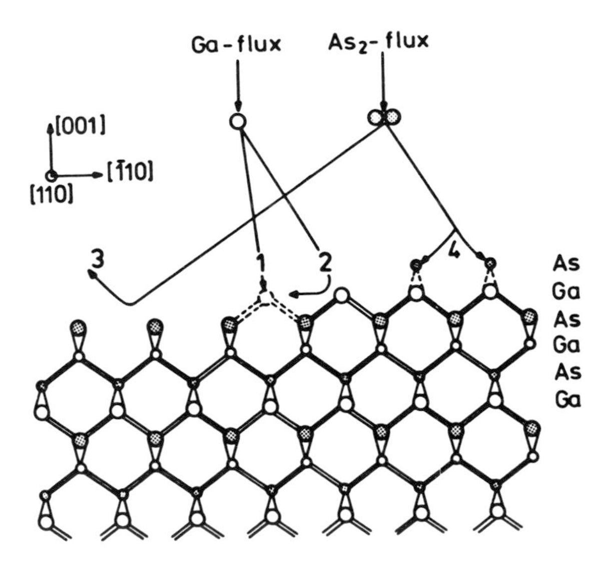

Figure 8: Schematic model of a GaAs (001), showing alternate layers of Ga and As build-up of surface stoichiometry during growth.

Surface crystallography is most conveniently monitored by electron diffraction techniques. Low-energy diffraction (LEED), which uses a back-scattering geometry, has been extensively used for work on cleaved surfaces, but is not really suitable in combination with MBE. For this, the forward scattering geometry of RHEED is more appropriate, since the electron beam is at extreme grazing incidence, whereas the molecular beams impinge almost normally on the substrate. Although the same basic information is obtained from both methods, a combination of RHEED and MBE enables the surface crystallography to be monitored even under dynamic conditions when the surface is growing.

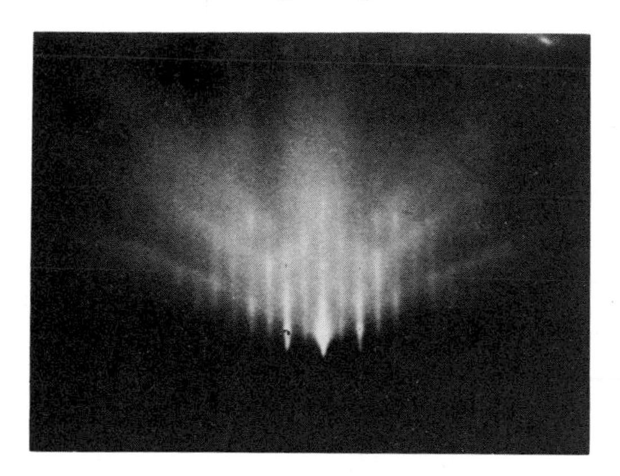

Figure 9: RHEED patterns from two orthogonal <110> azimuths of the GaAs (001) - 2x4 reconstructed surface

As an example of the information which can be obtained from RHEED we consider the GaAs(100) surface. Figure 9 shows the RHEED patterns in two-orthogonal <110> azimuths from a 2x4 reconstructed surface, one with a twofold and the other with a fourfold increase in periodicity. In addition to establishing the surface geometry, the RHEED pattern can also be used to evaluate the surface morphology (surface steps, facets, antisite disorder, etc.).

5. GROWTH DYNAMICS

The in-situ investigation of thin film growth dynamics has been made possible recently by the discovery and development of the RHEED intensity oscillation technique (Neave et al 1983, Van Hove et al 1983, Joyce et al 1988), which has proved extremely valuable for MBE and to a lesser extent for gas source beam techniques (Chiu et al 1987).

The primary observation is that damped oscillations occur in the intensity of all features of the RHEED pattern immediately following the start of growth. A typical data set for the growth of a GaAs(001) oriented film is shown in Figure 10 taken using the specular spot on the 00 rod. The steady state period corresponds precisely to the growth of a single molecular layer, i.e. a complete layer of Ga(Al) + As atoms, equivalent to $a_o/2$ in the [001] direction. Oscillatory response to a surface sensitive probe during thin film deposition is generally considered to be the manifestation of a two dimensional layer-by-layer growth mode. Similar results have been reported from a number of techniques, including LEED (Horn and Henzler 1987), helium atom scattering(Gomez et al 1985), and AES (Namba et al 1981), but will not be discussed here.

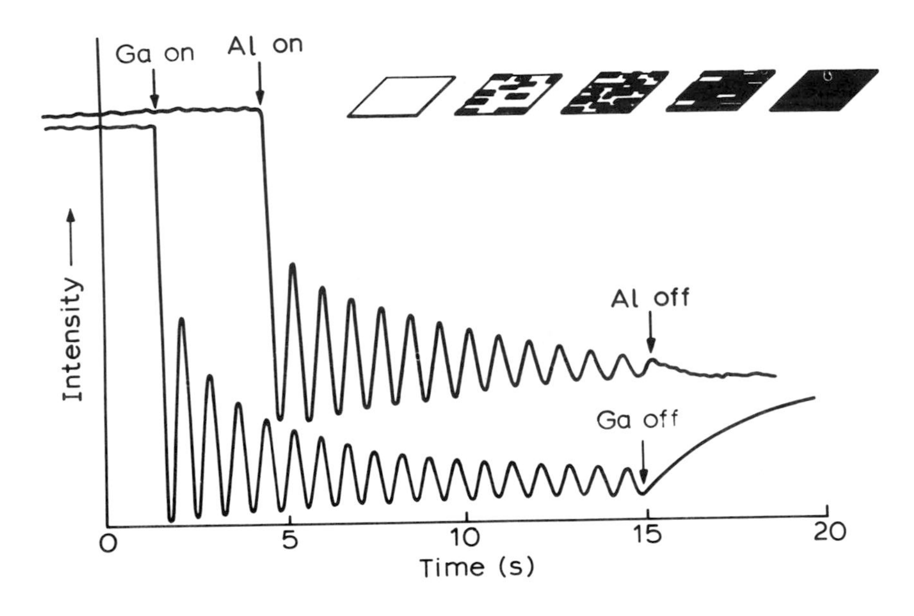

Figure 10: Intensity oscillations of the specular beam in the RHEED pattern from a AlAs and GaAs (001) - 2x4 surface during film growth.

Common to all models proposed to explain the oscillatory intensity response is the assertion that it occurs as a direct result of the changing surface topography associated with a layer-by-layer growth mode (Neave et al 1983). To examine this concept in more detail, we need to establish the nature of the diffraction processes contributing to the measured intensity, most usually made at the position of the specular beam on the 00 rod. Diffraction under the conditions used is a multiple scattering process and the elastic component of the specular intensity is attributed mainly to multiple diffraction and surface resonances. In addition, however, there is a very significant contribution from inelastic and/or incoherent processes, with the proportion of each (elastic and inelastic) depending on polar and azimuthal angles and beam energy. The primary and elastically diffracted beams undergo stronger diffuse scattering in the surface layer as it becomes disordered (assumes a higher step density) during growth. Diffusely scattered electrons which penetrate into the bulk can also be subsequently Bragg scattered and emerge as Kikuchi features, which are observed more strongly during growth than from a static surface.

Oscillations recorded under different diffraction conditions will therefore show the combined effects of a number of different diffraction processes. This is illustrated in

Figure 11, which shows oscillations from the specular spot on the 00 rod as a function of polar angle for [110] and [010] azimuths at a constant energy of 12.5 keV. The growing film was GaAs on a GaAs(001)-2x4 reconstructed surface, which was maintained throughout by using a temperature of 580°C, a Ga flux of $1x10^{14}$ atoms cm^{-2} s^{-1} and an arsenic (As_2) flux of $2x10^{14}$ molecules cm^{-2} s^{-1}. There is clearly a wide range of oscillation waveforms, but the growth conditions were invariant, so the differences must be attributed to diffraction, not growth effects. The most important

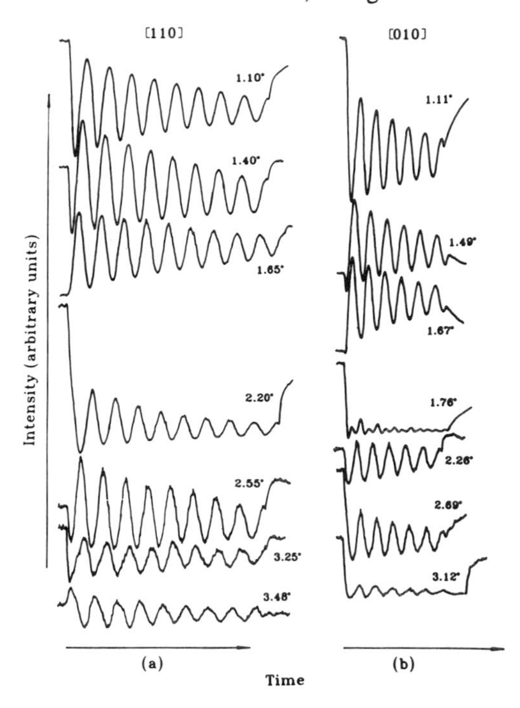

Figure 11: RHEED intensity oscillations of the specular spot on the 00 rod in [110] and [010] azimuths from a GaAs(001)-2x4 reconstructed surface at different polar angles. Primary beam energy = 12.5keV. Constant growth conditions throughout. T_s=580°C; $J_{Ga}=1x10^{14}$atoms $cm^{-2}s^{-1}$; $J_{As2}=2x10^{14}$ molecules $cm^{-2}s^{-1}$.

consequence is that the oscillations have a variable phase as a function of polar and azimuthal angles as shown in Figure 12. Data points were obtained by measuring the time to the second minimum, $t_{3/2}$ and normalising with respect to the period at steady state, T, to allow for any minor growth rate variations. The choice of $t_{3/2}$ is purely arbitrary and simply enables the phase of the oscillation to be assessed in relation to the start of growth. The steady state period is independent of diffraction conditions (except when double periodicity occurs) and dependent only on growth rate. If the oscillations are (damped) sinusoids, an ordinate value of 1.5 indicates the correspondence of oscillation maxima with layer completion. It is clear from Figure 12 however, where most of the data points were taken from near-sinusoidal oscillations, that this is seldom the case, even allowing a significant error range. This is a crucial factor when the technique is used to study growth dynamics.

Provided that due regard is paid to diffraction effects, it is nevertheless possible to obtain considerable insight into growth behaviour from RHEED observations. Absolute growth and evaporation rates, alloy composition, growth modes, adatom migration, surface relaxation and interface formation can all be investigated.

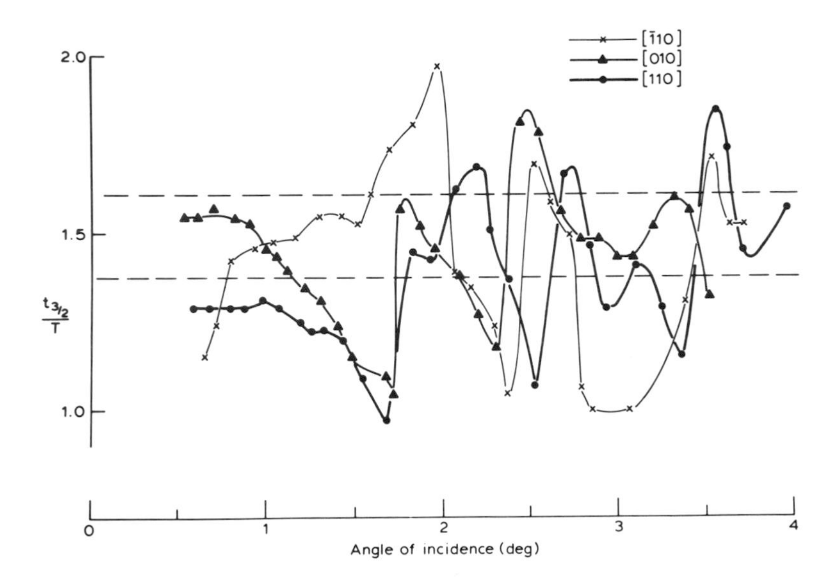

Figure 12: Phase relationships of RHEED oscillations as a function of the polar angle for different azimuths from a GaAs(001)-2x4 reconstructed surface. The growth conditions were constant as for Fig. 9. Phase is defined at the time taken to reach the second minimum normalized by the time of a complete period.

Growth and evaporation rates: the layer-by-layer mode

These measurements depend only on the steady state oscillation period, which is a high precision absolute rate monitor for layer-by-layer growth (Neave et al 1983, Van Hove et al 1983) or evaporation (Kojima et al 1985, Van Hove and Cohen 1985). Alloy composition ($III_A III_B V$) can also be determined in situ and this is especially important for high temperature growth where sticking coefficients of the Group III atoms are not always unity.

Apparently identical RHEED intensity oscillations have also been observed during the growth of GaAs from TEG and AsH_3 (Chiu et al 1987) and of InP from trimethyl indium and phosphine (Morishita et al 1988). Their presence certainly implies a two-dimensional layer-by-layer growth mode, but there must be doubt whether electron beam stimulated dissociation of the alkyls influences the actual growth rate.

Adatom migration during growth

The occurrence of a layer-by-layer growth mode implies that there must be substantial adatom surface migration. It is also well established (Foxon and Joyce 1975, 1977) that the Group III element flux is growth-rate controlling and this enables its surface migration behaviour to be determined from RHEED intensity oscillations (Neave et al 1985). The principle is a very simple extension to vicinal plane growth, illustrated in Figure 13.

The substrate is slightly misoriented in a specific direction by a know amount from a a low index plane (singular surface) and by assuming that this vicinal surface will adopt a minimum energy configuration by breaking up into low index terraces separated bymonatomic steps, an average step-free width l is defined. If the migration length of the Group III element is λ, for $\lambda > l$ the step edge will act as the major sink for migrating adatoms and there will be no two dimensional (2-D) growth on the terrace. Since RHEED intensity oscillations only occur as the result of 2-D growth, they will be absent for this growth mode. If, however, growth conditions are varied to the point where $\lambda < l$, 2-D growth will occur on the terraces and oscillations will be observed. It is therefore comparatively simple to study λ as a function of growth conditions.

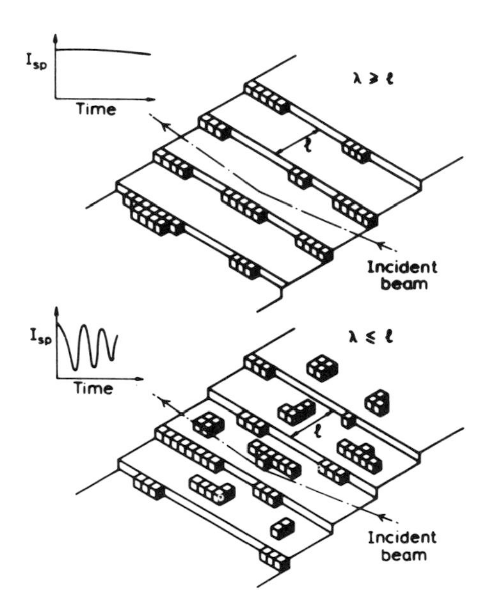

Figure 13: Schematic illustration of the principle of the vicinal plane method, showing the change in RHEED information as the growth mode changes from step propagation to 2-D growth.

It must be emphasized that in this context, surface migration refers to the transport of Ga adatoms in a weakly bound precursor state before final incorporation in a lattice site. Once such a site is reached, migration is assumed to have ceased. Incorporation is not a reversible process (at normal growth temperatures).

Ideally, the measurements should be performed by systematically varying the extent of misorientation and hence terrace width, but in practice it is simpler to use a single substrate with a known misorientation and vary the migration length by changing growth parameters. A typical data set obtained at fixed As_2 and Ga fluxes by varying the substrate temperature is shown in Figure 14.

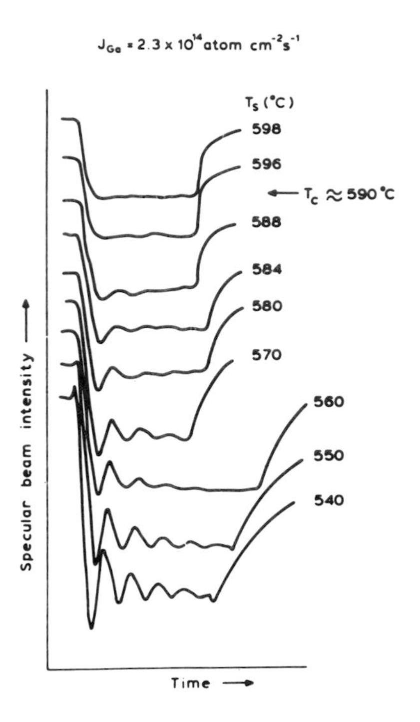

Figure 14: Data set showing the transition from oscillations to constant response as a function of substrate temperature, with constant J_{Ga}

If we assume that surface migration is isotropic, it is in principle possible to determine values of the surface migration coefficient D_o and the activation energy of surface migration E_D using the relationships:-

$$x^2 = 2 D\tau \tag{2}$$

and

$$D = D_o \exp(-E_D/kT) \tag{3}$$

Here χ is the mean displacement (terrace width), D the migration coefficient at a temperature T and τ is the surface lifetime in the precursor state. The choice of a value for τ is unfortunately not a simple matter.

Conventionally (Neave et al 1985, Van Hove and Cohen 1987) it has been equated to the monolayer deposition rate ($\tau = N_S/J$, where N_S is the number of surface sites per unit area and J the Group III atom flux) but this is certainly an oversimplification and at best represents an upper limit. Nishinaga and Cho (1988) have proposed an alternative approach, based on classical nucleation theory, which therefore requires large critical nuclei appropriate to a thermodynamic treatment involving surface free energy concepts. For typical MBE growth conditions however, the critical nucleus is likely to be a single adatom, so the reliability of a thermodynamic approach must also be questioned.

By equating τ with monolayer deposition rate as a working hypothesis, it is possible to determine an energy value, E_Q, from the temperature dependence of the disappearance of oscillations at different Group III atom fluxes. A value of $\approx$1.3eV was obtained by Neave et al (1985). This is not the migration energy alone, but includes a term representing bonds made by the atom with its nearest neighbours, E_n.

$$\text{Then } E_Q = E_n + E_D, \tag{4}$$

so E_D, the activation energy of migration, will be considerably less than 1.3eV, which seems reasonable given the essential role of surface migration in 2-D growth. It is important to realise, however, that migration parameters are not fundamental constants, they are involved in the kinetics, not thermodynamics of crystal growth and are pertinent only to adatom migration under specific growth conditions.

In spite of these quantitative difficulties, there is good evidence for a model in which growth is two-dimensional until the adatom migration length becomes less than the steady state mean terrace width, when step propagation becomes the dominant mode. It is implicit in this model that the observed damping of the oscillations is a measure of the transition to steady state, which is assumed to have been reached when the oscillation amplitude becomes zero.

Surface relaxation and growth technique modifications

If growth is stopped by shutting of the Group III flux whilst maintaining constant the substrate temperature and Group V flux, the commonly observed effect (Neave et al 1983) is for the intensity of the specular spot in the RHEED pattern to "recover" almost to its pre-growth value in a manner dependent on the precise point in the intensity oscillation at which the shutter is closed (Lewis et al 1985). The increasing intensity has been equated with "smoothing" of the surface, i.e. terrace widths increase. In general terms the "recovery" or relaxation occurs in two stages (fast and slow) and obeys an expression of the form:

$$I = A_o + A_1 \exp(-t/\tau_1) + A_2 \exp(-t/\tau_2) \text{ -} \tag{5}$$

where τ_1 and τ_2 are the temperature dependent time constants of the fast and slow stages respectively. For GaAs(001)-2x4, Neave et al (1983) reported an activation energy of $\approx$2eV associated with the rapid initial stage, possibly related to Ga-As bond dissociation at step edges. From their Monte Carlo simulation, Clarke and Vvedensky (1989) have proposed that during the fast stage two dimensional islands lose any "dendritic" structure acquired from the non-steady state growth phase by dissociation from sites with the lowest co-ordination. The slow step then results from adatom clusters evolving to form the maximum number of nearest neighbour bonds.

The development of the final surface morphology achieved by relaxation will depend critically on the stage in the monolayer deposition at which growth is stopped. In studying the phenomenon using RHEED, therefore, it is clearly essential to choose diffraction conditions where maxima correspond to monolayer completion (Joyce et al 1988).

The fact that termination (or interruption) of growth can lead to surface smoothing by a process of dissociation of two-dimensional islands, followed by surface diffusion and reformation in a lower energy configuration has given rise to a number of alternative forms of MBE growth. The modifications were introduced with the idea of growing more "perfect" interfaces in heterojunctions, quantum wells and superlattices, i.e. maximizing areas of single composition in the interface plane, or minimizing the number of layers normal to this plane in which compositional variations occur. They include interrupted growth (Madhukar et al 1985, Sakaki et al 1985), atomic layer epitaxy, ALE (Briones et al 1987) and migration enhanced epitaxy, MEE (Horikoshi et al 1988). In essence, however, they all rely on some form of interruption of one or other or both (anion and cation) fluxes.

In the simplest version of growth interruption, the shutter of the Group III cell, but not the Group V, is closed briefly (up to 60s) to allow the specular RHEED beam intensity to recover to its initial value. It is almost certain that this will result in a lower step density and it is likely to be more successful for lower binding energy, faster diffusing species, so, for example a considerably lower temperature is required for a GaAs layer than for AlAs or (Al,Ga)As (Tanaka and Sakaki, 1988). Its efficacy will also be critically dependent on the point in the monolayer when growth is interrupted, so correct choice of diffraction conditions is vital. Both ALE and MEE involve periodic interruptions of both fluxes during growth. ALE is based on an alternate supply of constituent elements to the substrate surface, regulated so that a single complete layer is provided with each pulse. This is not a problem for the Group V species, since their adsorption is self-regulated by the available Group III atoms in the surface. This is not so for the Group III atoms, however and an excess can accumulate as the free element on the surface if the amount per pulse is too large. This can be obviated by the use of RHEED, and provided correct diffraction conditions are used, exact monolayer quantities can be supplied. The results appear to indicate that "high quality material" can be prepared at rather lower temperatures than by conventional growth, but it should be remembered that the overall growth rate is low in ALE and it would be more valid to make a comparison with material grown conventionally at this rate and temperature. This information is not yet available. The basic mechanism seems to be a rapid relaxation process as the result of briefly interrupting growth very close to monolayer completion.

The MEE process also involves the alternating supply of cation and anion fluxes and it is claimed that by reducing the surface concentration of the Group V element during the impingement of the Group III atoms the latter's surface migration rate is enhanced, resulting in smoother, higher quality growth at low temperatures. Again, a better description would probably be that only very short relaxation times are required near monolayer completion and the alternating supply allows this to happen.

One final point needs to be raised about MEE, however. It is demonstrated that the observed RHEED intensity oscillations persist throughout the whole growth, even for relatively thick films and they are essentially undamped. This is claimed to be indicative of high quality growth, in contradistinction to conventional MBE, where oscillations tend to damp relatively quickly.

Two factors need stressing: in conventional growth, damped oscillations merely indicate a change of growth mode from two-dimensional to step propagation, and are not necessarily associated with material quality, only with step distribution and migration lengths. Much more importantly, though, the origin of the oscillations recorded in MEE is quite different from that for conventional growth. In MEE the oscillations occur simply because there is a transient change in surface reconstruction from a Group III stable form to a Group V stable form with each alternate pulse and the specular intensity is a strong function of surface reconstruction (Larsen et al 1986).

Growth modifications are thus seen to be an interesting approach to producing smoother interfaces and lower temperature growth. The mechanisms of ALE and MEE probably relate more to enhanced relaxation from near-complete monolayers than to enhanced adatom migration and in MEE in particular the RHEED intensity oscillations are associated with reconstruction effects, not the growth mode.

6. GROWTH OF HIGH PURITY STRUCTURES BY MBE

Typical growth rates for MBE are about $1MLs^{-1}$ or $1\mu m$ per hour. This corresponds to a pressure of about $1x10^{-6}$ Torr for the group III species at the sample surface. To achieve high purity material, therefore, it is essential to keep the partial pressure of gases which stick to a minimum (for $N_d + N_a < 10^{15}$ cm^{-3} we need $P_{imp} < 10^{-13}$ Torr assuming a unity sticking coefficient). The way we achieve this is to outgas thoroughly the sources prior to loading, use the best possible start materials and then bake the system extensively to remove water vapour. Full details of the methods we use have been published elsewhere (Foxon and Harris 1986).

For both GaAs bulk films and for two-dimensional-electron-gases (2DEGs) we find a progressive improvement in the properties of the films grown by MBE as the growth sequence continues after the system loading and bakeout. This is because background impurities which limit purity are gettered by the growth process and their concentration is gradually reduced providing there is no external leak in the system.

Bulk GaAs

Using this technique we have grown several thick intentionally doped GaAs films with electron mobilities at 77K of $>1x10^5$ and a best value of $1.33x10^5$ cm^2 V^{-1} s^{-1}. The free electron concentrations were about $2x10^{14}$ cm^{-3}. This suggests the background acceptor concentration in our material is below $2x10^{14}$ cm^{-3}. Recently we have reported an undoped GaAs film, grown using a superlattice prelayer, which at low temperature has luminescence dominated by free exciton emission ('t Hooft et al 1987).

2DEGs

For 2DEGs, the structure we have used for our high mobility samples is shown in Figure 15. The GaAs buffer is grown starting at 580°C, increasing gradually to 630°C. The Si flux is calibrated to give a doping level in GaAs of 2×10^{18} cm^{-3}, which in (Al,Ga)As gives a Si concentration of 1.33×10^{18} cm^{-3}. The undoped GaAs cap is to aid the contact technology by preventing oxidation.

If we consider a realistic picture of the 2DEG band diagram then for narrow spacers, d<100Å, the mobility will be determined by ionised impurity scattering near the 2DEG. For wider spacers, ~400Å, the ionised impurities near the surface (10x the number near to the 2DEG) will have a significant contribution and reducing or moving these should improve the measured mobility. For very wide spacer samples, with d>800Å, the background impurity of the GaAs will be the key factor in determining the mobility.

Figure 16 shows the results of a systematic set of samples in which only the undoped spacer layer thickness was intentionally varied. The general form of the curve is as expected, with a peak mobility for d~600Å. The drop in mobility at large spacer thicknesses occurs because reducing the 2DEG density results in a lower Fermi velocity and hence mobility, for a fixed scattering rate. Increasing the thickness of the dopedregion of (Al,Ga)As and hence removing some of the ionised impurities near the surface resulted in a 50% improvement in peak mobility, as expected.

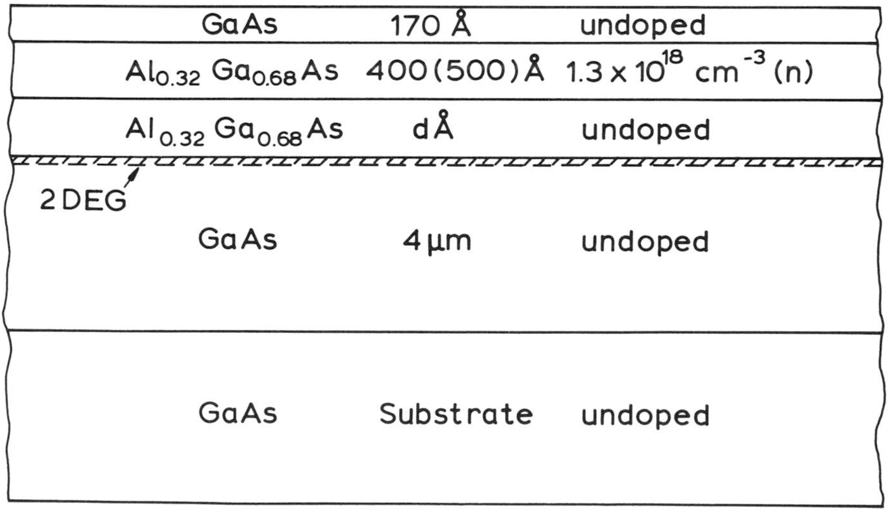

GaAs 170 Å undoped
$Al_{0.32}Ga_{0.68}As$ 400(500)Å 1.3×10^{18} cm^{-3} (n)
$Al_{0.32}Ga_{0.68}As$ dÅ undoped
2DEG
GaAs 4μm undoped
GaAs Substrate undoped

Figure 15: Structure of high mobility 2DEG samples

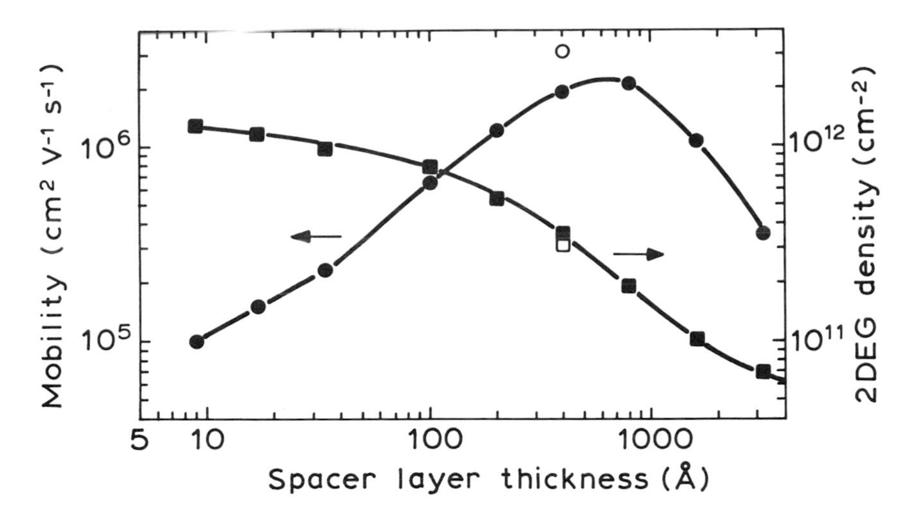

Figure 16: Effect of undoped spacer layer thickness on the density and mobility of 2DEG structures

For several years the best reported mobilities for 2DEG samples were approximately $2x10^6$ cm^2 V^{-1} s^{-1} at a sheet carrier density of about $5x10^{11}$ cm^{-2}. In our studies we were able to obtain similar mobility material at somewhat lower electron densities, as shown in Figure 16 (Foxon et al 1986) and by optimising the structure obtained mobilities of $3x10^6$ V^{-1} s^{-1} at 4K (Harris et al 1986, 1987). Using a prelayer to improve the background scattering in the GaAs English et al (1987) obtained a 2DEG with a mobility at <2K of $5x10^6$ cm^2 V^{-1} s^{-1} at a sheet carrier density of $1.6x10^{11}$ cm^{-2}. More recently even higher mobilities have been achieved at 4K over a wide range of electron densities (Foxon et al 1989). This has been achieved by combining the optimum structure for the doped regions (Harris et al 1987) with the use of a superlattice to trap impurities (t'Hooft et al 1987).

Qualitatively we can account well for the observed dependence of mobility on spacer layer thickness but quantitively theories still predict lower mobilities than are observed in experimental measurements.

For the structures discussed above, where the (Al,Ga)As layers are grown on GaAs (the so called "normal" structure) very high mobility 2DEGs can be obtained, but if the layer sequence is reversed and GaAs is grown on doped (Al,Ga)As (the so called "inverted" structure) much lower mobilities are observed. Various factors have been suggested for this inferior performance including interface roughness (Alexandre et al 1986), impurity build-up during the growth of (Al,Ga)As which is deposited in the GaAs near the 2DEG (Alexandre et al 1986) Si migration from the doped (Al,Ga)As into the GaAs either by diffusion or segregation (Inoue et al 1985, Gonzales et al 1986), (this point will be discussed in more detail below) and scattering from band-edge discontinuity fluctuations at the interface (Cho et al 1987). We have recently identified a new factor, electron localisation at the interface (Airaksinen et al 1988).

The structure of the samples grown for this study are shown in Figure 17. The 2DEG is formed in a narrow QW doped only on the surface side to avoid problems of Si migration. In sample C, GaAs is grown on bulk (Al,Ga)As but in sample D the alloy is replaced by a short period SL of equivalent band-gap. In this experiment any difference in mobility can only relate to the quality of the inverted interface.

For a conventional "normal" 2DEG structure we observe a monotonic increase in mobility as the sample is cooled from 300 to 4K with the carrier concentration being essentially constant. For sample C we observe a strong decrease in both mobility and carrier concentration as we go to low temperatures, but for sample D we observe behaviour typical of a 2DEG with a 22Å spacer.

Gold (1986) has recently predicted that a metal-insulator transition can occur in quantum well structures due to electron localisation by a randomly fluctuating potential. Strong localisation arises when the magnitude of the potential fluctuations exceeds the Fermi energy of the carriers as shown in Figure 18. By reducing the size of the fluctuations (by replacing the alloy with a superlattice) we were able to recover in sample D normal 2DEG behaviour. To test this idea further, additional samples of type D but with wider undoped spacers (lower 2DEG density) were grown. For a 40Å spacer sample, conducting behaviour was still observed after illumination but the 100 and 200Å spacer samples were completely insulating. We were thus able to estimate the size of the potential fluctuations to be about 30meV.

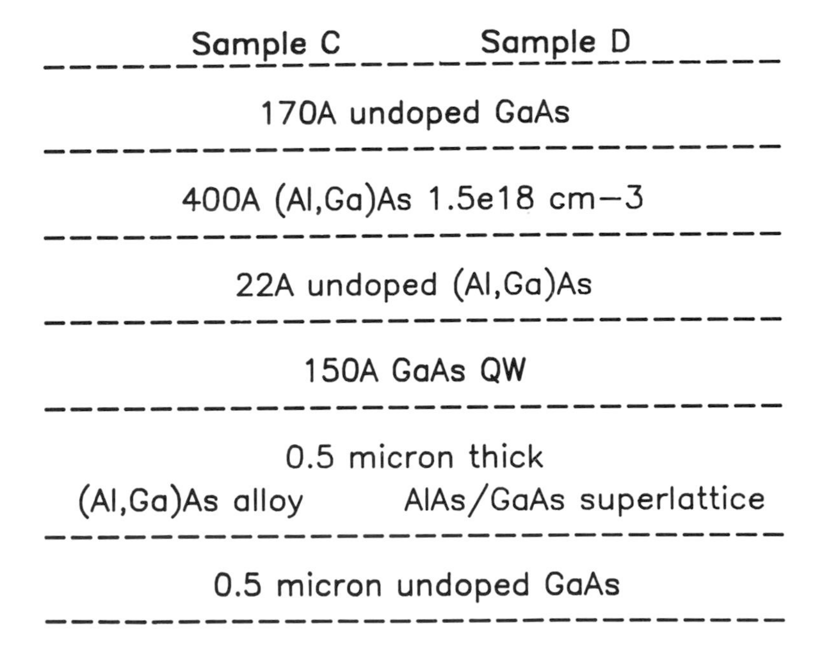

Figure 17: Structure of QW samples used to investigate the quality of the inverted interface

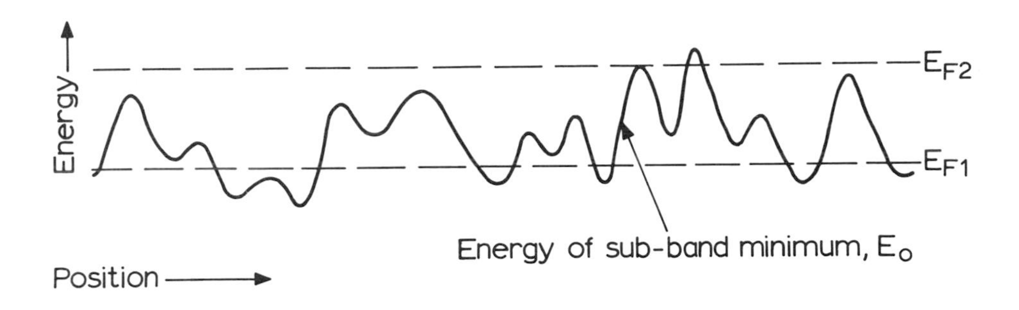

Figure 18: Illustration of how strong localisation occurs due to local potential fluctuations

The improved properties of the inverted interface obtained by replacing the alloy with a superlattice still do not given an interface of comparable quality to the "normal" interface under the growth conditions we have used.

7. QUANTUM WELLS (QWs) AND SUPERLATTICES (SLs)

Further insight into the quality of interfaces can be obtained by studying the physical properties of undoped QWs and SLs grown under essentially identical conditions. Figure 19 shows a cross-sectional TEM micrograph of a GaAs-AlAs SL comprising alternate layers 3 monolayers thick. Similar migrographs of a 2 + 2 ML SL can also be observed but at present 1 + 1 ML SLs cannot be resolved. Two features are evident from such micrographs; the first is that averaged over the depth of the film, interfaces abrupt on a scale of less than 1 monolayer can be produced; and secondly there is a general tendency for the wavy interfaces initially obtained to smooth out asgrowth proceeds. The mechanism responsible for this gradual improvement is not clear at present but the reduction of interface energy may be a factor.

By modelling the integrated intensities of the satellite reflections, X-ray diffractometry from MQW and SL structures can also give us valuable information on the lateral extent of the interfaces. From such studies estimates of interface roughness from 1 to 4ML have been proposed, with one interface (probably the GaAs on AlAs or (Al,Ga)As) extending over a larger distance in the growth direction.

Most information on the quality of the interfaces has been obtained from low temperature photoluminescence (excitation) (PL/PLE) studies of QWs and SLs. Here the key point to remember is that the spacing of interface steps Λ in relation to the diameter, D, of the exciton determines what will be observed in PL/PLE. When $\Lambda << D$ a pseudo-smooth interface will be obtained and when $\Lambda >> D$ we have a truly-smooth heterojunction. If $\Lambda \sim D$ excitons will be trapped at the interface and the energies of specific recombination peaks in PL will be lower than the corresponding

Figure 19: TEM micrograph of a GaAs(3μm)/AlAs(3μm) superlattice (grown by D. Hilton, TEM micrograph by J.P. Gowers).

positions in PLE, the so-called Stokes shift. In PL/PLE studies of QWs there are often examples of multiple peaks. In some cases they correspond to monolayer differences in well width at the upper "normal" interface (Deveaud et al 1984), but there are other examples in the literature (Bajaj et al 1986) where the authors correctly point out that the fine structure observed cannot correspond to QWs differing in width by an integer number of monolayers. It is important to point out that in the case of MQWs interwell width fluctuations due to variations in growth rate for example can give rise to such multiple peaks but studies of the temperature dependence of the PL/PLE rule out this explanation in the case referred to above.
The structural integrity of SLs can be studied using PL/PLE to give further information on the quality of heterointerfaces. We have recently shown that for GaAs-AlAs SLs grown at 630°C alloy like behaviour is observed for n = m = 1 or 2ML but for n = m >3ML true superlattices are formed with electronic properties quite different from those observed in $Al_{0.5}Ga_{0.5}As$ films grown under identical conditions (Moore et al 1988). This is entirely consistent with the X-ray data presented above which suggests an inverted interface extending over a few MLs. To produce "true" 1 + 1 or 2 + 2 ML SLs modified growth procedures will be needed.

8. DOPANT INCORPORATION

The choice of suitable p and n dopants is of crucial importance in exploiting MBE as a production technique for devices such as HEMTs, lasers etc..

The group VI elements suffer from a tendency to surface segregate when supplied in elemental form, which may be related to the fact that they evaporate in part as dimers. This problem can be overcome by using compound sources such as PbTe which dissociates to give Te atoms (readily incorporated at low temperatures) plus Pb atoms which are weakly bound to the surface and lost by re-evaporation. Above about 600°C, however, the group VI elements are also lost by re-evaporation and therefore not appropriate for most device structures.

The group II elements fall into two categories, Zn and Cd have too high a vapour pressure to be retained at typical growth temperature. Be and Mg do not suffer from this problem, but Mg is so highly reactive that it was difficult to obtain electrically active material in early MBE equipment and Be therefore has been almost universally used as the p type dopant in MBE.

The group IV element Ge is amphoteric and is therefore not a suitable n type dopant, Sn suffers from kinetically limited surface segregation which leaves Si as the most universally used n type dopant.

Synthesis of particular doping profiles can in principle be achieved by changing the temperature of the Si or Be cell to give the required flux as a function of time, but the thermal response time of the furnace limits the rate of change to about a decade in 100Å. Profiles of arbitrary shape can instead be obtained using "aomic-plane" doping

(Wood et al 1980). Here dopant is deposited during a growth interruption, the group III shutters are closed but the As flux is used to maintain the surface stoichiometry. By appropriate choice of dopant flux and interval any arbitrary profile can be synthesised.

This idea was extended to include the concept of "delta" doping (Zrenner et al 1984) in which the dopant atoms are deposited on a single plane to form a novel 2DEG system. This is different from the 2DEG systems formed at heterojunctions in that there is no large spatial separation of electrons and ionised impurities so that in this system extremely high mobilities are not obtained. The system is nevertheless interesting because very much higher sheet carrier concentrations can be obtained ($\sim 10^{13}$ cm^{-3} cf $\sim 10^{11}$ cm^{-3} in (Al,Ga)As/GaAs 2DEGs) and many more subbands can be occupied in the potential well formed by the dopant atoms (4 and 7 subbands at $4x10^{12}$ and 10^{13} cm^{-2} respectively, Ploog 1987).

In both "atomic-plane and "delta" doping the absence of any arriving group III element means that the number of Ga dangling bonds in the surface is minimised. For amphoteric dopants like Si and Ge this has the effect of reducing the number incorporated onto the As sublattice and reducing their tendency towards compensation. For Si, choosing alternative substrate orientations can increase the incorporation on the As sublattice and in fact p type material has been obtained on 211A or 311A Ga terminated surfaces (Wang et al 1985).

Both Si and Be suffer from a number of additional problems and neither is an ideal dopant source. Both show enhanced diffusivity at high doping levels (Gonzales et al 1986, Devine et al 1987). Here above a critical level the diffusion mechanism changes and as a result MQW or SL structures can become disordered. It is possible to see this directly in TEM micrographs of heavily doped samples.

For both Si and Be it has also been suggested that surface segregation is a problem. This has recently been studied using delta doped samples (Beall et al 1988). Ultra-high resolution SIMS measurements on Si delta doped GaAs samples have shown that there is evidence of surface segregation at temperatures above 550°C with a characteristic segregation length of around 50 to 100Å (close to conventional SIMS resolution limits). This is occasionally masked by very high diffusivity observed at doping levels > $2x10^{18}$ cm^{-3}. Key evidence in favour of segregation came from experiments where in conventionally doped films the substrate temperature was suddenly decreased, resulting in a doping spike as predicted by the segregation model, and a corresponding dip on increasing the temperature as expected. So far no convincing evidence for Be segregation has been obtained.

Despite the difficulties associated with Si and Be they can be employed to provide suitably doped films as evidenced by their use in MBE production of HEMTs (Fujitsu), lasers (ROHM) and high speed microwave devices (Philips).

9. ACKNOWLEDGEMENTS

We would like to thank our many colleagues at both Philips Research Laboratories and the Interdisciplinary Research Centre for Semiconductor Materials for their contributions to the work outlined in this article.

REFERENCES

Airksinen J M, Harris J J, Lacklison D E, Beall R B, Hilton D, Foxon C T and Battersby S J 1988 J. Vac. Sci. Technol **B6** 1151

Alexandre F, Lievin J L, Meynadier M H and Delalande C 1986 Surface Sci. **168** 454

Aspens D E, Colas A, Stedma A A, Bhat R, Koza M A and Keramidas V G 1988 Phys. Rev. Lett. **61** 2782

Bajaj K K, Reynolds D C, Litton C W, Singh J, Yu P W, Masselink W T, Fischer R and Morkoc H 1986 Solid State Comm. **29** 215

Beall R B, Harris J J and Clegg J B 1988 Semi. Sci. Technol. **3** 612

Briones F, Gonzdez L, Recio M and Vaquez M 1987 Japan J. Appl. Phys. **26** L1125

Chang C A, Ludeke R, Chang L L and Esaki L 1977 Appl. Phys. Let.. **31** 759

Chiu T H, Tsang W T, Cunningham J E and Robertson A 1987 J. Appl. Phys. **62** 2302

Cho N M, Ogale S B and Madhurkar A 1987 Appl. Phys. Lett. **51** 1016

Clarke S and Vvedensky D D 1989 J. Crystal Growth in the press

Deveaud B, Emery J Y, Chomette A, Lambert B and Baudet M 1986 Appl. Phys. Lett. **45** 1078

Devine R L S, Foxon C T, Joyce B A, Clegg B A and Gowers J P 1987 Appl. Phys. **A44** 195

English J H, Gossard A C, Stormer H L and Baldwin K W 1987 Appl. Phys. Lett. **50** 1826

Foxon C T, Boudry M R and Joyce B A 1974 Surface Sci. **44** 69

Foxon C T and Harris J J 1986 Philips J. Res. **41** 313

Foxon C T, Harris J J, Hilton D, Hewett J and Roberts C, Semi. Sci. and Technol. to be published.

Foxon C T, Harris J J, Wheeler R G and Lacklison D E 1986 J. Vac. Sci. Technol., **B4** 511

Foxon C T and Joyce B A 1975 Surface Sci. **50** 434

Foxon C T and Joyce B A 1977 Surface Sci. **64** 293

Foxon C T, Joyce B A and Norris M T 1980 J. Crystal Grow **49** 132

Gold A 1986 Solid State Comm. **60** 531

Gomez L J, Bourgeal S, Ibanez J and Salmaron M 1985 Phys. Rev. **B31** 2551

Gonzales L, Clegg J B, Hilton D, Gowers J P, Foxon C T and Joyce B A 1986 Appl. Phys. **A41** 237

Harbison J P and Farrell H H 1988 J. Vac. Sci. Technol. **B6** 733

Harris J J, Foxon C T, Barnham K W J, Lacklison D E, Hewett J and White C 1987 J. Appl. Phys. **61** 1219

Harris J J, Foxon C T, Lacklison D E and Barnham K W J 1986 Superlattices and Microstructures **2** 563
Heckingbottom R 1985 Molecular Beam Epitaxy and Heterostructures, ed. Chang L L and Ploog K Martinus Nijhoff, Holland pp 71-104
Horn M and Henzier M 1987 J. Crystal Growth **81** 428
Horikoshi Y, Kawashima M and Yamagodi H 1988 Japan J. Sppl. Phys. **27** 169
Inoue K, Sakaki H, Yoshino J 1985 Appl. Phys. Lett. **46** 973
Joyce B A, Neave J H, Zhang J and Dobson P J 1988 Reflection High Energy Diffraction and Reflection Electron Imaging of Surfaces ed. Larsen P K and Dobson P J, Plenum. pp 397-417
Kawai N J, Kojima T, Sato F, Sakamoto T, Nakagawa T and Ohta K 1986 12th Int. Symp. on GaAs and Related Compounds IOP Conf. Ser. 79 Adam Hilger pp 433
Kojima T, Kawai N J, Nakagawa T, Ohta K, Sakamoto T and Kawashima M 1985 Appl. Phys. Lett. **47** 726
Larsen P K, Dobson P J, Neave J H, Joyce B A, Bolger B and Zhang J 1986 Surface Sci. **169** 176
Larsen P K, Van der Veen J F, Mazur A, Pollman J, Neave J H and Joyce B A 1982 Phys. Rev. **B26** 3222
Lewis B F, Greenhaver F J, Madhurkar A, Lee T C and Fernandy R 1985 J. Vac. Sci. Technol. **B3** 1317
Madhukar A and Ghaisas S V 1988 CRC Critical Reviews in Solid State and Materials Science **14** 1
Madhukar A, Lee T C, Yen M Y, Chen P, Kim J Y, Ghaisas S V and Newman P G 1985 Appl. Phys. Lett. **46** 1148
Moore K H, Duggan G, Dawson P and Foxon C T 1988 Phys. Rev. **B38** 5535
Morishita Y, Maruno S, Gotada M, Nomeva Y and Ogata H 1988 Appl. Phys. Lett. **52** 42
Namba Y, Vook R W and Chao S S 1981 Surface Sci. **109** 320
Neave J H, Joyce B A, Dobson P J and Norton N 1983 Appl. Phys. **A31** 1
Neave J H, Dobson P J, Joyce B A and Zhang J 1985 Appl. Phys. Lett. **47** 100
Nishinaga T and Cho K 1988 Japan J. Appl. Phys. **27** L12
Ploog K 1987 J. Crystal Growth **81** 304
Robertson A, Chiu T H, Tsang W T and Cunningham J E 1988 J. Appl. Phys. **64** 877
Sataki H, Tanaka M and Yoshino J 1985 Japan of Appl. Phys. **24** L417
Tanaka M and Sakaki H 1988, Superlattices and Microstruct. **4** 237
'tHooft G W, Van der Poel W A J A, Molenkamp L W and Foxon C T 1987 Phys. Rev. **B35** 8281
Tsao J Y, Brennan T M and Hammons B E 1988 Appl. Phys. Lett. **53** 288
Van Hove J M and Cohen P I 1985 Appl. Phys. Lett. **47** 726
Van Hove J M, Lent C S, Pukite P R and Cohen P I 1983 J. Vac. Sci. Technol. **B1** 741
Van Hove J M and Cohen P I 1987 J. Cryst. Growth **81** 13
Wang W I, Mendex E E, Kuan T S and Esaki L 1985 Appl. Phys. Lett. **47** 826
Wood C E C, Metze G, Berry J and Eastman L F 1980 J. Appl. Phys. **51** 383
Woodbridge K, Gowers J P and Joyce B A 1982 J. Crystal Growth **60** 21
Zrenner A, Reisinger H, Koch F and Ploog K 1984 Proc. 17th Int. Conf. on Phys. of Semi. ed. Chadi J D and Harrison W A Springer Berlin 1985 325

Scanning Electron Microscopy (SEM) Microcharacterisation of Semiconducting Materials and Devices

D B HOLT

Characterisation refers to the complete range of techniques used in determining the structure and properties of semiconducting materials, multilayer structures and devices. It therefore includes (i) the diffraction techniques used for determining crystal structures, (ii) the microscopic techniques for determining defect microstructure and (iii) the physical measurement techniques for determining (volume averaged) values of physical properties.

The SEM (scanning electron microscope) techniques provide measurement with a micron scale resolution and so can determine defect microstructures and local values of physical properties. The term microcharacterization has come into use to emphasise this dual character of SEM techniques.

SEMs outsell TEMs (transmission electron microscopes) by 3 to 1 worldwide despite the fact that TEMs have much better (lateral, spatial) resolution and have highly developed interpretive theories available. Clearly, therefore, SEMs must have great advantages to offset this. One practical advantage is that SEMs can be used to examine macroscopic specimens with little or no special preparation, e.g. whole wafers or electronic devices can be inspected. Moreover, SEMs offer not only lateral, spatial resolution but also depth resolution, spectral (signal) resolution, time resolution, etc.

However, the big advantages of SEMs are those outlined in the next section. The advantages of the scanning principle are so great that many further types of scanning microscope are now appearing such as scanning laser and acoustic microscopes and scanning facilities have been incorporated into TEMs to produce "analytical" TEMs or TEMSCANS.

Scanning Electron Microscopy

The scanning electron microscope (SEM) is powerful and versatile because

(i) it has six modes of operation which provide information about many different groups of properties of solid objects
and
(ii) the information is obtained essentially as electrical signals suitable for eletronic data processing to provide quantitative values of many different properties as well as for presentation in various types of micrographs.

The SEM is, therefore, not only a family of different kinds of microscopes, it is also a family of microanalytical measuring systems. Its performance parameters depend on the basic subsystems of the SEM and the physical character of the signals underlying the six modes of operation.

Here we will introduce the operating principles of the SEM and the physical foundations of the six modes to lay the basis for a general understanding of SEM techniques with emphasis on those of greatest interest for semiconductor analyses: the EBIC (electron beam induced current) and CL (cathodoluminescence) techniques.

The Scanning Electron Microscope

The SEM is composed of two sub-systems (Figure 1):

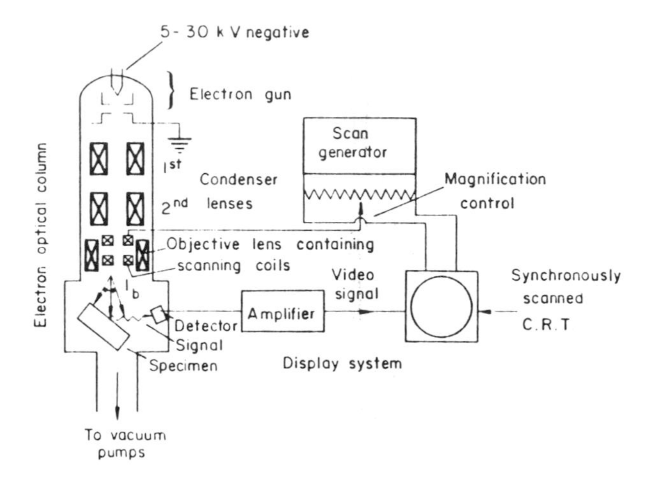

Figure 1. The SEM consists of an electron optical column and a detection–amplification–display system. The latter takes six different forms for the six modes of operation and may include a microcomputer for data processing.

(i) the electron optical column which produces a finely focussed beam of electrons that is scanned in a television–type raster over the specimen surface and

(ii) a signal detection, processing and display or read–out system. This display system detects one of the forms of energy emitted by the specimen under electron bombardment, i.e. transduces it into an electrical signal, amplifies, processes and displays or "prints it out" in some form. The SEM as a microscope is thus rather like a closed–circuit TV system. The electron optical column provides the "illumination" and the detection and display system produces the television–like picture.

The SEM is not closely analogous to traditional microscopes that use lenses to produce "images" in the sense of that term used in optics. The performance of image–forming microscopes (light microscopes and transmission electron microscopes) is ultimately limited by diffraction effects. This is not the case with the SEM.

Magnification

The scanning of the electron beam over the specimen and the scanning of the display CRO screen are driven by the same scan generator, as shown in Figure 1, so they are in synchronism and there is a one–to–one correspondence between the two areas scanned. These are not of the same size however, so the CRO provides a magnified dislay of the signal strength variation from point to point. Let 1 be the side of the area scanned on the specimen and L that of the display screen, then the magnification is simply:

$$M = L/1 \qquad (1)$$

The drive currents fed to the scan coils in the electron optical column are produced by a fraction of the output voltage of the scan generator which drives the CRO scan. Varying this fraction by means of a potentiometer as shown in Figure 1, therefore, provides continuously variable or "zoom" magnification.

The Six Modes of Operation of the SEM

The power of the beam, of current I_b and accelerating voltage V_b, is dissipated into several other forms of energy by interactions of the incident electrons with solid specimens. The five forms obtained by continuous beam bombardment are shown in Figure 2. These five forms of energy can be "detected" or transduced into electrical signals. They provide the information employed in the five older modes of operation of SEMs. These types of signal and the modes they underlie are:

Signal	Mode
1) emitted electrons	the emissive mode
2) X-rays	the X-ray mode or electron probe microanalysis (EPMA)
3) cathodoluminescence (visible and near-visible light)	the CL mode
4) currents or voltages	the charge collection (CC) mode (the signals are due to electrons collected by one contact and holes by the other)-includes EBIC (electron beam induced currents)
5) transmitted electrons	scanning transmission electron microscopy (STEM)

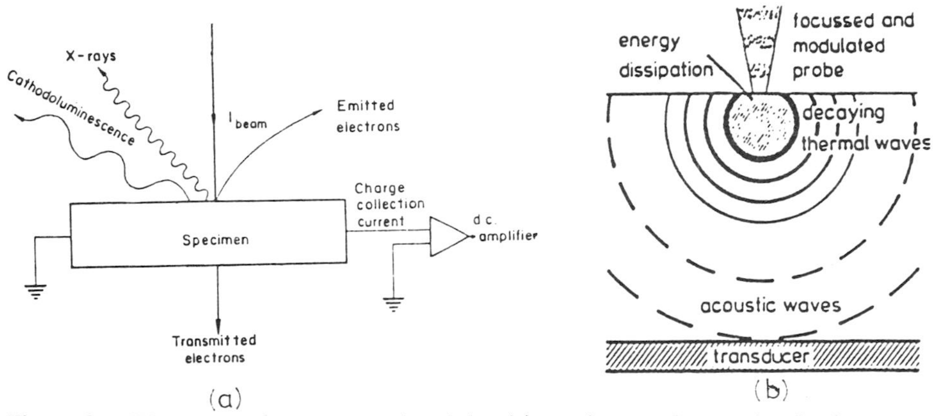

Figure 2. The types of energy produced by (a) continuous electron bombardment and used as signals in the first five modes are (i) emitted electrons, (ii) x-rays, (iii) cathodoluminescence (CL), (iv) charge collection (CC) signals, i.e. currents (known as EBIC electron beam induced currents) or voltages (known as EBIV) and (v) transmitted electrons. (b) Chopped beam bombardment can generate ultrasonic waves which are the signals of the new scanning electroacoustic microscopy (SEAM).

The Electroacoustic Mode

Recently a sixth mode has been introduced for which signals can only be generated by chopping the beam of electrons at a high frequency. Such pulsed electron bombardment produces intermittent heating and thermal expansion and contraction so ultrasonic waves result. GaAs and InP, for example, also generate waves piezoelectrically. The ultrasonic waves are detected and form the basis for the sixth mode, SEAM (scanning electro-acoustic microscopy).

The physical properties that can be microcharacterised by SEAM are only beginning to be explored. The obvious effects are the detection of cracks or delaminations of surface layers like metal contacts by the detection of premature reflections of the ultrasonic waves as in sonar. In materials such as the III–V compounds the waves couple piezoelectrically to space charges so depletion regions at junctions can be seen.

Developments in SEAM include moving to higher frequencies (MHz and above) to get shorter wavelengths and so higher resolution. The use of two or more ultrasonic detectors and the mixing of the detected signals,the use of phase and amplitude signals are also under investigation and this mode seems likely to be increasingly important in future.

Resolution

The resolution of SEMs is not essentially diffraction controlled. It is not, therefore, defined by Rayleigh's criterion of resolution nor can it be calculated by means of the Abbe theory of resolving power. It is determined by the physics of the electron optical column and of the detection and display system. It also depends on the physics of the signal generation process of the particular mode.

Lateral, Spatial Resolution

The (lateral, spatial) resolution is the diameter of the smallest area on the specimen from which a sufficiently strong signal can be derived to produce an acceptably noise-free picture point in the display cathode ray oscilloscope (CRO) screen. This is determined by:

(i) the minimum signal power required to give an acceptable signal/noise ratio in the detection-display system, and

(ii) the minimum area or volume in which the necessary beam power can be focussed or dissipated respectively as shown in Figure 3. The three-dimensional energy dissipation volume appears as an area on the two-dimensional micrographic CRO display.

Electron Beam Diameter and Current

The beam current that can be focussed into a spot of a diameter, d, on the specimen surface is limited by the electron optics of the column so the minimum diameter of electron probe increases as the 3/8ths power of the required beam current I_b. This is because of the severe aberrations of electromagnetic lenses which mean that as the beam is focussed down to smaller beam diameters, more and more electrons are lost at the necessary apertures in the column.

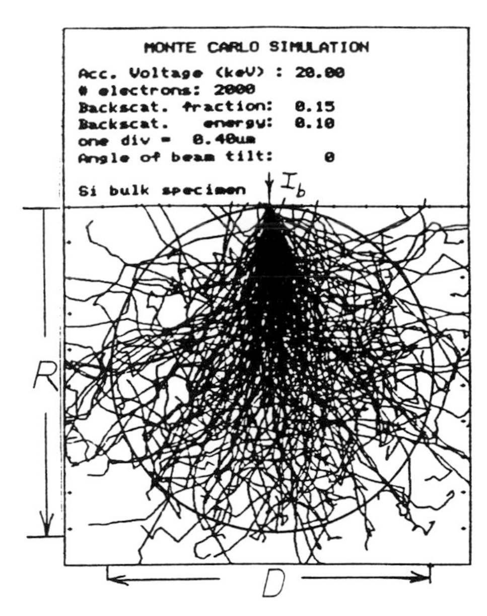

Figure 3. The incident beam of diameter d dissipates its energy throughout a volume of diameter D = R, the beam penetration or Gruen range. The 'bulk mode' signals, x-rays, CL and CC signals arise throughout this volume and have a resolution ≅ D. The secondary electrons escape only from a shallow surface layer so they give a resolution not much larger than d, the diameter of the incident electron beam. This Figure is a graphic display of a set of two-dimensional projections of Monte Carlo simulations of electron trajectories in silicon.

The Energy Dissipation Volume

The diameter of the energy dissipation (i.e. generation or cascade) volume is determined by electron scattering processes in the solid. It can often be approximated (depending on the atomic number of the material, Z and the beam voltage V_b) by a hemisphere or a sphere approximately tangential to the surface as shown in Figure 3. Even for small beam diameters, of the order of 10nm, the energy dissipation volume is of the order of 1μm in diameter for V_b = 20 - 30 KV. The performance characteristics of the detectors and display systems for the five modes require different minimum beam currents and hence diameters as listed in the Table below. Moreover, there is a basic difference between STEM which requires such thin specimens that there is virtually no beam spreading, the emissive mode in which beam spreading results in roughly a two-fold increase in the resolution as compared with the beam diameter and the 'bulk modes' X-ray, CL and CC in which the signal is obtained from the whole energy dissipation volume. In the latter modes the energy dissipation volume diameter is the resolution, almost independently of beam diameter, for small values of the latter.

Mode	I_b (A)	d	Resolution
Emissive	10^{-11}	~ 3nm*	~ 5nm*
X-ray	10^{-6}	~ 1μm	1-3μm
CL	"	"	"
CC	"	"	"
STEM	10^{-11}	0.5nm	0.5nm
SEAM			frequency dependent

* - These values depend on electron optical design and vary from instrument to instrument and have fallen considerably over the years.

In the new SEAM (electro acoustic mode) there is some controversy. To the extent to which the resolution arises in the thermal wave region, it depends on the thermal properties of the material. To the extent that it depends on ultrasonic wave interaction processes, it depends on their wavelength and so it is determined by the beam chopping frequency (since the velocity of sound is constant in this frequency region).

The Emissive Mode

The electron emission spectrum consists of primary or backscattered electrons, secondary (low energy) electrons and tertiary (intermediate energy) electrons as shown in Figure 4.

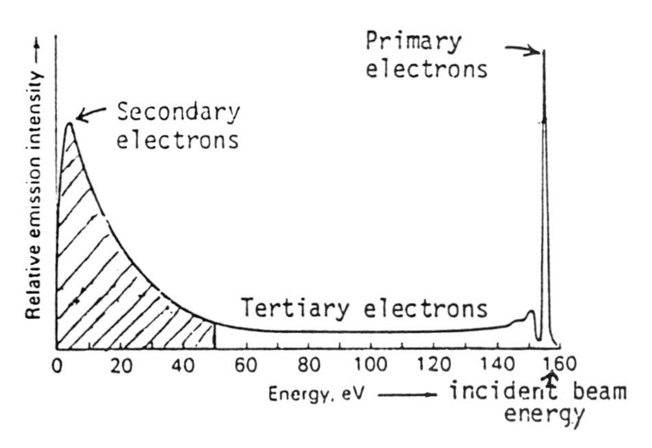

Figure 4. The electron emission spectrum. Secondary electrons are defined to be those emitted with energies less than 50eV.

The important ones are those in the large low energy (<50 eV) peak. These provide information about the surface orientation (topographic contrast). It is the secondary electron emissive mode SEM micrographs which resemble optical (light) views of the object but with far better resolution, and hence useful magnification than is available in light microscopes.

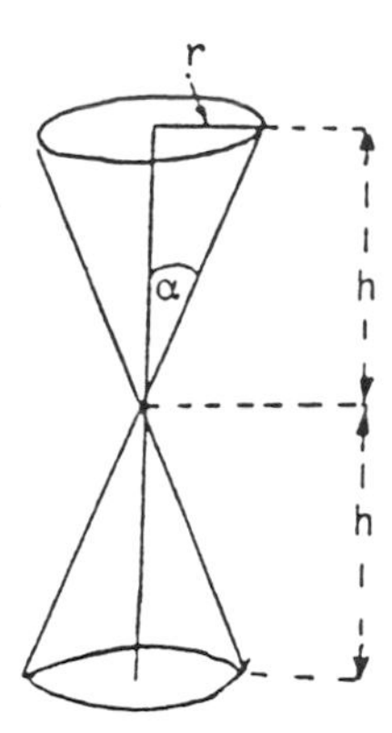

Figure 5. The depth of field in secondary electron SEM micrographs, 2h, is determined by the variation of the beam radius r, above and below the beam cross–over point, due to the beam convergence with semi–angle α.

The depth of field in the SEM is the depth in object space (i.e. on the specimen surface) for which the resolution remains adequate, so the picture on the display screen is sharp i.e. "in focus". Since the numerical apertures (Figure 5) of electromagnetic lenses have to be kept small because of their severe spherical aberrations (compared to glass lens systems for light) the semiangle of convergence of the electron beam, α, is very small, typically 10^{-2} to 10^{-3} radians. One radian is about 57°, so 10^{-3} radians is about 0.06°. Let δ be the smallest distance that can be resolved by eye on the display CRO screen or the pixel size and let the magnification be M. Then the resolution on the screen will not degrade so long as the radius of the electron probe, r remains

$$r \leqslant \frac{\delta}{2M}$$

but in Figure 5 it can be seen that

$$r = \alpha h$$

so that the depth of field, i.e. the depth within which points on the object surface can lie and still appear sharply in focus on the viewing screen is

$$2h = \frac{\delta}{\alpha M} \qquad (2)$$

The "standard eye" can resolve 10^{-1} mm so that we set δ equal to this value. Then for, say, M = 1,000x and $\alpha = 10^{-2}$ radian, 2h = 10μm. This value is about two orders of magnitude greater than the depth of field of light microscopes of the same magnification. If M = 100x and $\alpha = 10^{-3}$ radians, 2h = 1 mm. Figure 6 is a well–known crystal growth morphology picture exhibiting such a depth of field.

The reason that SEM emissive mode pictures look like visual observations is illustrated in Figure 7. There is a general correspondence of bright and dark areas, where many and few electrons are detected, respectively, with the appearance of the same areas viewed by eye. N.B. the micrograph appears as if viewed in the direction of the electron beam, not as if viewed from the detector, as one would intuitively expect.

Figure 6. Large depth of field secondary electron micrograph of hopper–morphology whiskers of $Pb_xSn_{1-x}Te$. They became wider as they grew up from the furnace tube, but they were broken off, inverted and glued down on a mounting stub for examination. The width and depth of field in this micrograph are about a mm. (Micrograph courtesy of Mr N S Griffin, reproduced by kind permission of the Plessey Co Ltd).

The detailed contrast in the two cases is in fact quantitatively similar due to the fact that the number of secondary electrons emitted depends on the angle of incidence of the beam in a way that is inversely related to Lambert's Cosine Law for the diffuse scattering of light. That is

$$I_{sec.\ electrons} = C\ I_b \sec\theta \qquad (3)$$

whereas the diffusely reflected light intensity, L_r, is related to the incident intensity, L_i by Lambert's Cosine Law:

$$L_r = C'\ L_i \cos\theta \qquad (4)$$

where θ is the scattering angle. Since L_r is the signal that is physically analogous to $I_{sec.\ els.}$, when we write the expression for the case of light in the form analogous to equation (2), therefore:

$$L_i = (1/C') \times \text{signal} \times \sec\theta \qquad (5)$$

This is the algebraic expression of the interchange of illumination and viewing direction between the light and secondary electron pictures that we noted in the analogy in Figure 7.

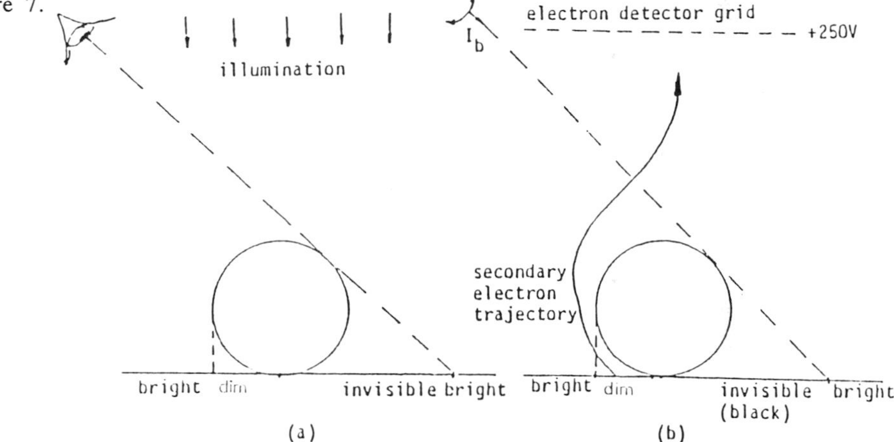

Figure 7. Resemblance between the pictures obtained by direct viewing by eye (a) and secondary electron SEM viewing (b).

Voltage and Magnetic Contrast

The electron detector can be modified to give a greater ability to select electrons of a particular energy and/or angle of emission from the surface. That is, instead of using a simple positively charged grid to attract all the numerous, low-energy secondary electrons as in the standard Thornley-Everhart detector (Figure 8), an electron spectrometer may be used. It is then possible to observe or measure small differences of electrostatic potential over the surface. This is referred to as voltage contrast. Stroboscopic voltage contrast enables one to observe the pulses of low and high voltage that constitute the information in digital circuits running around operating circuits (Figure 9).

This is the best means for observing the operation of VLSI circuits to verify the design logic, etc. The beam can, alternatively, be left stationary on a selected node of the circuit during operation to record the voltage waveform. The voltage contrast techniques (both stroboscopic microscopy and waveform point analysis) are commonly

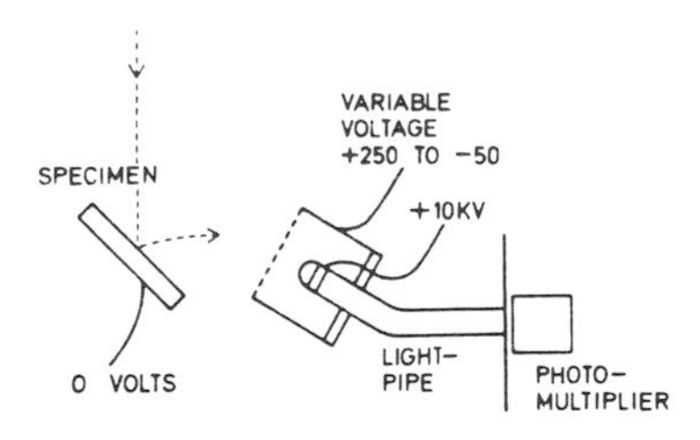

Figure 8. In the Thornley–Everhart emitted electron detector a potential of about +250 V on the grid, relative to the specimen at earth potential (on average) attracts the numerous low–energy secondary electrons. These are accelerated by about +10kV on the aluminium coating on the scintillator. These electrons now have a high energy and each produces numerous photons in the scintillator. The perspex light pipe guides them to the photomultiplier tube. There the photons cause photoelectric emission of electrons which are multiplied by secondary electron emission along a chain of dynodes. The high sensitivity of this detector first made the emissive mode SEM a practical instrument. By turning off the attractive grid voltage the detector is made sensitive to the high energy primary (backscattered) electrons only, so this detector can act as a simple electron spectrometer.

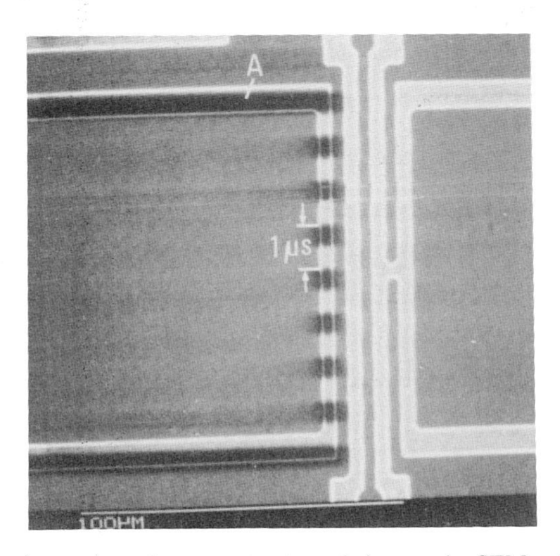

Figure 9. A stroboscopic voltage contrast, emissive mode SEM micrograph of a 1 MHz square wave on an interconnection in an integrated circuit. The light areas are at a more negative potential than the dark areas. The time interval between the "time resolution" is indicated. (After H P Feuerbaum, Scanning 5, (1983) pp14).

known as 'e–beam testing' by VLSI circuit workers. E– beam testers are increasingly integrated into CAD (computer aided design) systems for design validation as well as being used for inspection for process validation and VLSI circuit failure analysis.

The fringing magnetic fields at ferromagnetic domain walls can similarly be made visible. This is referred to as magnetic contrast.

Channelling Patterns

A form of diffraction pattern can be obtained arising from channelling of the electrons between the atomic planes in crystals. These electron channelling patterns (ECPs) provide information on the crystallographic orientation and perfection with a spatial resolution of the order of a micron. An example is shown in Figure 10.

Figure 10. Electron chanelling pattern (ECP) of Si in (100) orientation.

The X-ray Mode (Electron Probe Microanalysis - EPMA)

High energy electrons incident on solids result in the emission of x-ray spectra consisting of broad continuous bands of 'white' x-rays or Bremstrahlung which are not useful and sharp lines that are. The emission lines arise from transitions between inner shell energy levels in the atom and their wavelengths are characteristic of the element.

In SEMs it is customary to use "energy-dispersive" x-ray spectrometers. These employ Li drifted Si (or sometimes Ge) detectors. These are blocks of high-resistivity material with a large electric field applied. Incident x-ray photons produce current pulses in a manner very like the action of the older Geiger-Muller tubes. The height of the pulse is proportional to the energy of the photon and since $E = h\nu = hc/\lambda$ it is also inversely a measure of the wavelength, λ. A pulse-height analyser is used to count the pulses. The energy (wavelength) of a peak in the x-ray spectrum indicates the element concerned and the height of the peak is a measure of the concentration of the element in the energy dissipation volume. "Point analyses" can be carried out to determine the chemical composition of the material at the point of impact of the electron beam or one wavelength can be selected and the variation of the count rate can be displayed to produce an x-ray mode SEM micrograph which is a map of the distribution of the selected element over the area scanned. Quantitative work requires the application of "ZAF corrections" for atomic number (Z), self-absorption (A) and fluorescence (F) effects. Proprietary software for these iterative corrections is supplied by the manufacturers of computerised EDS (energy dispersive spectrometer) x-ray systems.

EDS microanalysis is quick and convenient and at least semi-quantitative. Hence it is very widely used for materials identification e.g. checking whether the metallization on a device is Al or Ag based. However detection limits are at best one part in 10^4 and light elements are less easily detected than heavy ones. Thus it can be used to check the composition of, e.g., III-V alloys but not to detect dopants in Si.

The Cathodoluminescence Mode

The total intensity of cathodoluminescence (cathode ray bombardment induced light) (CL) emitted is small due to the low power of the beam and the low luminescent efficiency of most materials. For spectrometry, efficient collection of the light is therefore necessary and is obtained, e.g. by means of the semi-ellipsoidal mirror, lens and fibre optic light guide shown in Figure 11. The light then passes through a monochromator and is detected by a photomultiplier and photon counting system. The spectrum obtained from a point can be recorded, displayed and printed out by the microcomputer. Such point analyses are obtained with the beam stationary or scanning a small area. One wavelength can then be selected and displayed while scanning to give a micrographic map of the emission of light by a particular luminescence mechanism.

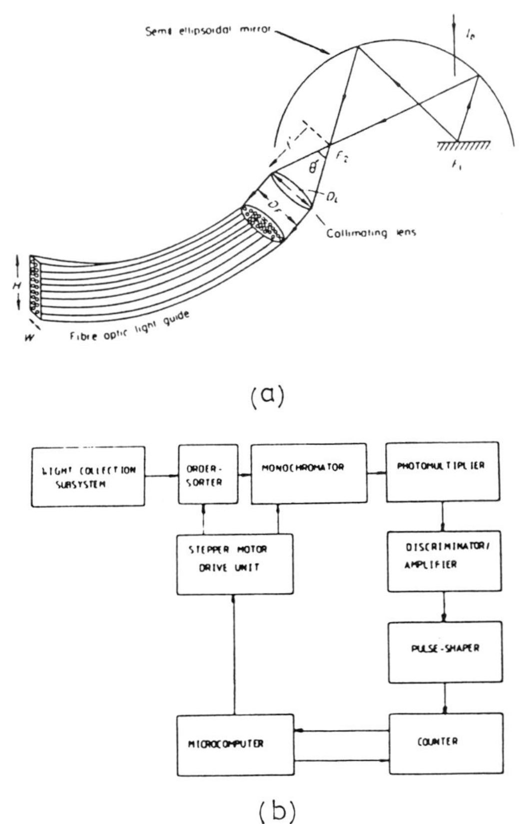

Figure 11. Components of a computer controlled spectroscopic CL detection system. (a) The light collector is based on a semi-ellipsoidal mirror. (b) A block diagram of the entire system.

The luminescent emission bands from specimens at near room temperature used for such "monochromatic" CL microscopy provide information of two kinds. Intrinsic CL arises from electron-hole recombination between states in or near the conduction and valence band edges so this 'near bandgap' radiation has a peak photon energy approximately equal to E_g. Variations of composition in III-V alloy layers and temperature distributions in operating devices can be measured in this way, for example. Extrinsic CL arises from transitions involving localised energy states often associated with impurities. It can, therefore, give information about impurity distributions in the material. At low temperatures, the technique can detect down to doping levels 5 or 6 orders of magnitude lower than the x-ray mode in favourable cases, i.e. down to dopant concentrations of $10^{14}cm^{-3}$ or below.

SEM-CL is important for the study of optoelectronic materials and light-emitting devices such as semiconductor lasers. It provides much the same information about energy states as does photoluminescence (PL), which however has better spectral resolution.

There are two reasons for the higher spectral resolution in the case of PL. Firstly, whole macroscopic specimens can be immersed in cryostats which gives lower temperatures than can be attained even in liquid helium cooled SEM CL systems. Emission spectra sharpen up at lower temperatures so more detail can be resolved in PL. Secondly, laser illumination can be used to excite the whole specimen volume so PL is not limited in spectral resolution by low intensities. In the case of CL only the energy dissipation volume (about one micron cubed) is excited so intensities are low. The compensating advantage of CL is that it also provides spatial resolution, both lateral and depth resolution. Depth resolution is attained by varying the SEM beam accelerating voltage and hence the electron penetration range. In this way greater or lesser depths can be excited. By using monochromatic CL microscopy with a wavelength emitted by one material only, individual layers in epitaxial multilayer III-V structures can be examined through overlayers. This and the effect of low temperatures, which often increase emission intensities and so improve signal-to-noise ratios and hence spatial resolution, is illustrated in Figure 12.

TEM-CL. TCL and ECL.

The technique described above can be termed scanning electron microscope cathodoluminescence (SEM-CL) to distinguish it from other CL techniques.

CL signals are also emitted and can be displayed or spectroscopically analysed in scanning transmission electron microscopes. As will be discussed below, these are of two types: "analytical TEMs (transmission electron microscopes) or TEMSCANs which are basically TEMs with added scan coils and "dedicated STEMs" or FEG-STEMs (field emission gun-scanning transmission electron microscopes). In both cases, to get electron transmission, specimens are examined that have been thinned to between a few hundred nm and about $1\mu m$. As the specimens are so thin little electron beam spreading takes place (compare Figure 3), i.e. the excited, energy dissipation volume is small. This means that spatial resolution is better than in an SEM but CL intensities are much lower. Thus an improvement in spatial resolution is obtained at the expense of a reduction in spatial resolution. This trade-off between improved spatial resolution and lower signal strengths occurs in all modes. This technique is difficult but some impressive results have been produced (Steeds 1989).

TCL and ECL (transmission and emission CL) are simple "panchromatic" (all wavelengths) micrographic techniques. In TCL a Ge or Si photodetector diode is placed immediately below a slice of material and in ECL the detector is placed above and to one side but in direct line of sight. The CL reaching the detector produces a video signal which is synchronously displayed in the usual way. These techniques are relatively simple but can be quite useful for III-V epitaxial multilayer materials. The use of the two types of detector which have different near infrared response ranges and of the transmission and emission positions gives some spectral separation. (The substrate and epilayers are generally transparent in different wavelength ranges).

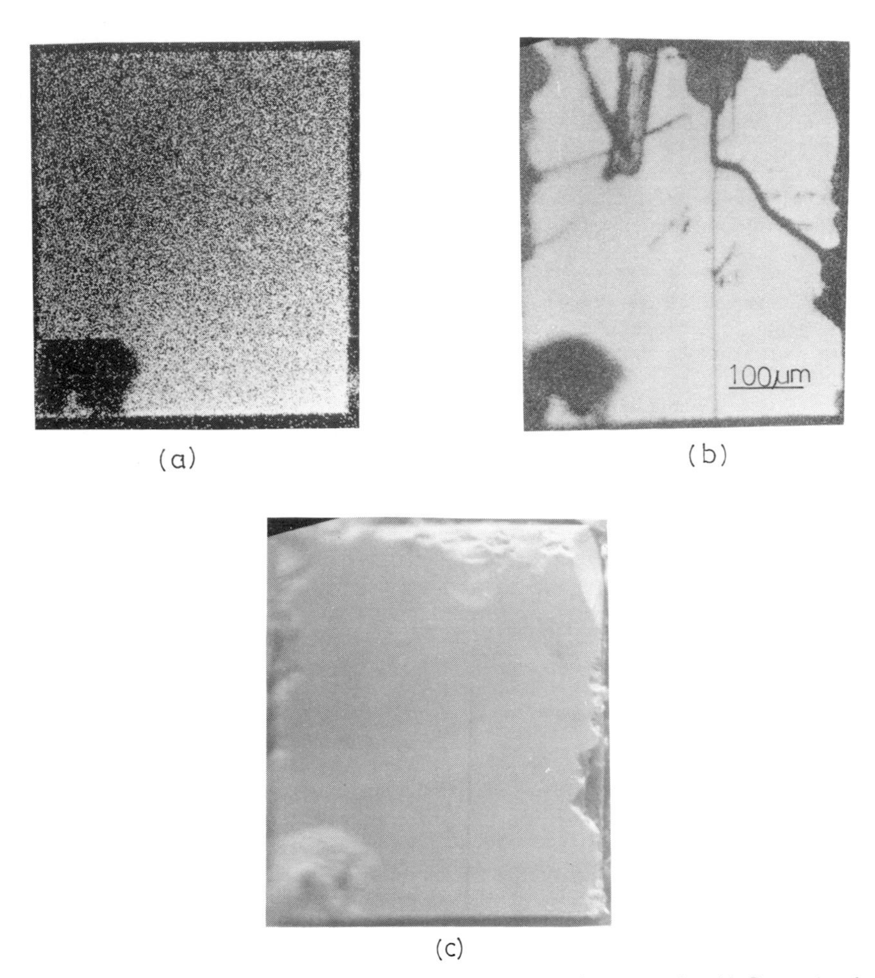

Figure 12. CL micrographs of the low x, active layer in an early $Al_xGa_{1-x}As$ double heterostructure laser taken through the high x guiding layer above it (a) at room temperature and (b) at liquid helium temperature. (c) Emission mode micrograph of the top of this device, showing that no surface features correspond to the main dark lines and areas in the internal active layer.

The Charge Collection Mode

This mode is based on the "charge collection" (CC) signals (currents or voltages) obtained from two contacts applied to the specimen. If no external bias is present, then an "electron voltaic effect" (EVE) is necessary to generate an e.m.f. that can drive a signal through the external detection circuit as shown in Figures 13 (a) and (b). The analogous photovoltaic effect is applied in semiconductor radiation detectors and in solar cells. The EVE causes electrons to be collected by one contact and holes by the other, hence the name of the mode.

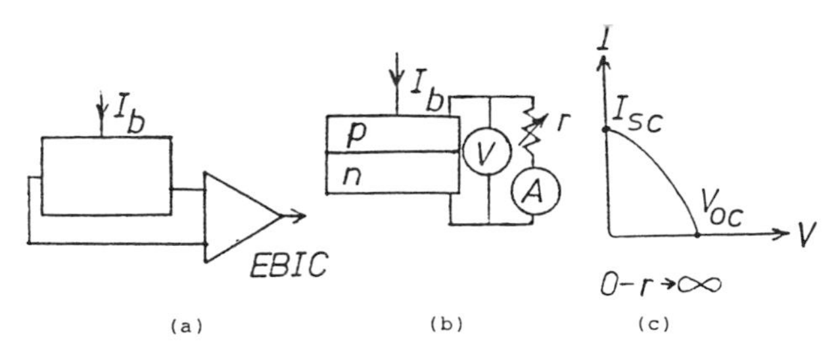

Figure 13. Charge collection due to the electron voltaic effect (EVE) (a) schematic circuit and (b) circuit for a p–n junction used to record (c) the barrier EVE current voltage characteristic.

The most important of these phenomena is the "barrier electron voltaic effect" (BEVE). This generates strong signals only when the beam is incident on or near an electrical barrier such as a p–n junction, Schottky barrier (contact) or heterojunction between two different semiconductor materials. It reveals their positions micrographically and the signals can be analysed to give the values of the electrical properties involved. These barriers are essential for solid state electronic devices so the CC mode is important for electronic materials and device characterization.

Special detection systems are required for this mode so that either a well–defined short circuit current Isc, or open–circuit voltage Voc, is obtained (Figure 13). Only then can the signals be quantitatively interpreted. The use of Isc is often referred to as EBIC (electron beam induced current) microscopy. The use of Voc signals is similarly referred to as EBIV. A more quantitative treatment of EBIC will be given below.

Scanning Transmission Electron Microscopy.

In order to obtain resolution in STEM comparable with that in a conventional TEM (transmission electron microscope) a completely different SEM design is necessary. "Dedicated" STEMs are ultra–high vacuum (UHV) instruments employing field emission (electron) guns (FEGs) instead of the thermionic emission guns used in "conventional" SEMs. They are coming to be known as FEG STEMs. As the specimen is in a UHV environment its surface can be kept atomically clean for extended periods. FEG STEMs can therefore also usefully employ surface analytical methods such as Auger emission spectroscopy (AES). Consequently a scanning Auger microscopy (SAM) mode has been developed in FEG STEMs.

Because the specimens have been reduced to thicknesses of the order of hundreds of nm for electron transparency, beam spreading (Figure 3) is made negligible. The high brightness of the FEG makes it possible to get the necessary 10^{-11} A of beam current into a spot of diameter 5nm which is the STEM resolution. The EBIC and CL modes can be applied in FEG STEMs and, due to the absence of beam spreading, the (lateral, spatial) resolution in these modes can also be impressive. This is usually limited by the inverse relation of spatial to signal (spectral) resolution. In STEMs the excited volume is small so e.g. the intensity of the CL emission is low which limits the wavelength resolution attainable and the doping–concentration sensitivity of the technique. This problem is so severe that for many materials it is necessary to increase the electron beam diameter in order to increase the beam current and excite a larger volume to get a detectable signal.

This inverse relation of the two forms of resolution is not confined to STEMs. It applies in (conventional) SEMs to 'bulk' specimens, too. If a signal is unacceptably weak, the beam power must be increased. This can be done by increasing the beam current i.e. by increasing the beam 'spot' size which makes the spatial resolution worse. It can also be done by increasing the 'beam energy' i.e. the energy per incident electron in KeV but increasing the accelerating voltage which increases the penetration range R (Figure 3) and hence D, the resolution for bulk modes.

Beam electron collisions knock electrons from atomic inner shell levels to lead to the emission of the characteristic x-ray photons used in x-ray microanalysis (EPMA). These scattering events involve losses of the same characteristic quanta of energy by the beam electrons. Transmitted electron energy loss spectroscopy (EELS) has recently emerged as a technique for identifying the elements present in the excited volume, with the advantage that the losses exciting both characteristic x-ray and Auger emission appear in the EELS spectrum. EELS is applied in both FEG STEMs and in conventional i.e. thermionic gun combined scanning and TEM instruments. These are the so called analytical electron microscopes or TEMSCANs. (The latter is a trade name). STEM observations can also be carried out in these instruments.

It is hoped that this introductory account has given an indication of the still rapidly growing family of scanning electron microscope techniques and some idea of the types of information obtainable by their use. We turn next to a more detailed discussion of the interpretation of the information produced by one important mode, EBIC, and the means for extracting quantitative values of physical properties from the EBIC signal variations.

The Barrier Electron Voltaic Effect

The operation of the barrier electron voltaic effect (BEVE), the source of EBIC signals, is illustrated in Figure 13. Physically it involves the separation ("collection") of hole-electron pairs generated by the incident electrons by the built in field in the depletion region of an electrical barrier, a p-n junction or an heterojunction.

The incident beam generates, per second, a number of pairs, given by:

$$\Delta N = G\ I_b \qquad (6)$$

where I_b is the beam current in electrons per second and G, the generation factor , is the number of pairs generated per incident electron given by:

$$G = \frac{(1-f)\ E_b}{e_i} \qquad (7)$$

where E_b is the beam energy per electron in KeV, f the fraction of energy lost on average with the backscattered electrons and e_i is the "ionization" energy to create a hole electron pair. It is found that:

$$e_i \simeq 3\ E_g \qquad (8)$$

The charge collection short circuit current can then be written as

$$I_{cc} = \eta_{cc}\ G\ I_b \qquad (9)$$

where η_{cc}, the charge collection efficiency, is a property of the barrier (and a function of the beam voltage etc.). η_{cc} is the fraction of the carriers generated (GI_b) that are collected.

Consider the case of plan view scanning of planar technology devices such as bipolar Si integrated circuits (Figure 14.) Scanning a p-n junction at a constant depth will produce a constant signal (I_{cc}) if η_{cc} is constant. Contrast (variations in brightness) will arise only if the junction varies in e.g. depth, depletion region width etc. Thus, a good junction will appear uniform and any processing imperfections or electrically active defects passing through the junction will be visible.

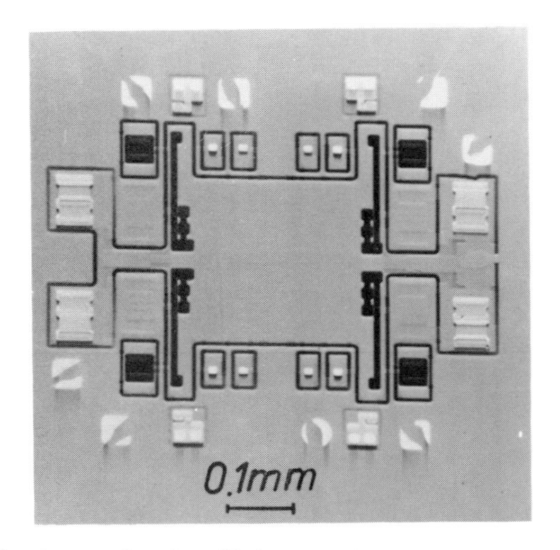

Figure 14. EBIC micrograph of a Si integrated circuit. This is a quad NAND gate with a similar logic gate in each corner. Due to the way charge is collected to the inverting input of the first op amp in the detecting circuit p–n junctions appear bright while n–p junctions appear dark. Not all circuit components were active under the operating conditions used for this micrograph.

This technique reveals all the barriers, the 'working parts' of devices, that are present plus electrically active defects only. (Electrically inactive defects are generally of no interest or may be desirable for gettering etc). EBIC is widely used for inspection for quality assurance, process validation, yield improvement and failure analysis.

To extract quantitative data we must treat charge collection mathematically, following Donolato (1985, 1988)

The Phenomenological Theory of EBIC Contrast

This theory is described as phenomenological because it simply assumes defects are volumes in which the lifetime is reduced. That is, defects are assumed to be sites of enhanced recombination and so reduce the number of carriers collected, producing dark contrast. First, however, we must consider charge collection in the absence of any defects.

Charge Collection by a Schottky Barrier.

Consider the simple case of a Schottky barrier on the beam entry surface and a defect lying below (Figure 15.) Schottky barriers are often evaporated onto as–grown material to provide charge collection and make EBIC microscopy possible.

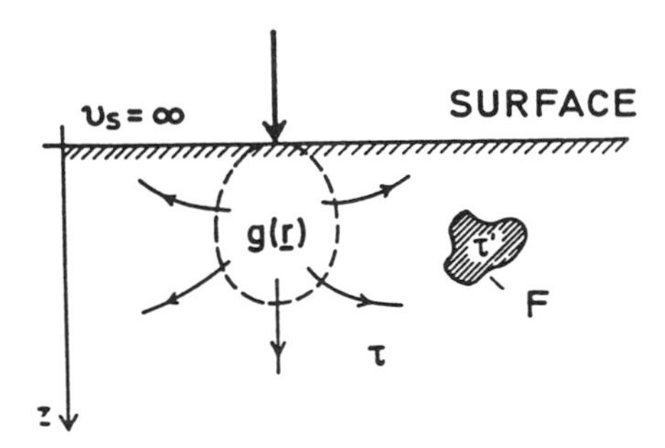

Figure 15. The simplest model of a Schottky barrier on the beam entry surface is an infinite surface recombination velocity. Electron–hole pair production is represented by the generation function $g(\underset{\sim}{r})$ and a defect is modelled by a volume F in which the lifetime is reduced to τ'. (after Donolato 1985).

Under an electron bombarded Schottky barrier in the absence of any defects the EBIC current can be written:

$$I_o = \int_V g(\underset{\sim}{r})\ \Phi(\underset{\sim}{r})\ dV \qquad (12)$$

where $g(\underset{\sim}{r})$ is the generation rate at point $\underset{\sim}{r}$, $\Phi(\underset{\sim}{r})$ is the charge collection probability (fraction of the charges generated that are collected) for that point and V is the volume under the contact. In this case $\Phi(\underset{\sim}{r})$ – exp (–z/L), and the situation is effectively one dimensional so one can write:

$$I_{cc} = I_o = \int_V g(z)\ \exp(-z/L)\ dz \qquad (13)$$

That was the result for the assumption of Figure 15, that the barrier can be modelled by assigning an infinite recombination velocity to the surface. This amounts to assuming zero carrier density there due effectively to unity collection of minority carrier charges reaching the surface. In fact, Schottky barriers consist of a metal layer which absorbs some of the beam energy and so reduces the number of carriers generated, and a depletion region of width W. W depends on both the Schottky barrier height Φ_B and the doping density in the semiconductor.

The charge collection efficiency in depletion regions is generally assumed to be unity, so equation (13) can be rewritten (neglecting the thickness of the metal layer)

$$I_o = \int_o^W g(Z)dZ + \int_W^R g(z)\ \exp\{-(Z-W)/L\}\ dz \qquad (14)$$

Consider the effect of increasing the beam voltage (Figure 16a), which produces an increase in the penetration range R. At first R < W and the charge collection efficiency is one. The charge collection efficiency η_{cc} is the integral of the charge collection probability $\Phi(\underset{\sim}{r})$. It can be seen from equations (9) and (7) that as V_b and hence E_b increases for a constant beam current I_b, I_{cc} will increase linearly as shown in Figure 16(b).

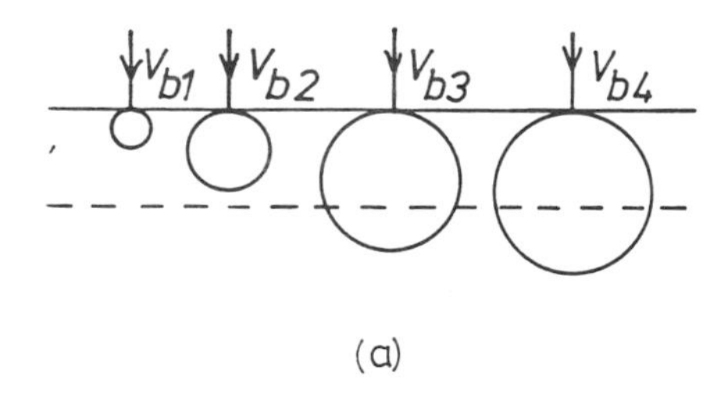

(a)

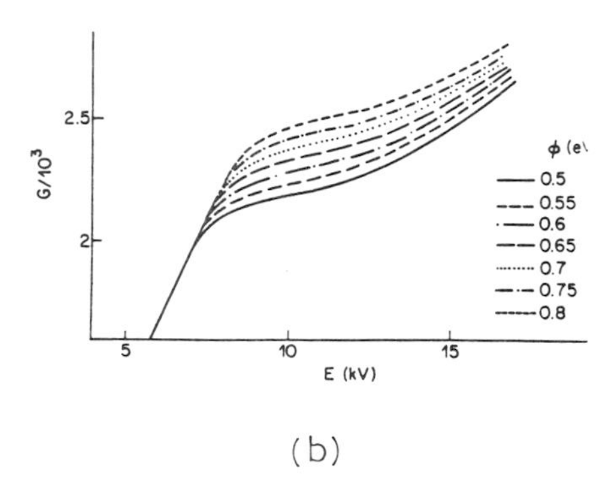

(b)

Figure 16. (a) Increasing the beam voltage (energy) V_{b_1} - V_{b_4} increases the beam penetration range and energy dissipation volume until it penetrates through the Schottky barrier depletion region of width W. (b) The EBIC gain (I_o/I_b) varies with beam voltage as shown for a series of values of Schottky barrier height Φ (and W.) (After Gibson et al, 1987).

Once the range R > W carriers begin to be generated below the depletion region. Such carriers, to be collected must diffuse up to the edge of the depletion region. Their charge collection probability is $\exp\{-(z-W)/L\}$. The result is that the cc current gain then increases more slowly with V_b as shown in Figure 16(b). The depletion region width W increases with Φ_b so a family of curves is obtained as shown for any particular material and doping density. By fitting experimental data to a set of such curves Schottky barrier heights can be evaluated.

Monte Carlo simulations run on microcomputers can be used to obtain the distribution of carriers under the beam $g(\underset{\sim}{r})$ and this in turn can be used to compute families of curves like those in Figure 16(b) (which was in fact calculated in this way). This method is making EBIC an increasingly practical measurement tool. Of course, the curves depend also on the values of metal thickness and minority carrier diffusion length so information can also be obtained on such additional parameters.

Other barriers, such as a p-n junction at a certain depth, with a depletion width W, can be modelled similarly. Again EBIC curve fitting can be used to evaluate materials and junction parameters.

EBIC Defect Contrast Theory

Returning to the simple model of Figure 15 and equations (12) and (13), consider the influence of the defect. This is modelled by Donolato (1985,1988) as a volume F in which the minority lifetime is reduced from τ to τ'. The Donolato first order theory assumes further that the carrier density p_o is not significantly reduced in the volume F so the net carrier generation rate can be written

$$G(\underset{\sim}{r}) = g(\underset{\sim}{r})\ (V) - (1/\tau')\ p_o(\underset{\sim}{r})\ (F) \qquad (15)$$

Hence the EBIC current becomes:

$$I_{cc} = \int G(\underset{\sim}{r})\ \Phi(\underset{\sim}{r})\ dv = I_o - \frac{1}{\tau'} \int p_o(\underset{\sim}{r})\ \Phi(\underset{\sim}{r})\ dV \qquad (16)$$

$$= I_o - I^*$$

Donolato defines the contrast profile across a defect to be (Figure 17)

$$i^*(x) = \frac{I^*(x)}{I_o} \qquad (17)$$

Often contrast is taken to be the maximum value

$$C = \frac{I_o - I_D}{I_o} \qquad (18)$$

where I_D is the minimum value of the EBIC signal on scanning across the defect.

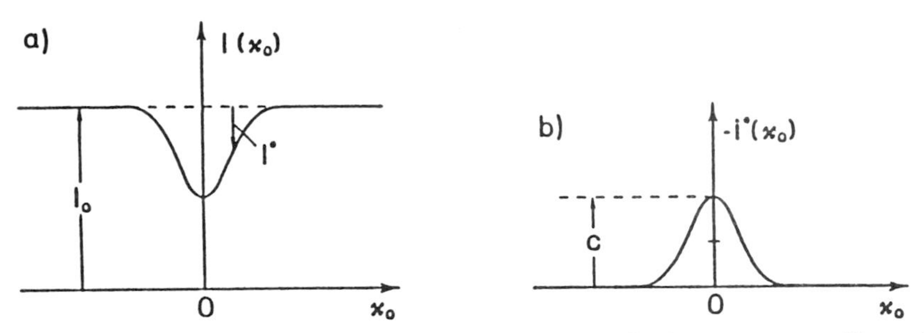

Figure 17. (a) An EBIC profile across a defect and (b) the contrast profile. (After Donolato 1985).

Information about defect properties is obtained from such contrast profiles.

A major attraction of this treatment is that it applies to point, line and surface defects. Only the form taken by the final defect term depends on the geometry of the defect. For a dislocation, the defect volume is a cylinder of radius r_d around the dislocation line within which the lifetime is reduced from τ to τ' (Figure 2). The integration becomes not dV but $\pi\ r_d{}^2$ dl. By taking the constant area of the cylinder out in front, the expression for a dislocation becomes:

$$I^* = \gamma \int po(\underset{\sim}{r})\ dl \qquad (19)$$

where $\gamma = \pi\ r_d{}^2\ /\ D\tau'$ is the dislocation strength or line recombination velocity. Donolato's theory leads to expressions of the form

$$C = \gamma F\ (R,\ L,\ \text{specimen-defect geometry}) \qquad (20)$$

where F is a complicated expression that has to be found explicitly for each experimental situation. However, having obtained an appropriate expression, from measured EBIC line scan profiles, knowing the set of geometrical factors involved such as the dislocation depth, values for γ can be obtained. The few values measured all lie in the range 0.01 to a few times 0.1

Again Monte Carlo simulations to obtain $g(\underset{\sim}{r})$ can be used in programs to compute EBIC profiles and curve fitting can then be used to evaluate γ. By curve fitting the 'perfect crystal' EBIC signal I_o many or all the parameters needed for the evaluation of γ can be obtained. This is making the method much more useful.

Although this field is very new the results already show that extended dislocations have much greater recombination strengths than constricted ones and a physical model has been proposed to treat the temperature dependence of dislocation EBIC contrast.

In the case of surface defects such as grain boundaries in polycrystalline Si solar cells for example, the defect volume becomes a slab of thickness h and the second integral in equation (16) is over the surface. The defect strength obtained is then the grain boundary (or other area defect) surface recombination velocity. It has been shown that EBIC can be used to measure the changes in grain boundary surface recombination velocities with passivation treatments. This opens the possibility of applying EBIC to evaluate the effectiveness of alternative processing procedures.

The EBIC method has spatial resolution which not only makes it possible to measure the electronic properties of defects but also to measure the variations in materials or barrier properties in non-uniform materials and devices. This is a characeristic it shares with the CL mode for which a similar phenomenological contrast theory has recently been developed (Yakubowicz 1987).

There are analogous techniques in which a scanning light beam provides the excitation (carrier generation). They are known as OBIC or LBIC (optical or laser beam induced current) and SRPL (spatially resolved photoluminescence). The first papers on interpretive theory for these techniques have just appeared as have the first commercial scanning laser microscopes.

The amount of interest these electronic (and optoelectronic) microcharacterization techniques have aroused is such as to leave no doubt that development will continue to be rapid and that much will be learned by their application to materials and device problems.

REFERENCES AND

Suggestions for Further Reading:

Books

Holt D B and Joy D C (Editors) 1989 *SEM Microcharacterization of Semiconductors.* (Academic press, London)

Newbury D E , Joy D C, Echlin P, Fiori C E and Goldstein J I 1986 *Advanced Scanning Electron Microscopy and X-ray Analysis* (Plenum Press, New York).

Review Articles

Donolato C 1985 *in Polycrystalline Semiconductors. Physical Properties and Applications* (G Harbeke, Ed.) (Springer–Verlag, Berlin). pp 138–154. "Beam–Induced Current Characterization in Polycrystalline Semiconductors."

Donolato C 1988 *Scanning Microscopy* **2**, pp 801–811 "Recovery of Semiconductors and Defect Properties from Charge–Collection Measurements."

Jakubowicz A *Scanning Microscopy* **1**, pp 515–533 "Theory of Electron Beam Induced Current and Cathodoluminescence Contrasts from Structural Defects of Semiconductor Crystals; Steady–State and Time–Resolved Problems."

Steeds J W 1989 *in Proc. BIADS 88. J. Phys. Applique.* (to be published) "Performance and Application of a STEM–Cathodoluminescence System."

Yacobi B G and Holt D B 1986 *J. Appl. Phys.* **59** R1–R24. "Cathodoluminescence Scanning Electron Microscopy of Semiconductors."

Depth Profiling of Semiconductor Materials by Secondary Ion Mass Spectrometry

J.B. CLEGG

1. INTRODUCTION

It is apparent from the articles in this volume that considerable progress has been made in the growth of layered semiconductor structures which are required for the fabrication of advanced electronic devices. The evolving epitaxial growth technologies have stimulated the development of a broad range of characterisation techniques which have been systematically applied to these new materials in order to gain a clearer insight into their chemical and physical properties. Many questions are posed during the development of the crystal growth technologies and within the frame of reference of this article, these questions have centred around the nature and distribution of dopant, impurity and matrix atoms within the layered structures. Questions often asked are (1) what is the form of the dopant profile (i.e. concentration as a function of depth), (2) does the dopant profile show abrupt step changes between different concentration levels, (3) how sharp is the interface at a hetero-junction and (4) are spurious electrically active impurity elements present? We are also now seeing the additional requirement for accurate studies of lateral dopant distributions; a topic which is relevant in the development of small scale devices.

Many of these problems have been addressed by the application of SIMS but the reader should remember that other techniques are available which provide complementary information in particular areas. The principle aim with materials characterisation is to establish a reliable data base giving materials properties and to this end a multi-technique approach is always necessary. However it is fair to say that for high sensitivity depth profiling SIMS is the first choice method.

It is widely known that positive and negative ions sputtered from surfaces during keV primary ion bombardment form the basis of SIMS. The historical development of SIMS can be traced back to J.J. Thomson's (1910) experiments with "Kanalstrahlen". These rays striking against a metal plate in a discharge tube caused secondary rays to be emitted in all directions, uncharged for the most part, with only a small fraction carrying a positive charge. The development of these initial observations into a practical analytical technique took many years and the first commercial SIMS instrument was developed by the GCA Corporation in 1967, interestingly for the extra-terrestrial micro-analysis of lunar material. The initial application of SIMS to semiconductor analysis was disappointing due to poor instrument design and performance. However in the late 1970's, a new generation of SIMS instruments, much better suited to the demanding requirements of semiconductor work, became available. Since then SIMS has played a major role in semiconductor characterisation studies and is now regarded as a mature and established technique.

In this article we shall consider the basic principles and methodology of SIMS together with a description of the latest instrumentation available. This is followed by case studies, many from our own work, which show the application of the method thereby identifying its advantages and disadvantages. Finally some concluding remarks are made regarding future SIMS developments in this area. One of the aims of this article is to give some guidance to the materials scientist on the reliability and interpretation of SIMS data. For a fuller description of SIMS and its application the reader is referred to the book by Benninghoven et. al. (1987).

2. PRINCIPLE OF SIMS

In SIMS the sample is bombarded in vacuo by an energetic beam of primary ions (1 to 20keV). As a result, particles are sputtered from the sample surface, some of which are in the form of secondary ions. The positive or negative ions are extracted into a mass spectrometer and separated according to their mass to charge (m/e) ratio. This is shown schematically in Figure 1, together with the three basic types of signal output. With primary beam rastering, the sputter erosion causes the sample surface to recede in a very controlled way. By monitoring the intensity of one or more mass peaks as a function of bombardment time, an in-depth concentration profile is obtained. Generation of secondary ion images provides information concerning the lateral distribution of elements across the sputtered surface at a specific depth. With modern instrumentation the ion images from successive sample layers may be stored thus enabling 3-D element analysis within the sputtered volume.

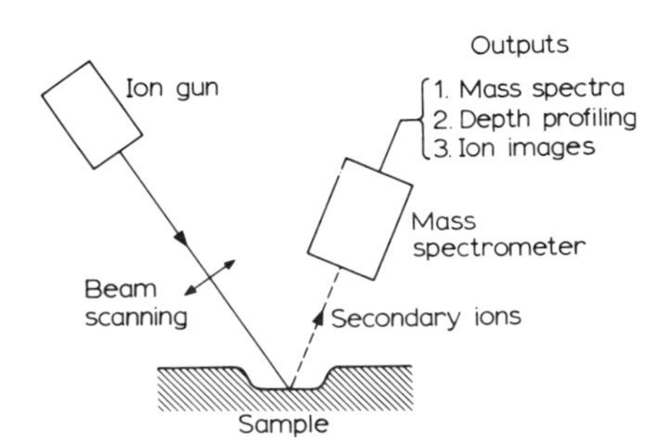

Figure 1: Principle of dynamic SIMS with beam scanning to give a flat bottomed crater in the target. Crater sizes are generally in the range $5x5\mu m^2$ to $500x500\mu m^2$. With modern instrumentation outputs 2 and 3 are combined.

3. PROJECTILE-TARGET INTERACTIONS

The interaction of (primary) projectile ions with the target host atoms is of importance as it gives rise to both removal and ionization of the sample atoms. We consider here some key aspects of these coupled processes some of which are shown schematically in Figure 2. In a SIMS experiment most of the projectile ions incident on the target penetrate (as neutral species) into the near surface region. According to the linear cascade theory (Sigmund 1969) the projectile shares its energy with target atoms initially at rest in a series of binary collisions, a process in which fast recoils are created. These in turn set other target atoms in motion and give rise to an isotropic collision cascade. The cascade continues to develop until the transferable energies become less than the displacement energy (ca 10eV) and the primary species comes to rest in the host lattice (ion implantation). Under the conditions prevailing in SIMS the life-time of the cascade is about 10^{-12} sec. and its dimensions are of the order of 10nm. When a collision sequence intersects the surface an atom or group of atoms may receive sufficient energy in a suitable direction to overcome the surface binding forces and are sputtered from the target. The most probable energy of these emitted atoms and ions is about 10eV and the depth of origin in the SIMS signal is restricted to about 2 to 3 atom layers.

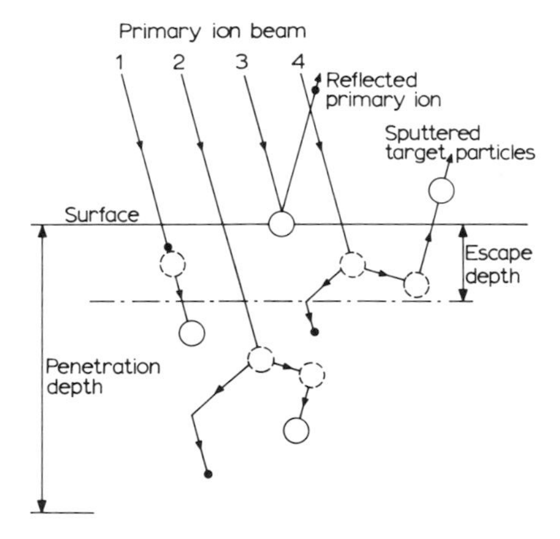

Figure 2: Schematic of projectile-target interactions occurring in SIMS. Process (1) shows recoil implantation (2) cascade mixing (3) primary ion reflection and (4) cascade mixing with sputtering.

There are three important points which arise from Figure 2:-

1. The sputtering process in dynamic SIMS leads to a loss in bonding information between atoms.

2. Whilst the escape depth of ca. 1nm suggests a high depth resolution it should be remembered that the target atoms are intermixed within a region

corresponding to the penetration depth of the projectile ion (see Figure 2 for primary recoil mixing and cascade mixing). These effects which can give rise to broadening of the depth profile may be minimised by employing low beam energies and oblique bombardment.

3. The target surface composition is changed due to the implantation of the projectile species. For high sensitivity SIMS, oxygen and cesium beams are employed and these give rise to the formation of an oxidised chemical phase in the near surface region. It has been shown that the bombardment of Si with O gives an altered layer with a composition of $Si0_x$, where $x \leq 2$ depending on the analysis conditions (Augustus et. al. 1988). The thickness of this layer (20nm for 4keV normal incidence O_2^+ bombardment) is thicker than expected from ion implantation theory and suggests that some of the implanted oxygen diffuses deeper into the target. We have also observed that O bombarded GaAs surfaces are converted into an oxidised As deficient phase with a thickness somewhat thicker than with Si targets (Gale and Gowers 1989). Thus the SIMS signal is actually originating from a region which has a different composition from the bulk material. Additional perturbing process which occur in this altered layer (e.g. radiation enhanced diffusion and segregation and preferential sputtering) must be considered when interpreting high resolution depth profiles. As a guide we can say that all these process limit the best attainable depth resolution to about 3nm with 10nm being more general.

The processes of secondary ion formation which occur during sputtering are complex and there is no unified theory available which adequately describes these processes. Here we shall briefly consider some aspects of ion emission which are important to high sensitivity analysis.

It has been known for some time that the presence of electronegative elements such as oxygen at the surface of a bombarded sample enhance the degree of ionization, β_i (the fractional amount of element i sputtered as i^+ or i^-), of secondary ions by up to three orders of magnitude compared to emission from oxygen free samples under noble gas bombardment (Wittmaack 1980a). A comparable effect is observed with negative ion emission when the sample is bombarded with an electropositive element e.g. Cs. The yield enhancement, which is thought to be associated with changes in the work function (Williams and Evans 1978), is critically dependent on the concentration of O or Cs in the surface region. For example Wittmaack (1981b) has shown that with O bombardment of Si, the enhanced signal intensity is proportional to $C_o^{3.7}$ where C_o is the near surface oxygen concentration. In a simple model Deline et. al. (1978) proposed that C_o was inversely related to the matrix sputter yield (Y_m, the number of target atoms sputtered per incident projectile ion). Thus the SIMS enhancement of any element is controlled by both the matrix (as Y_m is matrix dependent) and the bombardment conditions which also affect Y_m. Generally the greatest enhancement is obtained with near normal incident bombardment. However such conditions can result in a degraded depth resolution as considered previously and hence we see an interplay between sensitivity and resolution.

Even with conditions of maximum yield enhancement β_i shows a variation of about 4 orders of magnitude over the periodic table (Storms et. al. 1977). Thus we see that β_i is both element and matrix dependent (the so-called SIMS matrix effect). This obviously leads to problems with quantification as the secondary ion intensities in the mass spectrum (after correction for the isotopic abundance) do not directly relate to the composition of the target. The SIMS spectrum of GaAs leads one to believe that the target was arsenic deficient. This point is continually being emphasised by the protagonists of the more quantitative sputtered neutral mass spectrometry (SNMS)!

It is useful here to give two simple relationships which assume the ionization processes are thermal in character (Jurela 1973):

$$\beta_i^+ = K_i^+ \exp(-E_i/kT) \tag{1}$$

$$\beta_i^- = K_i^- \exp(EA_i/kT) \tag{2}$$

where k is the Boltzmann constant, E_i is the first ionization potential, EA_i is the electron affinity and K_i and T are effectively adjustable parameters which take into account the matrix effect. These relationships only give a semi-quantitative estimate of β_i and for accurate work one must use standard samples for SIMS quantification.

4. INSTRUMENTAL ASPECTS

There are two types of instrumentation which are used in depth profiling studies (a) the ion microscope and (b) the ion microprobe (see Figure 3). Ion microscopes or direct imaging system retain the spatial distribution of the extracted secondary ions as they pass through the mass spectrometer. The mass resolved secondary ions can be

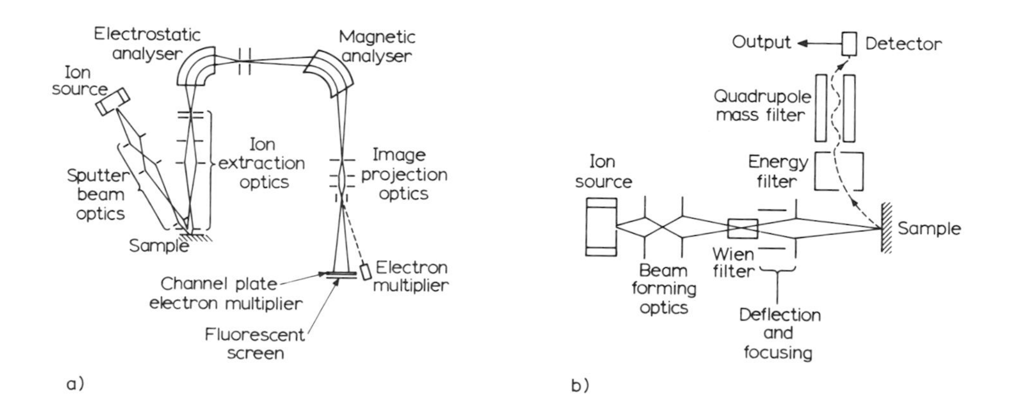

Figure 3: Schematic diagrams of the basic SIMS systems; (a) the direct imaging ion microscope and (b) the raster scanning ion microprobe.

imaged onto a micro-channel plate to observe the ion image or counted with a multiplier. A lateral resolution of about 1μm is achieved within a $\approx$200μm imaged field. The very precise secondary ion optics allow physical apertures to be placed in the column which restrict the field of view of the spectrometer. Thus analytical signals which originate from the crater wall can be largely gated out thereby increasing the dynamic range of the depth profile. This type of crater edge rejection is known as optical gating. Although this instrument can use a broad static beam for sputtering, most depth profiling work requires the beam to be raster scanned to achieve a flat crater base (see Figure 1).

Ion microprobes (Figure 3b) use a finely focused ion beam which is digitally rastered in order to obtain lateral and in-depth information from the sample. The lateral resolution of these systems is controlled by the quality of the primary beam focus. It should be remembered that even though lateral resolutions of about 0.5μm are attainable, this is at the expense of analytical sensitivity. This point will be considered in more detail in a following section. With the microprobe systems crater edge effects are reduced by electronic gating of the detected signal so that only ions which have originated from central regions of the crater are counted. Most microprobe systems also exhibit some form of optical gating and this gives further edge rejection. In the main, the microprobe systems have employed quadrupole mass filters rather than the more expensive sectored electrostatic/magnetic analysers. It should be pointed out that the instrument type (a) in Figure 3 can operate in the probe mode when equipped with a micro-focused gun.

The depth profiling performance of both instrument types has been enhanced recently by improvements to the methods employed for ion intensity data acquisition and storage. With the microprobes, ion intensity data for each sputtered area or pixel are stored sequentially by a computer while automation of either TV camera-based or resistive anode encoder based imaging systems serves to acquire the data in the microprobe mode. Thus the analyst has available after the analysis, the digitised secondary ion images for each rastered frame taken during the experiment. The depth profile can then be obtained from any particular area or volume of the sputtered target (image depth profiling). Arbitrary gated areas can be defined after the analysis and any local surface contaminants which would otherwise influence the profile shape, can be gated out (contamination blanking). With older instruments the occurrence of spurious contaminants within the fixed gated area often required the profile experiment to be repeated in an unaffected sample region.

5. DEPTH PROFILE QUANTIFICATION

The measured secondary ion intensity is dependent on the product of a series of parameters:-

$$I_i = \dot{z}\, \beta_i\, \eta_i\, A\, N\, c_i \qquad (3)$$

where

I_i = count rate of element i corrected for the isotopic abundance (c/s)
$\dot{z}$ = sample erosion rate (cm/s)
β_i = fractional amount of element i sputtered as i^+ or i^-
η_i = detected/emitted ions from area A
A = analysis or gated area (cm^2)
N = sample atomic density ($atoms/cm^3$)
c_i = atomic concentration of element i

The sample erosion rate depends on:-

$$\dot{z} = I_p\, Y_m/A_b\, N = j_p\, Y/N \qquad (4)$$

where

I_p = primary ion flux (ions/s)
A_b = bombarded or rastered area (cm^2)
j_p = primary ion flux density ($ions/cm^2/s$)
Y_m = target sputtering yield (atoms/primary ions)

Before considering some aspects of quantification we shall make some comments regarding the above equations. It is apparent from (3) that the analytical sensitivity (the count rate per unit concentration) is enhanced by increasing both A and $\dot{z}$. As most semiconductor depth profile studies are carried out on the as-prepared layers and slices prior to device fabrication, the actual sample size available is not generally restrictive ($\approx$6x6mm^2). In practice the maximum sizes of A_b and A are about 1x1mm^2 and 0.3x0.3mm^2. The sample erosion rate $\dot{z}$ given in equation (4) is not a freely variable parameter and is controlled in the limit by the useful ion flux at the target and by other factors such as the required profile data density (the number of measurement points per unit depth eroded). In contrast to the use of relatively large analysis areas there is the increasing demand for restricted area and on-chip analysis where the region of interest may only have an area of several microns square (Vandervorst 1988). However it can be seen from equation (3) that a decrease in A is accompanied with a corresponding reduction in the analytical sensitivity for a given sputter rate. The limits of detection are accordingly degraded by several orders of magnitude.

To obtain a quantitative depth profile the raw intensity versus time data must be converted into concentration versus depth i.e. c_i values are determined for each eroded step interval. The target erosion rate is simply obtained by measuring, with a surface profilometer, the total eroded depth (z) of the crater formed after an erosion time (t). With a sample of constant matrix composition, $\dot{z}$ is given by (z/t); this relationship is valid under steady state sputtering conditions, i.e. for eroded depths greater than about 10nm from the sample surface or a step-up or step-down interface. With equation (3) we need to know the product $\beta_i\ \eta_i$, known as the useful ion yield, to be able to calculate c_i. As mentioned previously there are no first principle methods of obtaining β_i and η_i is difficult to measure. Thus in practice one has to establish the useful yield by the analysis of a reference material containing impurity i generally at a higher level

of concentration (ca. $10^{19}cm^{-3}$) than in the unknown sample. The requirement in quantitative SIMS to calibrate each impurity in each matrix is clearly a restrictive feature of the technique.
Ion implantation provides a convenient means of preparing suitable standards. Almost any isotope can be implanted in any matrix with a known dose. Integration of the profile then allows the useful ion yield or a corresponding sensitivity factor to be determined for element i in the given matrix. To optimise the analytical accuracy it is preferable to carry out the calibration procedure just before or after the analysis of the unknown. Accuracies of about $\pm 10\%$ can be achieved at the $10^{17}cm^{-3}$ level with elements which show high yields. If these sensitivity factors are subsequently used over extended time periods then the concentration uncertainty is increased (10 to 50%). When calibration samples are not available one has to resort to the use of empirical expressions which predict the useful ion yield. In some instances, with semiconductor materials, the error may amount to a factor of 2 (Morgan and Clegg 1980a) but more commonly the error may be an order of magnitude. In this situation it is advisable to consider the use of other techniques which give more quantitative raw data.

6. DEPTH PROFILING APPLICATIONS

Ion Implanted Profiles

Ion implantation is widely used for the controlled doping of semiconductor materials. Originally the technique was used to produce rather deep ca.1μm implants but the emphasis has now changed towards much shallower implants giving junction depths (ca.10 to 100nm) much closer to the surface (Brotherton et. al. 1986). For the optimisation of the various semiconductor process technologies it is important to have a knowledge of the dopant profile. Whilst several ion implantation theories such as LSS have been developed which describe the profile in an amorphous or random target, they do not give an accurate representation of the full profile in crystalline semiconductors. The reasons for this are (1) ion channelling which extends the ion range and (2) dopant spreading during the anneal treatment required to remove crystal damage. SIMS has played an important role in studying such effects and as an example we first consider some ion channelling data in Si.

When an ion enters the single crystal in a direction of a low-indexed crystallographic axis, the penetration can be much deeper than expected because of the steering effect due to open channels between regular rows and planes of atoms. The channelled ion has infrequent collisions with the target atoms and electronic losses determine the path length. The ion is finally stopped by nuclear collisions at the end of the path. To minimise this effect the crystal slice is orientated in the implanter to give a reduced transparency (non-channelling or random orientation). Figure 4 shows some channelling measurements we have made with (100) Si using three particular slice orientations in the implanter. It can be seen that the conditions of 22° rotation and 10° tilt give rise to the sharpest profile. With the more commonly used condition

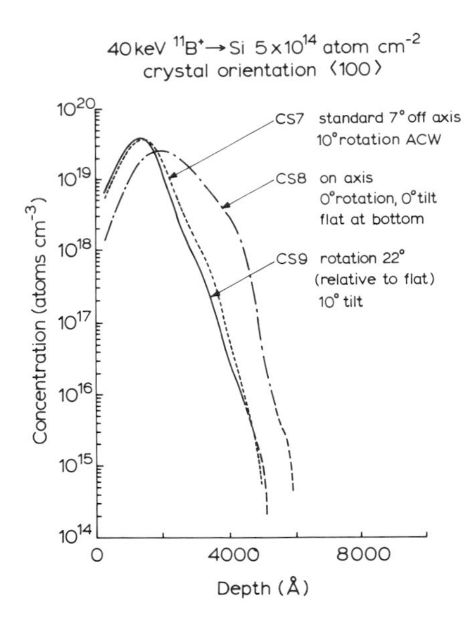

Figure 4: SIMS boron profiles showing the effects of ion-channeling. The slice orientations in the implanter are given. The SIMS data has been digitally smoothed.

there is seen to be a small profile shift due to planar channelling while on axis conditions give rise to a much broader profile (planar and axial channelling).

For these measurements we used the given ion implanted dose in CS9 to establish the B sensitivity factor. Inspection of Figure 4 shows that a profile dynamic range of 5 orders of magnitude is obtained with the detection limit for B being in the mid-$10^{14}cm^{-3}$ region. The precision of the measurements at this level of concentration, which is controlled by counting statistics amounts to about $\pm 50\%$ with the un-smoothed data. It should also be noted that the data in the near surface region has not been plotted. SIMS quantification in this region is difficult due to non-equilibrium sputter conditions and with these relatively deep implants it is permissable to omit these points from the profile.

The second example of the application of SIMS to the measurement of implanted profiles is taken from our work with hot electron structures where the doped layer thickness should be comparable to the carrier mean path length. We have investigated the use of low energy implantation to make such structures and have used SIMS to measure the implanted dopant profiles (Shannon and Clegg 1984). Our results from a 5keV As implant into crystalline Si (random orientation) are shown in Figure 5. To enhance the SIMS depth resolution, low energy bombardment was used at 45°

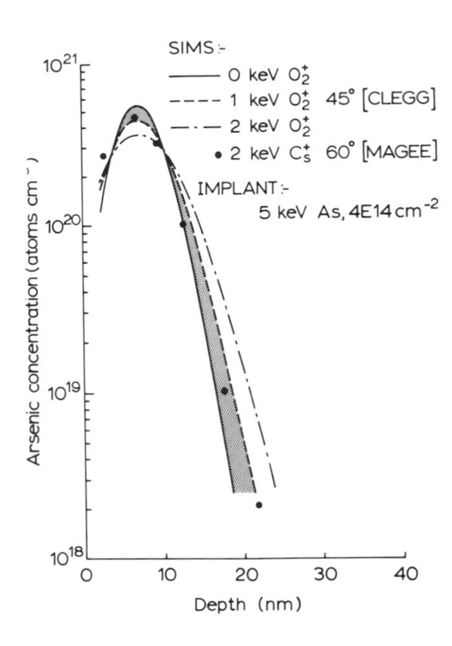

Figure 5: SIMS arsenic profiles as a function of probe energy. The points were taken from a profile measured by Magee.

incidence. The data show that as the probe energy is reduced the profiles become sharper, i.e. the measurements are limited by atomic mixing effects during sputtering. An extrapolation to zero probe energy gives rise to the continuous curve (Clegg 1987a). The most probable profile is thought to lie within the shaded region. Also note that with this sample, the first 2nm of the profile is not plotted due to an interference from the thin native oxide. It is encouraging to note that rather good agreement is shown with independent measurements made by Magee (1985). Of importance is the fact that his data with Cs bombardment extended down to the $10^{16}cm^{-3}$ region and the profile showed evidence of significant ion channelling in the tail region. The profiles given in Figure 5 represent state of the art performance at the limit of SIMS depth resolution.

Dopant profiles in MBE GaAs epitaxial layers

SIMS has been particularly useful for studying dopant incorporation in III-V compound semiconductors. Many device structures require rather abrupt dopant profiles and migrational processes which redistribute dopant atoms during growth are obviously undesirable. With this in mind we have investigated the incorporation of Sn in GaAs (Harris et. al. 1984). For a particular mixer diode structure we required an active undoped layer to be grown on a n^+ buffer layer. As can be seen from Figure 6 layer growth at the normal substrate temperature of 550°C gave rise to an exponentially decaying Sn profile in the active layer even though the Sn effusion cell had been shuttered. However when the active layer was grown at a reduced temperature (490°C) this effect is seen from Figure 6 to be largely suppressed. These observations have been explained in terms of a kinetically limited surface segregation process which occurs during MBE growth (Harris et. al. 1984). We note that Si rather than Sn is now widely used for n-type doping as the driving forces for segregation are much less.

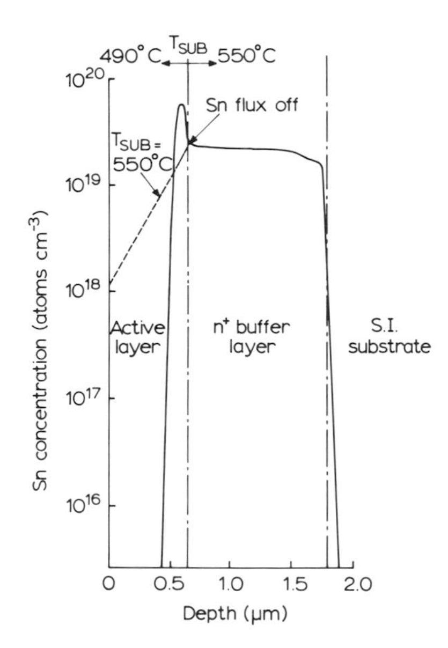

Figure 6: SIMS tin profiles of a MBE GaAs structure grown with different substrate temperatures.

Two points regarding these SIMS measurements should be mentioned. First we see that the profile is measured down to $5 \times 10^{15} cm^{-3}$ and second the interface sharpness may be quoted in terms of the depth over which the concentration changes by one order of magnitude; with the modified growth procedure this amounts to 22nm/decade.

Interest is currently being shown in the use of an interrupted MBE growth technique to give atomic plane or δ-doping. Before SIMS is applied to the analysis of such structures it is advisable to have a knowledge of the SIMS resolution and how it may vary with eroded depth. To obtain this information we have prepared a special test sample, containing a series of δ-spikes, using a reduced growth temperature (400°C) so as to minimise any Si migrational processes. The analysis of this test sample then gives a profile whose shape is determined entirely by sputter induced atom relocation effects. This profile therefore represents the form of the resolution function from which the limiting depth resolution can be obtained (Clegg and Beall 1989b. An interesting aspect arising from this work is the comparison we have made between the SIMS δ-spike profile and the predicted profile given by the IMPETUS model. This model, which predicts the sputtered flux of atoms during the SIMS experiment, takes into account the ballistic mixing of the target atoms and the accommodation of the implanted projectile ions (Armour et. al. 1988). The comparison, using an atomically sharp input δ-spike, is shown in Figure 7. It can be seen that the agreement is rather good (Badheka et. al. 1989). Again we emphasize that the SIMS profile, shown by the circles, is entirely limited by mixing effects. Measurement of the profile width at x0.61 of the peak maximum leads to a depth resolution of 2.5nm with our particular experimental conditions.

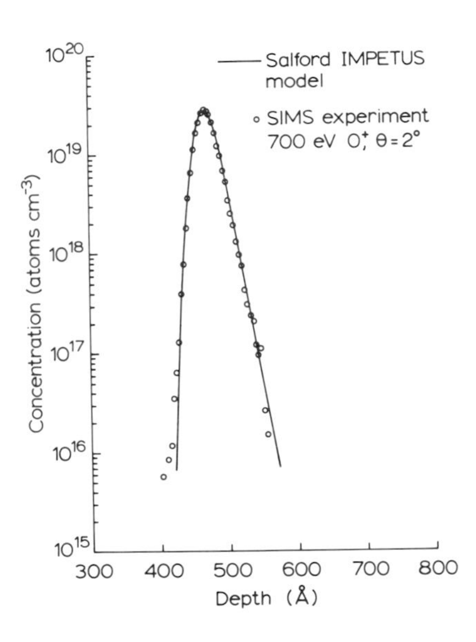

Figure 7: Predicted profile shape for a sputter mixed silicon δ-doped plane ($1x10^{13}cm^{-2}$Si) in GaAs using the IMPETUS code. The independent SIMS data are also given.

These limiting resolution measurements are important as they give a resolution base-line from which we can judge results obtained from the δ-doped samples grown at normal temperatures (Beall et. al. 1988). This is exemplified by the data shown in Figure 8. To enhance any time dependent Si migration, the growth time was deliberately extended by 30min interrupts between spike 1 and 3. These profiles show quite clearly that the longer each spike was held at temperature the broader it became. Note that this observation is not due to a loss of SIMS resolution with depth; with our experimental set-up we have shown (Clegg and Beall 1989b) that the resolution is essentially constant for eroded depths up to ca.500nm. If the profile broadening shown in Figure 8a was due to Fickian diffusion from a plane source of atoms then one would expect the profile to have a Gaussian form with a standard deviation σ given by:

$$\sigma = (2Dt)^{\frac{1}{2}} \tag{5}$$

where t is the anneal time and D is the diffusion coefficient. Inspection of an expanded plot of δ-spike 2 (see Figure 8b) actually shows that it has the form of two half Gaussians. The slight asymmetry which is caused by SIMS mixing can be removed

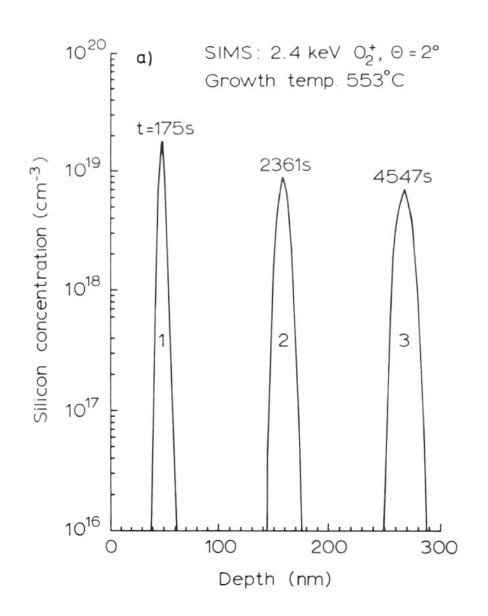

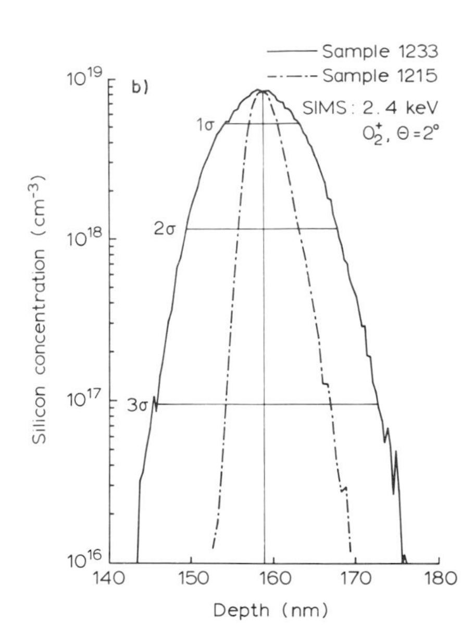

Figure 8: (a) SIMS profiles of δ-doped planes deposited at 553°C with interruption of the layer growth to give extended anneal times and (b) an expanded plot of δ-spike (2) with the chained profile showing the form of the resolution function.

by a cross correlation technique applied to the resolution function and the experimental profile. After such corrections we arrive at a value of D_{si} of $3.8x10^{-17}cm^2s^{-1}$ at 553°C (Clegg and Beall 1989b).

Profiling heterostructures

The profile examples given so far have involved layers of uniform matrix composition. When one attempts to measure the distribution across a heterostructure interface then the analytical task becomes more complicated as both the useful ion yield and the sputter rate are changing. Morgan and Maillot (1988) have investigated the measurement of dopant profiles across the Si/SiO_2 interface and have developed procedures to obtain quantitative data. Similar studies have been made by Galuska and Marquez (1986) on the $Al_xGa_{1-x}As/GaAs$ system while Gao et. al. (1988) have reduced the problem by selecting the analysis conditions so as to suppress the matrix effect. Thus there are available calibration procedures which aim to give quantitative profiles across heterojunctions but in the main most SIMS profiles still show raw data with their correspondingly systematic biases.

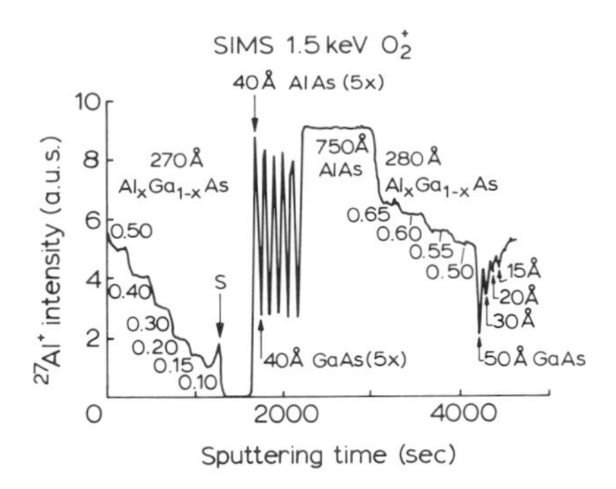

Figure 9: SIMS Al^+ profile obtained from a complex $Al_xGa_{1-x}As$ multilayer structure. The composition of the various layers is indicated in the figure. Taken from Boudewijn et. al. (1986) and reproduced with permission of John Wiley & Sons Ltd.

An example of multilayer matrix element profiling is shown in Figure 9. The multilayer was grown by MOVPE and the expected x values are indicated in the figure. From a comparison with the growth scheme and cross-sectional TEM results it was shown by Boudewijn et. al. (1986) that all the untended features were observed in the Al profile and even the presence of 15Å of GaAs was just detectable. Furthermore, quantitative SIMS analysis with standards showed that the actual x content of the 270Å thick AlGaAs layers agreed within 5% of the intended value. These SIMS measurements also highlighted a problem of mass flow control during growth as is shown by the unexpected spike marked S in Figure 9. SIMS matrix element profiles like the one shown in this figure are frequently presented as this information is available together with the impurity data in one single experiment. However one should always consider the use of other more quantitative techniques such as Auger depth profiling or SNMS (Reuter 1986) to determine the matrix element distributions.

Restricted area analysis

The increasing demand for depth profile analysis in smaller areas has resulted in the development of image depth profiling (see section 4). Here we consider the use of the checkerboard technique to control the actual area analysed (Frenzel et. al. 1988). Part of the rastered area is divided into a 16x16 pixel array and intensity data in each pixel in each frame are stored. Post analysis data processing then allows depth profiles to be constructed from various regions of the sputtered sample contained within different sub-gates. Results from the analysis of a Fe implanted GaAs substrate are shown in Figure 10. Taking the total gated signal, a shoulder is seen to be present on the side of the implant profile due to the presence of a buried inhomogeneity containing Fe. Using the appropriate gated area as displayed on the ion image, this feature was gated out so as to reveal the original implant profile. It is worth noting that this particular checkerboard size does not give an excessive amount of data for storage and is compatible with likely probe sizes (1 to 30μm) and raster widths (5 to 500μm) now used in profiling studies.

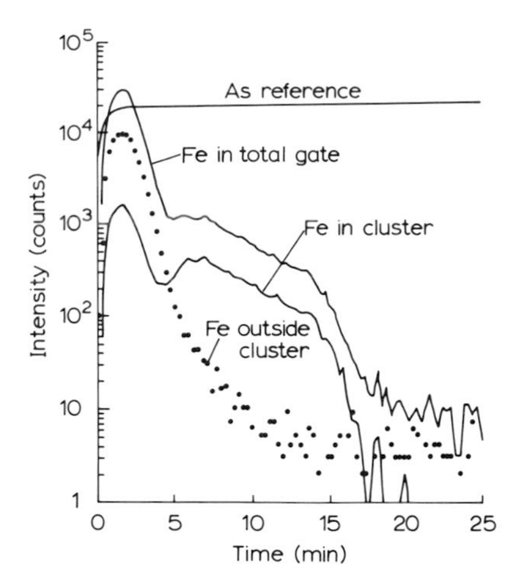

Figure 10: SIMS iron profile in an implanted GaAs sample. The gated area was selected using the checkerboard technique. Maximum gated area 125x125μm^2, detection limit ca. $1 \times 10^{14} cm^{-3}$. Reproduced with permission of Atomika.

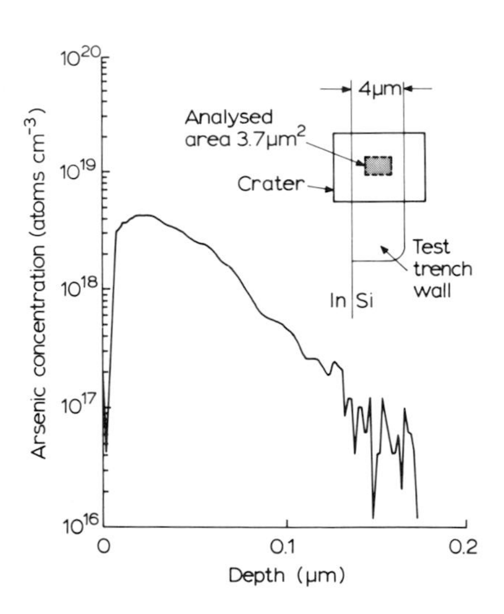

Figure 11: SIMS profile of arsenic measured in a test Si trench wall (15keV Cs^+, 10^{-11}A, smoothed data). Taken from von Criegern et. al. (1988) and reproduced with permission of John Wiley & Sons Ltd.

An example of restricted or device area analysis is shown by the work reported by von Criegern et. al. (1988) involving the analysis of Si trench walls used in 4 megabit memories. The trench walls are typically 1μmx1μm wide and 4μm deep and are doped with an SiO_2/As source which is subsequently removed. A knowledge of the dopant incorporation in the side walls is important for device optimisation. These workers developed a special sample preparation technique whereby longer trenches (200μm) were sectioned and supported in an In mount prior to SIMS analysis. A 0.6μm Cs microbeam (10^{-11}A) was raster scanned over an area not greater than 6x6μm^2 in order to keep the analysis time to about 1 hour. The As profile obtained from an electronically gated area of 3.7μm^2 is shown in Figure 11. The profile is followed over 2 orders of magnitude with the limit in the mid-$10^{16}cm^{-3}$ region or 1ppma. Interestingly these measurements confirmed other observations which indicated a reduced peak doping. This work is a good example of careful SIMS optimisation to solve a particular analytical problem.

7. FINAL REMARKS

This article shows that considerable progress has been made with the development of the SIMS technique and its application to semiconductor problems. The method now has a good quantitative base for depth profiling studies in most semiconductor systems. The interaction between the crystal grower and the SIMS experimentalist to produce, for example, test samples has been particularly beneficial in many areas. Further benefits regarding the improvement in our knowledge of SIMS have also been obtained by comparisons with other characterisation techniques. The basic problem associated with the variability of the secondary ion yields still causes difficulties and with this in mind much effort is now being directed towards the analysis of the sputtered neutral flux. Matrix effects are still present to some extent in these measurements and the best solution may be to combine both SIMS and SNMS in one analytical system.

ACKNOWLEDGEMENTS

The author would like to thank Dr. C.W. Magee and Dr. J. Maul for supplying the Si(As) and GaAs(Fe) profile data.

REFERENCES

Armour, D G, Wadsworth M, Badheka R, van den Berg J A, Blackmore G, Courtney S, Whitehouse C R, Clark E A, Sykes D E and Collins R, 1988, Proc. SIMS VI (France) 1987, eds. A Benninghoven, A M Huber and H W Werner (John Wiley) pp399-407.

Augustus P D, Spiller G D T, Dowsett M G, Kightley P, Thomas G R, Webb R and Clark E A, 1988, Proc. SIMS IV (France) 1987, eds. A Benninghoven, A M Huber and H W Werner (John Wiley) pp485-488.

Badheka R, Wadsworth M, van den Berg J A, Armour D G and Clegg J B, 1989, to be published in Surf. Interface Anal.

Beall R B, Clegg J B and Harris J J, 1988, Semicond. Sci. Technol. **3** 612

Benninghoven A, Rudenauer F G and Werner H W, 1987, Secondary Ion Mass Spectrometry (John Wiley)

Boudewijn P R, Leys M R and Roozeboom F, 1986, Surf. Interface Anal. **9** 303

Brotherton S D, Gowers J P, Young N D, Clegg J B and Ayres J R, 1986, J. Appl. Phys. **60** 3567

Clegg J B, 1987a, Surf. Interface Anal. **10** 332

Clegg J B and Beall R B, 1989b, to be published in Surf. Interface Anal.

Deline V R, Katz W, Evans C A and Williams P, 1978, Appl. Phys. Lett. **33** 832

Frenzel H, Maul J L, Martens H, Raab R and Scholze Ch, 1988, Proc. SIMS VI (France) 1987 eds. A Benninghoven, A M Huber and H W Werner (John Wiley) pp219-224

Gale I G and Gowers J P 1989, private communication.

Galuska A A and Marquez N 1986, Proc. SIMS V (USA) 1985 eds. A Benninghoven, R J Colton, D S Simons and H W. Werner (Springer Verlag) pp363-365

Gao Y, Godefroy and Mircea A 1988, Proc. SIMS VI (France) 1987 eds. A Benninghoven, A M Huber and H W Werner (John Wiley) pp761-764

Harris J J, Ashenford D E, Foxon C T, Dobson P J and Joyce B A, 1984, Appl. Phys. **A33** 87

Jurela Z, 1972, Rad. Effects **13** 167

Magee C W, 1985, private communication.

Morgan A E and Clegg J B 1980a Spectrochimica Acta 35B 281

Morgan A E and Maillot P, 1988b, Proc. SIMS VI (France) 1987 eds. A Benninghoven, A M Huber and H W Werner (John Wiley) pp709-716

Reuter W 1986, Proc SIMS V (USA) 1985 eds. A Benninghoven, R J Colton, D S Simons and H W Werner (Springer Verlag) pp94-102

Shannon J M and Clegg J B 1984 Vacuum **34** 193

Sigmund P 1969 Phys. Rev. **184** 383

Storms H A, Brown K F and Stein J D, 1977, Anal. Chem. **49** 2023

Thompson J J 1910 Philos. Mag. **20** 752

Vandervorst W 1988, Proc. SIMS VI (France) 1987 eds. A Benninghoven, A M Huber and H W Werner (John Wiley) pp409-418

Von Criegern R, Zeininger H and Rohl S, 1988, Proc. SIMS VI (France) 1987 eds. A Benninghoven, A M Huber and H W Werner (John Wiley) pp219-224

Williams P and Evans C A 1978, Surf. Sci. **78** 324

Wittmaack K 1980a, Nucl. Instr. and Meth. **168** 343

Wittmaack K 1981b Appl. Surf. Sci. **9** 315

Localised Vibrational Mode Spectroscopy of Impurities in Semiconductor Crystals

R.C. NEWMAN

1. INTRODUCTION

Semiconductor crystals are doped during their growth to make them n or p-type and they may also become contaminated inadvertently with other impurities. Examples of the latter process are the introduction of oxygen into silicon crystals grown from melts contained in a silica crucible: the incorporation of carbon into gallium arsenide: the introduction of hydrogen into III-V compounds, etc. To understand the electrical and optical properties of crystals it is necessary to know the concentrations of the impurities present and also their distribution amongst possible lattice locations. As an illustration it is known that silicon impurities in GaAs are amphoteric and may occupy either gallium lattice sites Si(Ga) where they act as donors, or arsenic lattice sites Si(As) where they act as acceptors. At high concentrations it might be expected that near neighbour Si(Ga) - Si(As) donor acceptor pairs would be present and even larger clusters of impurities may occur because of the limited silicon solubility. All of these defects are in fact produced, but in addition pairing with other impurities may take place. As the processing of the material proceeds, involving heat treatments either with or without prior ion-implantation, the distribution of the impurity lattice sites may change. The purpose of this chapter is to show how material may be characterised by means of the infrared vibrational absorption induced by the presence of the impurities and their complexes. Further details may be obtained from Fan (1956), Spitzer (1971), Newman (1973) and Barker and Sievers (1975).

2. THEORETICAL BACKGROUND

2.1 Perfect Crystals

It is not possible to describe the vibrational modes relating to impurities without first discussing the modes of the perfect crystal. This task is in itself extremely demanding but insight sufficient for the later discussion can be obtained by considering simple models. The basic crystallographic unit cells for the diamond and zinc blende structures are rhombohedral and contain two atoms. It follows that the vibrational modes may be classified into three acoustic and three optic branches. A simplification occurs for a one-dimensional model as there is then only one branch of each type (Dekker 1958). It is assumed that the masses of the two atoms in the unit cell are M and m, where M > m and that they are linked only to nearest neighbours by springs of force constant f. The frequencies of the normal modes are given by:

$$\omega^2 = f\left(\frac{1}{M} + \frac{1}{m}\right) \pm f\left[\left(\frac{1}{M} + \frac{1}{m}\right)^2 - \frac{4\sin^2(qa)}{Mm}\right]^{1/2} , \tag{1}$$

where q is the wavevector and a is the distance between the atoms. The amplitudes of vibration A and B of alternate atoms along the chain satisfy the relation:

$$\frac{A}{B} = -\frac{2f\cos(qa)}{M\omega^2 - 2f} . \tag{2}$$

For the optic branch (Fig.1), corresponding to the positive sign in Equation 1, A and B have opposite signs and neighbouring atoms vibrate out of phase. For the acoustic branch with the negative sign in Equation 1, A and B have the same sign and neighbouring atoms vibrate in phase. The maximum lattice frequency ω_{Max} which occurs at q = 0 (vibrational wavelength $\lambda = \infty$) is given by:

$$\omega^2_{Max} = 2f\left[\frac{1}{M} + \frac{1}{m}\right]^{1/2} . \tag{3}$$

The two sub-lattices vibrate out of phase and for any "diatomic" pair the force constant is 2f and the reduced mass is $(M^{-1}+m^{-1})$. At the zone boundary where $q = \pm \pi/2a$, it can be shown that B = 0 for the acoustic branch: all the atoms with mass m are at rest and $\omega_A^2 = 2f/M$. For the optic branch A = 0: the heavy atoms are at rest and $\omega_o^2 = 2f/m$. If m = M, as for silicon, the optic and acoustic branches are degenerate at the zone boundary but otherwise there is a "gap" between the two bands of frequencies.

Absorption of incident radiation can only occur if both energy and momentum (crystal momentum) are conserved. As the momentum of a photon $\hbar\omega/c$ is so small because of the large value of the velocity of light c absorption would only be possible at (very near to) q = 0 for the optic modes ($p = \hbar\omega/c$ is essentially a vertical line on Fig.1). For silicon there is no dipole moment because of the inversion symmetry of each "di-silicon molecule" and there is no coupling to the radiation. As a consequence there is no first order absorption. However for III-V compounds the two atoms in the unit cell, for example Ga and As in GaAs, are different and there is a dipole moment leading to strong absorption around ω_{Max} called the Reststrahl band. In this spectral region the crystal is essentially opaque.

Additional absorption for both types of structure arises from changes in the vibrational states of two or more lattice modes (phonons), provided the sum of their wavevectors is zero (Johnson 1959). If two modes are excited the strength of the resulting absorption decreases as the temperature of the crystal is lowered but reaches a minimum usually near 77K. On the other hand, the excitation of one phonon and the de-excitation of a second leads to absorption the strength of which tends to zero as $T \rightarrow 0K$. Clearly this process cannot occur once modes have fallen into their ground states. Thus at low temperature (4K) a III-V compound shows little absorption below the Reststrahl, although there is strong two phonon (summation bands) absorption between ω_{Max} and $2\omega_{Max}$. Unfortunately most of the impurity-induced absorption of interest falls in the latter spectral region. To achieve optimum sensitivity for detecting the impurity absorption it is necessary to cool the host crystal to a temperature of 77K or lower. It is noted in passing that three and four phonon intrinsic processes also occur but the associated absorption becomes progressively weaker with the increasing order of the process.

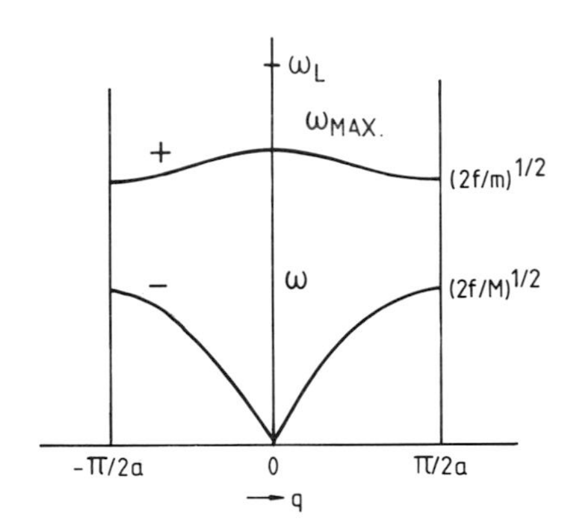

Figure 1
The optical and acoustical frequencies of a diatomic linear chain as a function of the wavevector q showing the maximum lattice frequency ω_{Max} and a localised mode frequency ω_L.

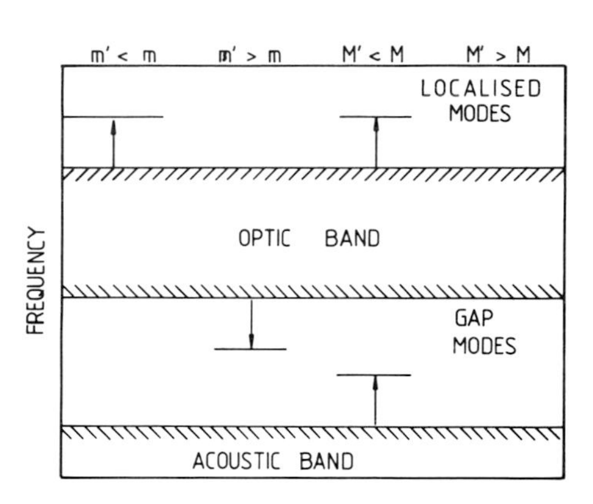

Figure 2
The occurrence of localised and gap modes for impurities of mass m' or M' substituting for host lattice atoms of mass m or M in a diatomic linear chain.

2.2. CRYSTALS CONTAINING IMPURITIES

When impurities are introduced into a crystal the translational symmetry is destroyed, leading to a relaxation of the selection rule ($\Delta q = 0$) for optical absorption, and the normal modes of vibration are modified. For atoms with a mass greater than those that they replace in the host, the modified modes will usually lie within the bands of the optic or acoustic modes of the perfect crystal. Incident radiation will produce excitations because of impurity induced dipole moments but the absorption will be spread over a range of energies. Consequently, little detailed information can be obtained about the identity, or lattice location, of the impurity unless independent data of a different type are available. Exceptions can occur in compound semiconductors with M>>m, as for GaP where M = 69 (^{69}Ga, 60% abundant) or M = 71 (^{71}Ga, 40% abundant) while m = 31 (^{31}P, 100% abundant). If ^{31}P atoms are replaced by ^{75}As there is a reduction in the zone boundary frequency ω_o corresponding to the optic branch and a so-called "gap mode" is produced between ω_o and ω_A(Gledhill et al 1981). Similar effects could occur in InP. However, the width of the gap (ω_o-ω_A) is usually small and it is not certain that the new mode will fall in this region.

The presence of impurities with low masses m_{imp} are of most interest. Before proceeding with the main discussion it is noted that gap modes may again occur if the impurity replaces a heavy host lattice atom as there is then an increase in the frequency of the zone boundary acoustic mode ω_A (Fig.2) Examples of gap mode absorption in GaP are shown in Fig.3. There may also be relatively sharp absorption features (band modes) at other energies where there is a low density of perfect lattice modes as the coupling to the lattice is then necessarily small (Gledhill et al 1981). A good example is afforded by B(Ga) in GaP (Fig.3) where a mode at 156 cm^{-1} occurs just above the maximum frequency of certain acoustic modes of the lattice (see histogram in the figure).

The most important effect of the presence of a light impurity is the emergence of a vibrational mode at a frequency ω_L greater than ω_{Max} (Fig.2) (Dawber and Elliott 1963, Newman 1973). To first order, the local interatomic forces f′ may be assumed to be little changed from those of the host crystal and we may write:

$$\omega_L^2 = 2f'\left[\frac{1}{m_{imp}} + \frac{1}{\chi M_{nn}}\right], \tag{4}$$

where M_{nn} is the mass of the nearest neighbours to the impurity and χ is a parameter that depends on the local angle bending and bond stretching force constants. Since $\omega_L > \omega_{Max}$, vibrations cannot propagate through the lattice and the mode is said to be localised. For an impurity with

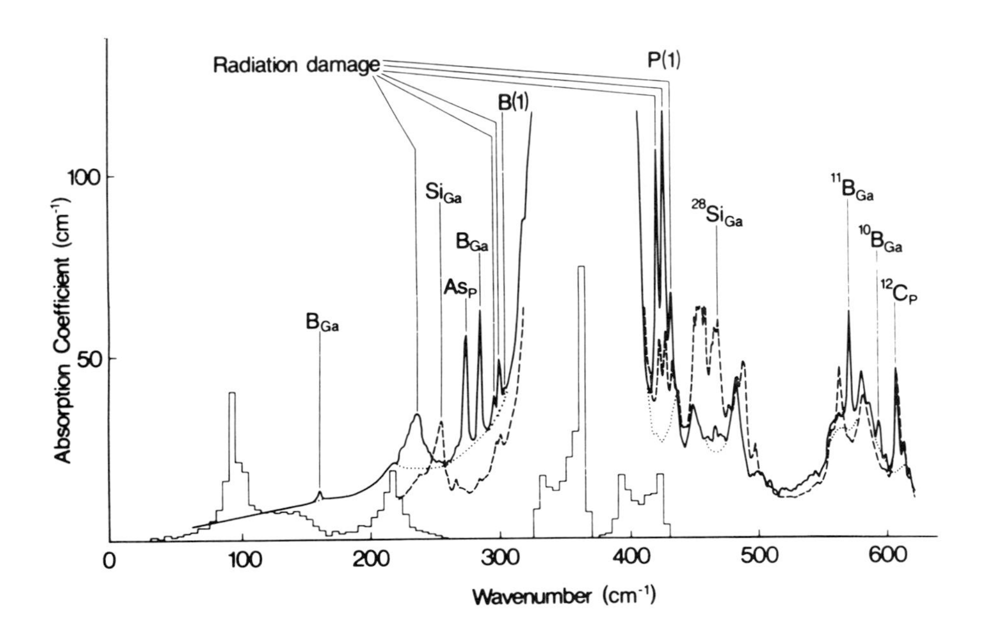

Figure 3
The one-phonon density of states for GaP (Histogram) showing a gap around 300cm^{-1}. Superposed are IR absorption spectra at 80K and 1cm^{-1} resolution of (a) pure, unirradiated GaP (......), (b) GaP containing B(Ga), As(P), C(P) and Si(Ga) impurities following 2MeV electron irradiation to a high dose of 10^{19} electrons cm^{-2} (———) and (c) GaP containing high concentrations of Si(Ga) and C(P) (-------). Note the gap modes from As(P), B(Ga) and Si(Ga) but not C(P): cf Fig. 2. Also note the resonant mode from B(Ga) at 156cm^{-1} in the region of the acoustic modes, and the many lines from radiation damage complexes (Gledhill et al. 1981).

$M_{imp} \to 0$, only the impurity itself would have a vibrational amplitude but in any practical case the motion of the nearest neighbours has to be taken into account. Thc motion of more distant neighbours is usually negligible and ω_L is well described by Equation 4.

For a substitutional atom in the zinc blende structure the symmetry of the impurity is tetrahedral Td and the first excited state of the localised oscillator is triply degenerate: displacements of the impurity in the x, y or z directions are clearly equivalent. If impurity complexes are present, the symmetry will be lowered and the degeneracy of the excited state will be partially or wholly removed. A light impurity paired with a different impurity in a nearest neighbour site has only trigonal (C_{3v}) symmetry. Longitudinal $\omega_{\parallel}$ and $\omega_{\perp}$ modes involving impurity displacements parallel and perpendicular to the pair axis will have different frequencies. For second neighbour pairs (e.g. Si(Ga)-Cu(Ga) in GaAs) there is only one reflection plane, the corresponding point symmetry is C_s and three non-degenerate localised modes of vibration occur. The magnitudes of the splittings of the originally triply-degenerate mode will depend on the identity of the perturbing atom. It follows that the pairing of atoms allows the characterisation of heavy atoms even though these atoms themselves do not give rise to localised vibrational modes (LVM). The technique is particularly valuable when pairs such as Zn(Ga) - H are produced in GaAs because of the high sensitivity of detecting the absorption due to the hydrogen. Comparisons can then be made with analogous Be(Ga) -H pairs, where the light ^{9}Be(Ga) atom gives an LVM line as well as the hydrogen atom, enabling the structure of the defect to be determined in a definitive way (Nandhra et al 1988).

The full width at half height Δ of an LVM absorption line depends on the lifetime of the oscillator in its excited state, strains in the host crystal and the presence of mixed isotopes of the crystal (Dawber and Elliott 1963). The lifetime can be increased by reducing the temperature of the sample leading to sharper lines. It should be remembered that in general $\omega_{MAX} < \omega_L < 2\omega_{MAX}$ and cooling the sample is also advantageous in reducing the intrinsic two phonon absorption in this spectral region. Measured values of Δ are typically about one wavenumber (1 cm^{-1}) at 77K, but in high quality crystals sharper lines can be observed for some impurities. For Si(Ga) in GaAs, $\Delta \simeq 0.4 cm^{-1}$, while linewidths as low as 0.03 cm^{-1} have been reported for hydrogen paired with certain impurities in InP (Clerjaud et al 1988). The presence of various isotopic combinations of ^{69}Ga and ^{71}Ga surrounding an impurity located on an arsenic lattice site in GaAs lead to fine structure of the LVM line as a result of the second term in Equation 4. Such observations allow unambiguous determinations of the site occupied since no such splitting occurs for impurities occupying gallium lattice sites (^{75}As, 100% abundant and only the first neighbours have significant vibrational amplitudes).

Finally, it should be noted that the energy and width of an LVM line show measurable changes depending upon homogeneous and inhomogeneous strains in a crystal. Now that Fourier Transform Infra Red (FTIR) spectroscopy is commonly used the frequencies of LVM lines can be quoted with great precision (say 0.01 cm^{-1} or less), but such accuracy has to be used with discretion. The presence of defects or impurities leads to variations in the lattice parameter of crystals which in turn modifies the vibrational frequency and in addition the position of a line depends on the sample temperature (shifts of 4 cm^{-1} between 77K and room temperature are typical).

3. THE STRENGTH OF AN LVM LINE

Radiation incident on a harmonic oscillator will induce absorption to the next higher quantum level or stimulated emission to the next lower level. The net absorption is independent of the initial level occupied and as a consequence is independent of the sample temperature. The integrated absorption may then be written

$$\int \alpha \, d\omega = \frac{2\pi^2 N \eta^2}{n m_{imp} c} , \qquad (5)$$

where α is the absorption coefficient at the angular frequency ω (equal to $2\pi c\bar{\upsilon}$: $\bar{\upsilon}$ is the energy of the mode in wavenumbers), n is the refractive index of the host crystal and c is the velocity of light (Newman 1973). The left hand side of Equation 5 can be measured, while n m_{imp} and c are known, leaving the unknown quantities N and η. N is the concentration of impurities occupying the site which gives rise to the particular LVM line, e.g. Si(Ga) atoms in GaAs. The value of N cannot always be determined by chemical or electrical methods alone since some impurities can be present in several types of sites and complexes, as discussed elsewhere for silicon in GaAs (Spitzer 1971). The quantity η is called an apparent charge and is defined as the dipole moment per unit displacement in the particular mode. Values of η are usually close to the electron charge e, as might be expected, but the presence of a static charge on an impurity does not necessary mean that η is large nor vice versa. In fact for neutral $^{12}C^0(Si)$ $\eta \cong 2.5e$, whereas for negative $^{11}B^-(Si)$ in silicon $\eta \cong e$. It is therefore necessary to carry out a calibration for each impurity in each lattice site or complex in every host lattice if quantitative data are required. ASTM calibrations for oxygen and carbon in silicon exist but generally agreed calibrations for other systems are not yet available.

The calibration formula Equation 5 applies to a single LVM line arising from an impurity in a site of tetrahedral symmetry. If there is a perturbation from an adjacent atom or a lattice defect it is a simple matter to add the contributions to the integrated absorption from each of the new lines if they are all observed. However it should not be assumed that the value of η remains unchanged for the perturbed modes because there will be a redistribution of the local electron density.

4. THE EFFECT OF THE CHARGE STATE ON THE LVM FREQUENCY

In section 3 it was implied that some defects could exist in more than one charge state, e.g. B^0 or B^- in silicon. Another example would be carbon acceptors $C^0(As)$ or $C^-(As)$ in GaAs. Little effect on ω_L is expected if the charge state of a shallow impurity is changed as the "missing electron in the bonding" or hole is distributed over several neighbours up to a radius corresponding to the first Bohr orbit which is large because of the high dielectric constant of the crystal and the low effective mass. This expectation has been confirmed for carbon in GaAs. However for centres with deeper electronic levels this argument breaks down as the charge is more localised near the impurity. An interstitial oxygen atom paired with a lattice vacancy in silicon is such a defect. The complex is in effect an off-centre substitutional impurity with orthorhombic (C_{2v}) symmetry which may be either neutral or negatively charged with an energy level at (E_c- 0.17eV). The values of ν_L for $(O_i - V)^0$ and $(O_i - V)^-$ are 836 and 884 cm^{-1} respectively (Newman 1973). Similar results have been obtained very recently for centres involving oxygen impurities in GaAs.

5. EXPERIMENTAL DETAILS

The samples to be examined are usually in the form of a parallel slab with a thickness d in the range 1 cm to 0.2 mm depending upon the strength of the background two-phonon absorption. To obtain meaningful data the transmission though the sample at the peak of an absorption band must be close to or greater than 5%, otherwise stray light will impair the photometric accuracy: the sample thickness must be chosen to meet this requirement. Because semiconductors have a high refractive index (for silicon n = 3.5) there are large reflection losses at each of the polished slab surfaces. The sample will act as a Fabry Perot etalon resulting in oscillations in the transmitted intensity as the wavelength is changed. The simplest way to overcome this undesirable effect is to polish the sample to a wedge shape with a bevel angle of about 2^0. For thin samples there is a penalty as the thickness is not known accurately and there will be a corresponding uncertainty in the estimated absorption coefficient. In the absence of multiple internal reflections (no fringes) the impurity induced absorption coefficient α may be written as

$$\ln(T_s/T_r) = - \alpha\, d, \quad (6)$$

where T_s is the transmission through the sample at a particular wavelength and T_r is the corresponding transmission through a pure reference sample of the same thickness held at the same temperature. The conversion from transmission to absorption coefficient for a spectrum is usually carried out on a data station used in conjunction with the spectrometer. Either dispersive or FT spectrometers may be used. The former lack speed and resolution, but may have a better photometric accuracy. High resolution spectra can be obtained only with FTIR interferometers which also have a high speed, due mainly to parallel data acquisition.

In the discussion so far it has been assumed that the sample does not exhibit electronic free carrier or photo-ionization absorption. If such absorption is strong in the spectral region where LVM absorption is to be investigated methods have to found to remove the electronic component. An invaluable method for silicon, GaAs and GaP is to irradiate samples with 2MeV electrons at room temperature using a Van de Graaff generator. Host lattice atoms are displaced to form Frenkel pairs which may dissociate to form vacancies and self-interstitials which act as deep electron and/or hole traps and so deplete the conduction band (valence band) of electrons (holes), when shallow impurities are present. Such a procedure has been used to examine silicon impurities in GaAs because the Si(Ga) is itself a shallow donor (Spitzer 1971). The compensation process is not without problems as the radiation-induced defect may be mobile under the conditions of the treatment and may then be selectively trapped at sites adjacent to the impurities to be investigated. The complexes so formed have to be distinguished from the grown-in centres, a process that can be effected by the examination of samples following increasing doses of irradiation. In general, low doses of irradiation are desirable to minimise internal strains in the crystal. Another possibility is to use samples which are closely compensated during growth by the simultaneous incorporation of both shallow donors and acceptors such as Si(Ga) and Be(Ga) in GaAs, as very small irradiation doses are then required.

As an alternative to irradiation, a compensating impurity may be diffused into samples containing shallow donors or acceptors. In the past lithium and copper have been used for this purpose but a disadvantage is that impurity complexes form such as Si(Ga)-Li(Ga) and Si(Ga)-Cu(Ga) in GaAs. More recently, it has become clear that diffusion of atomic hydrogen derived from a plasma may lead to the passivation (neutralisation) of electrically active centres. Complexes with grown-in impurities may form and the perturbation of the hydrogen or deuterium modes may be used to characterise the impurity. The sensitivity of the method is high because of the small hydrogen mass (see Equation 5.)

6. EXAMPLES OF LVM ABSORPTION

A low resolution spectrum of GaAs containing $^{12}C(As)$ and $^{13}C(As)$ acceptors is shown in Fig. 4. Doping with such isotopically enriched impurities shows unambiguously that the LVM line at 582 cm^{-1} is due to

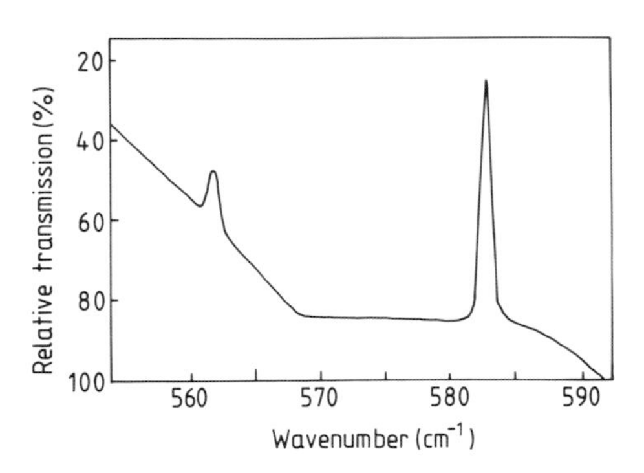

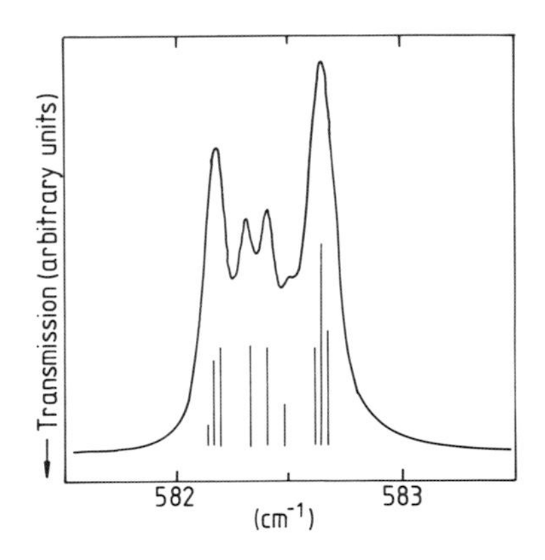

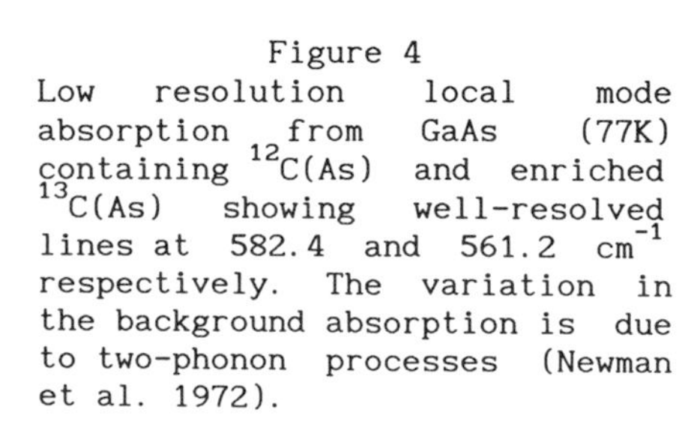

Figure 4
Low resolution local mode absorption from GaAs (77K) containing $^{12}C(As)$ and enriched $^{13}C(As)$ showing well-resolved lines at 582.4 and 561.2 cm^{-1} respectively. The variation in the background absorption is due to two-phonon processes (Newman et al. 1972).

Figure 5
High resolution transmission spectrum (0.06 cm^{-1}) of the LVM line from $^{12}C(As)$ in GaAs (Theis et al. 1982) and the predicted fine structure. The height of each bar is proportional to the corresponding fine structure line strength and the overall width has been fitted to the experimental data (Leigh and Newman 1982).

carbon ^{12}C and not some other light impurity (Newman et al 1972). From the separation of the two modes a value of χ = 2.65 is deduced (Equation 4). A high resolution (0.03cm^{-1}) spectrum of the $^{12}C(As)$ line is shown in Fig.5, together with vertical bars indicating the positions and relative strengths of the calculated fine structure components (Leigh and Newman 1982). The complexes and their symmetries are $^{12}C^{69}Ga_4$ (T_d), $^{12}C^{69}Ga_3{}^{71}Ga$ (C_{3v}), $^{12}C^{69}Ga_2{}^{71}Ga_2$ (C_{2v}), $^{12}C^{69}Ga^{71}Ga_3(C_{3v})$ and $^{12}C^{71}Ga_4(T_d)$, giving a total of nine components. Each of the end features shown in Fig.5 consists of three non-resolved components: more recently, the three intervening features have been more clearly resolved in less strained crystals. These measurements demonstrate that the carbon atom is bonded to four Ga nearest neighbours. Similar fine structure has been observed in LVM lines due to $^{11}B(As)$ (80% abundant) and $^{10}B(As)$ (20% abundant): these centres have been termed impurity antisite defects. Another example is the LVM line from $^{28}Si(As)$ although the fine structure is less well-resolved than for the lighter impurities. Correspondingly well resolved structure, although expected, has not been measured for $^{12}C(P)$ in GaP because of line broadening due to strains in the crystals.

Absorption from GaAs highly doped with silicon is shown in Fig.6. Most of the impurities in such n -type material are expected to be present as Si(Ga). The strong sharp lines ($\Delta\cong$0.4cm^{-1}) at 384, 379 and 373 cm^{-1}are due to $^{28}Si(Ga)$ (92.3% abundant), $^{29}Si(Ga)$(4.7% abundant) and $^{30}Si(Ga)$ (3% abundant). The line at 399 cm^{-1} showing structure arises from $^{28}Si(As)$: the shift in energy from 384 ($^{28}Si(Ga)$) occurs as a result of small changes in the local force constants (f'). The line at 393 cm^{-1} is due to the doubly degenerate transverse mode of $^{28}Si(Ga)$-$^{28}Ga(As)$ nearest neighbours donor-acceptor pairs where the two atoms have amplitudes which are out of phase. The corresponding longitudinal mode which splits into four components in material containing enriched ^{30}Si, namely $^{28}Si(Ga)$-$^{28}Si(As)$, $^{28}Si(Ga)$- $^{30}Si(As)$, $^{30}Si(Ga)$-$^{28}Si(As)$ and $^{30}Si(Ga)$-$^{30}Si(As)$ occurs near 464 cm^{-1} but is not shown in the Figure. Additional absorption from two complexes involving silicon occurs near 367-369 cm^{-1}. The defects labelled Si-X and Si-Y have unknown structures although Si-Y is now thought to be a Si(Ga)-V(Ga) complex. It is interesting that Si-X defects are produced in both MBE and MOCVD epitaxial GaAs and appear to act as electron traps which limit the maximum n-type carrier concentration.

The effect of the charge state of the (O_i-V) complex in silicon on the local mode frequency is shown in Fig.7. An n-type crystal was irradiated with 2 MeV electrons until the Fermi level was lowered to a position close to the electronic level of the defect at (E_c-0.17eV). By changing the sample temperature it was possible to "fine tune" the Fermi level and so change the electron occupancy of the centre in a fully reversible way. The shift in the positions of the two lines is attributed to small changes in the local force constants when the electron is captured. An anologous centre in GaAs, would be an oxygen atom bonded to two gallium atoms. Recently observed LVM lines each with a triplet fine structure corresponding to $^{16}O^{69}Ga_2$, $^{16}O^{69}Ga^{71}Ga$, $^{16}O^{71}Ga_2$ have been found at 730cm^{-1} and 715 cm^{-1}. Another line, possibly from interstitial oxygen bonded between one arsenic and one gallium atom, shows only the expected doublet splitting (Fig.8, Schneider et al 1989).

Finally, some results for n and p- type GaAs diffused with atomic hydrogen (or deuterium) are shown in Figs.9 and 10. A detailed

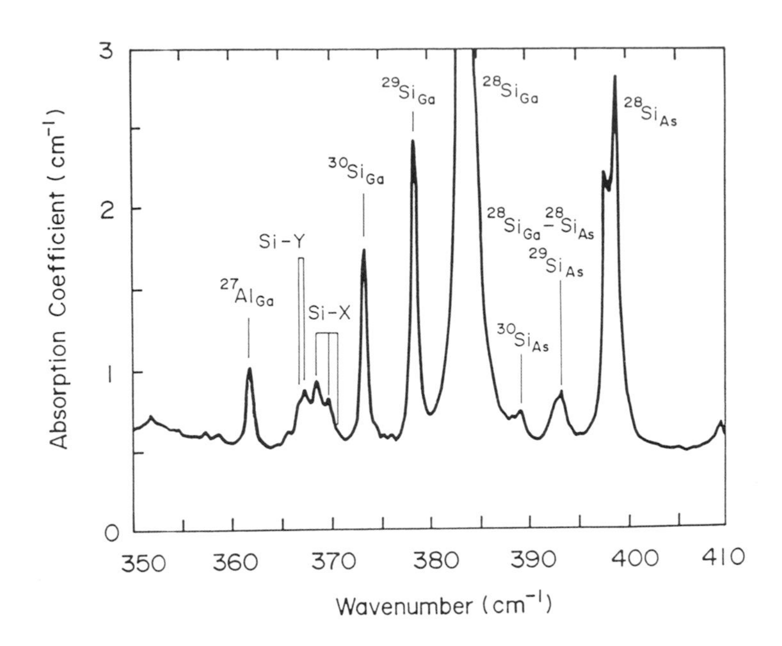

Figure 6
Infrared spectrum (0.1 cm^{-1} resolution, 6K) of bulk GaAs (Bridgman material) doped with silicon impurities showing LVM lines from the various complexes. Note the fine structure of the Si(As) line, the inadvertent contamination from Aℓ(Ga) impurities, and the absorption due to the complexes Si-X and Si-Y. It is now thought that Si-Y may be a Si(Ga) impurity paired with a second neighbour gallium vacancy (Ono and Newman, 1989).

discussion is beyond the scope of this chapter but the high frequency LVM lines are due to the vibration of Si-H bonds in the Si doped material (Pajot et al 1988), whereas the line in the Be doped sample arises from an H-As bond adjacent to the Be impurity. A full analysis of the structures can be made from the IR data, which also reveal (a) a bond bending mode for Si-H complexes, modified Si(Ga) and Be(Ga) modes and (c) isotopic analogues when deuterium is introduced instead of hydrogen.

7. SCOPE OF TECHNIQUE AND SENSITIVITY

The techniques outlined have provided significant information in many areas of investigation. Perhaps the most important topic from a technological view point is the examination of neutral substitutional carbon and interstitial oxygen in silicon. The former defects although benign as such, can capture self-interstitial atoms generated during device fabrication and are then ejected into interstitial sites. The C_i defect is mobile at room temperature and forms complexes with most other

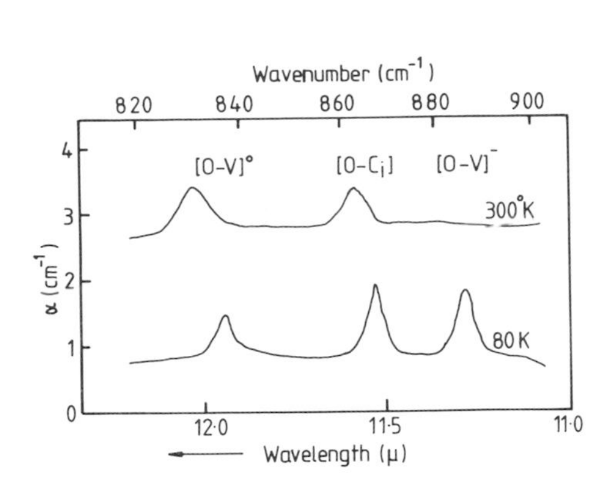

Figure 7
Absorption spectrum of n-type silicon containing oxygen impurities following electron irradiation to produce [O-V] pairs which have a localised electronic level at (E_c-0.17eV). The irradiation was terminated when the Fermi level ε_F reached this position at 80K. The lines at 836 cm^{-1} and 884 cm^{-1} are due to the vibrations of $[O-V]^o$ and $[O-V]^-$ complexes respectively. At 300K, ε_F is lowered slightly and only absorption from $[O-V]^o$ is observed. The line at 865cm^{-1} is due to an unrelated irradiation-induced defect involving an interstitial carbon atom trapped next to a grown-in interstitial oxygen impurity (Newman 1973).

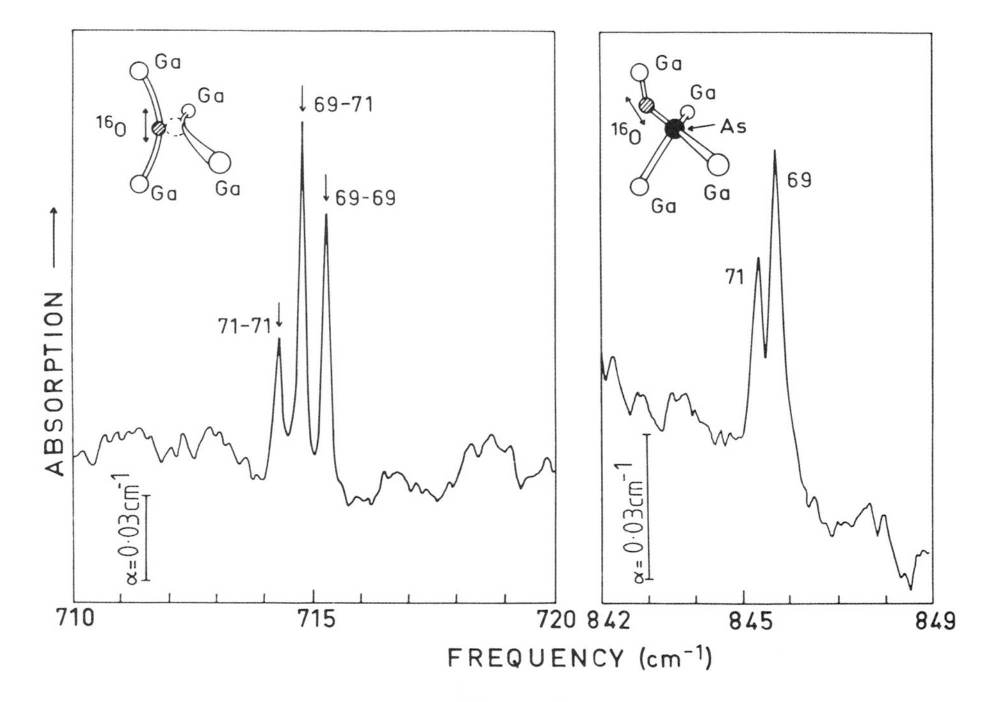

Figure 8
High resolution absorption spectra of GaAs (4.2K) containing oxygen impurities in an estimated concentration of about 3×10^{15} cm^{-3}, showing a triplet and a doublet feature due to mixed ^{69}Ga and ^{71}Ga nearest neighbours. Possible models for the defects shown in the inserts would correspond to interstitial oxygen and [O-V] defects found in silicon (Schneider et al. 1989).

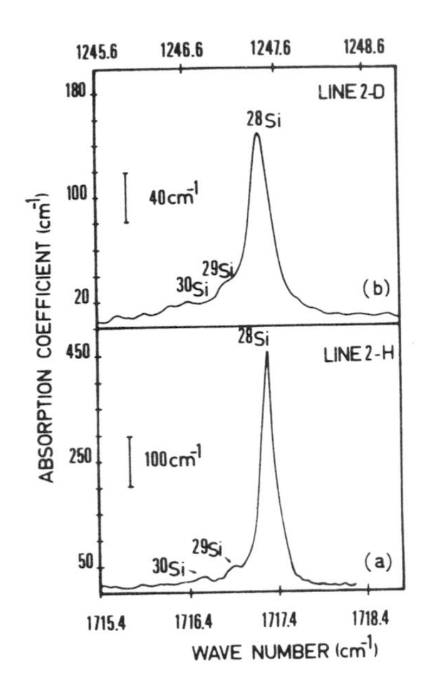

Figure 9
Absorption (4.2K) due to the stretching modes of Si(Ga)-H and Si(Ga)-D in passivated MOCVD silicon doped n-type GaAs. The fine structure on the side of the main Si-H line is due to the naturally occurring ^{29}Si(4.7%) and ^{30}Si(3%) isotopes. The linewidth is only 0.16cm^{-1} (Pajot et al 1988).

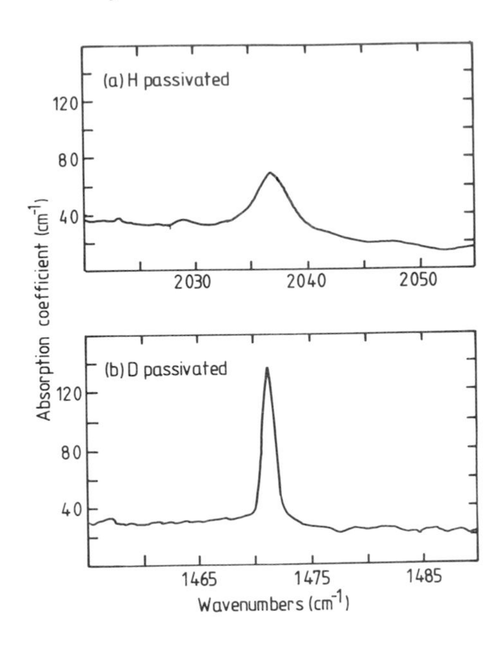

Figure 10
Absorption (4.2K) due to the stretching modes of As-H and As-D at nearest neighbour sites to Be(Ga) impurities in MBE GaAs after a passivation treatment (Nandhra et al. 1988).

impurities. It can also lead to the formation of particles of silicon carbide and can influence the growth of epitaxial silicon. Oxygen impurities form aggregates and large SiO_2 precipitates that are crucial for intrinsic gettering which enables complex integrated circuits to be fabricated. Carbon in semi-insulating GaAs crystals acts as an electron trap for the so-called mid-gap defect EL2, so pinning the Fermi level in this position. Silicon in GaAs has already been discussed, but an extension of previous work to special MBE material has indicated that the Si(Ga) defect itself may be the bistable DX-centre found in GaAs under high pressure. Other applications relate to diffusion measurements and the formation of complexes in silicon such as thermal donor centres which again involve oxygen impurities.

LVM absorption is now established as a spectroscopic technique. It received a considerable impetus with the availability of commercial high resolution interferometers. Not only can fine structure effects be measured but lower impurity concentrations can be detected in thinner layers because of the improved signal to noise ratio. It can be misleading to quote actual sensitivities as they are different for different impurities in the various hosts. For example, it has already been implied that the sensitivity for detecting ^{12}C in silicon is some six times (2.5^2) greater than that for ^{11}B but it has to be remembered that the intrinsic two phonon absorption underlying the ^{12}C line is relatively stronger. Nevertheless sensitivities are typically in the region of 10^{15} atom cm^{-3}, or lower, for a sample 1 mm in thickness. For sharp lines as for Si(Ga) in GaAs the sensitivity has recently been measured to be 3×10^{13} atom cm^{-2} in epitaxial layers with an instrumental resolution of 0.1 cm^{-1}. Thus for a layer 1μm in thickness the detectable Si(Ga) concentration would be 3×10^{17} cm^{-3}; these values are comparable with those used to grow test layers of doped material by MBE or MOCVD.

REFERENCES

Barker Jr., A.S. and Sievers A.J., 1975, Rev.Mod.Phys. 47, Suppl.No. 2, 51-180.

Clerjaud B., Côte D., Krause M. and Naud C., 1988, Mat.Res.Soc.Symp.Proc. 104, 341-4.

Dawber P.G. and Elliott R.J., 1963, Proc.Roy.Soc. A 273, 222: 1963, Proc.Phys.Soc. 81, 453-60.

Dekker A.J., 1958, Solid State Physics (London:Macmillan).

Fan H.Y., (1956), Rep.Prog.Phys. 19, 107-155.

Gledhill G.A., Kudhail S.S., Newman R.C., Woodhead J. and Zhang G.Z., 1981, J.de Physique, Colloq.C6, Suppl.No. 12, 42, C6-685-7.

Johnson F.A., 1959, Proc.Phys.Soc. 73, 265.

Leigh R.S. and Newman R.C., 1982, J.Phys.C:Solid St.Phys. 15, L1045-51.

Nandhra P.S., Newman R.C., Murray R., Pajot B., Chevallier J., Beall R.B. and Harris J.J., 1988, Semicond.Sci.Technol. 3, 356-60.

Newman R.C., 1973, Infrared Studies of Crystal Defects (Taylor & Francis : London) pp1-187.

Newman R.C., Thompson F., Hyliands M. and Peart R.F., 1972, Solid St. Commun., 10, 505-7.

Ono H. and Newman R.C., 1988, J.Appl.Phys. (in press).

Pajot B., Newman R.C., Murray R., Jalil A., Chevallier J. and Azoulay R., 1988, Phys.Rev.B 37, 4188-95.

Schneider J., Dischler B., Seelewind H., Mooney P.M., Lagowski J. Matsui M., Beard D.R. and Newman R.C., 1959 Appl.Phys.Lett., in press, April.

Spitzer W.G., 1971, Festkörper Probleme XI edited by O. Madelung (Pergamon : Vieweg), 1.

Theis W.M., Bajaj K.K., Litton C.W. and Spitzer W.G., 1982, Appl.Phys.Lett., 41, 70-2.

Point Defect Studies using Electron Paramagnetic Resonance

R.C. NEWMAN

1. INTRODUCTION

Many point defects in a semiconductor involve an unpaired electron and sometimes there is associated orbital motion. The electron spin produces a magnetic dipole moment and the orbit adds a further contribution. If such centres are present in a low concentration so that they are non-interacting the crystal becomes paramagnetic. When an external magnetic field B_o is applied there is alignment of the magnetic dipoles which have discrete energy levels determined by quantum mechanical principles. The crystal is then subjected to incident radiation of frequency ν and for particular values of B_o the differences in the energy levels ΔE of the magnetic dipoles will be given by $h\nu=\Delta E$, where h is Planck's constant. When this happens there will be absorption of the radiation, and the process is known as electron paramagnetic resonance, often abbreviated to EPR. It often happens that the orbital contribution to the magnetic moment is very small so that the resonance relates primarily to the energy levels of the spin dipole moment. For that reason, the process is frequently called electron spin resonance, or ESR.

Real resonance spectra can be exceedingly complex, but if they can be properly interpreted a wealth of microscopic detail is obtained about the defect under examination. More specifically, the symmetry is found, the value of the nuclear spin I of a central atom, the nuclear spins of neighbouring host lattice atoms, the extent of the localised wave function, the response to the system to external constraints such as a uniaxial stress, etc. Additional information can be obtained from more complicated resonance techniques involving both electron and nuclear spin flips (ENDOR), although a discussion is outside the scope of this chapter. Such resonances can also be detected by optical methods (Cavenett 1981, Spaeth et al. 1985) which have broadened the scope of the technique. As with other diagnostic methods, EPR is most usefully employed in conjunction with other techniques, particularly as defects can usually be present in more than one charge state and may not always be paramagnetic. For example, early work on a vacancy-oxygen complex in silicon failed to reveal the central oxygen atom directly and it was only the correlation of the EPR data with infrared measurements of the localised vibrational mode of the oxygen which allowed a positive identification to be made (Watkins and Corbett 1961). It is not intended to list identified spectra here nor to give a comprehensive discussion of the theory. An excellent overview of the physical concepts has been published by Watkins (1975) and a more detailed mathematical review has been given by Low (1960).

2. DIAMAGNETIC SEMICONDUCTOR CRYSTALS

An electron with charge -e in an orbit with angular frequency ω and area A_o constitutes a circulating current $I_o=-e\omega/2\pi$ giving rise to a magnetic dipole moment $\mu_o=I_oA_o$. The angular momentum **l** (in units of $\hbar$) is given by $m\omega r^2$, where r is the radius of the orbit and $A_o=\pi r^2$. It follows that

$$\underline{\mu}_o = -g_o\mu_B\mathbf{l}, \tag{1}$$

where $\mu_B=e\hbar/2m$ is the Bohr magneton, m is the mass of the electron, $\hbar$ is Planck's constant h divided by 2π and has the dimensions of angular momentum, and $g_o=1$. The direction of the dipole is along the normal of the area element A_o. Electrons also have an intrinsic spin angular momentum **s** leading to a second contribution μ_s to the total moment μ_T.

$$\underline{\mu}_s = -g_s\mu_B\mathbf{s} \tag{2}$$

where g=2.0023. The value of μ_T is the vector sum of μ_o and μ_s.

In a multi-electron atom the contributions to μ_T are obtained by summing the vector components from all the electrons. Stable configurations occur when electron shells, or certain sub-shells are full. The magnetic moment of an electron with orbital angular momentum **l** will be cancelled by that of another electron with **l** in the opposite sense. The clockwise and anticlockwise currents then cancel. Similarly, each spin moment is cancelled by that of a second electron. Thus $\Sigma l_i\equiv0$ and $\Sigma s_i\equiv0$, so that there is no permanent moment and the configuration is diamagnetic. This condition would also apply to a perfect ionic crystal of say $Na^+(1s^22s^22p^6)Cl^-(1s^22s^22p^63s^23p^6)$.

A covalently bonded crystal such as silicon is considered next. Each atom has an electron core with the configuration $(1s^22s^22p^6)$ and the four remaining electrons $(3s^23p^2)$ of the free atom participate in the bonding. When an assembly of atoms is brought together, the discrete 3s and 3p levels broaden into bands of levels which increase in width and then overlap. Ignoring normalisation, the wavefunctions become mixed, or hybridised, to form states of the type $\psi=|3s\rangle\pm|3p_x\rangle\pm|3p_y\rangle\pm|3p_z\rangle$ called sp^3 orbitals (Heine 1960). There are three p-components because of the three-fold degeneracy of the atomic state, whereas an s-state is non-degenerate. Four orbitals of each atom point along <111> directions towards its four nearest neighbours and overlap corresponding orbitals from those neighbours. Every tetrahedral bond thus formed therefore contains two electrons with spin "up" and spin "down" respectively. The spin magnetic moments cancel while the orientations of the "orbits" are fixed by the crystal geometry. There is no resultant orbital angular momentum which is said to be quenched, and there is no resultant magnetic dipole moment. The energy levels of these orbitals may be identified with the valence band of silicon. Antibonding orbitals (the remaining four wavefunctions) are directed away from the central atom along crystallographic $\langle\bar{1}\bar{1}\bar{1}\rangle$ directions where there are no nearest neighbours and overlap cannot occur: they have higher energies and are identified with conduction band states. At the equilibrium lattice spacing there is an energy gap E_g between the two bands and so a perfect crystal at the absolute zero of temperature will again be diamagnetic since all states in the valence band are filled and all states in the conduction band are empty. Similar considerations apply to III-V compound semiconductors with the zinc blende structure.

3. PARAMAGNETIC RESONANCE FROM AN UNPAIRED SPIN

When certain impurities or lattice defects are present in a crystal, unpaired electrons may be introduced. The simplest example is that of a shallow donor impurity with the electron occupying the ground state at a low temperature. Such a defect would appear to be analogous to a hydrogen atom: there is an unpaired spin, but no orbital angular momentum (s-state). When an external magnetic field **B** (derived from an electromagnet) is applied, there is an interaction energy given by $H=-\mu_s.\mathbf{B}$ corresponding to the result of classical physics for the alignment of a bar magnet. μ_s is given by Equation (2) so that H_{ez}, the electronic Zeeman energy is

$$H_{ez} = g_s\mu_B \mathbf{B.s} \qquad (3)$$

The scalar product $\mathbf{B.s} = B_xs_x+B_ys_y+B_zs_z$ may be simplified if the z-axis is chosen to be parallel to **B**, giving $B_x=B_y=0$ and $B_z=B_o$. Because of quantisation the only allowed values (eigenvalues) of the operator s_z are $m_s=\pm1/2$ (in units of $\hbar$ see section 1), so that Equation (3) can be written

$$E_{ez} = g_s\mu_B B_o m_s \;;\; m_s = \pm1/2. \qquad (4)$$

The originally doubly-degenerate spin state $|s,m_s=\pm1/2>$ is split into two Zeeman levels with an energy separation ΔE (see Fig. 1) given by

$$\Delta E = g_s\mu_B B_o, \qquad (5)$$

which is proportional to B_o. For $B_o\simeq0.35T$ (or 3500 Gauss), the value of ΔE is about $0.3cm^{-1}$ (wavenumber units) corresponding to radiation with a wavelength of 3cm (microwave, X-band).

Under equilibrium conditions, the number of centres N_1 in their ground state will be greater than the number N_2 in the upper state. The Boltzmann factor gives $N_2/N_1=\exp(-\Delta E/kT)$, where $N_1+N_2=N$. The ratio N_2/N_1 defines the "spin temperature" which will be the same as that of the crystal. The important quantity is (N_1-N_2), given by:

$$\frac{N_1-N_2}{N_1+N_2} = \frac{1-\exp(-\Delta E/kT)}{1+\exp(-\Delta E/kT)} \simeq \frac{\Delta E}{2kT}. \qquad (6)$$

At 4K, $kT\simeq3cm^{-1}$, while kT has larger values at higher temperatures. Since ΔE is small compared with kT, the right hand side of Equation 6 is small, implying that (N_1-N_2) is also small. Nevertheless it is that difference that is essential to observe magnetic resonance. The principle is to illuminate the sample with radiation of frequency ν and to detect the resonance that occurs when $h\nu=\Delta E$. The value of ν is fixed by the dimensions of a microwave cavity and so the magnitude of ΔE is varied by slowly changing the applied field B_o (Equation 5). At resonance, electrons in the lower state ($m_s=-1/2$) (Fig. 1) are excited to the upper state to give absorption, while electrons in the upper state ($m_s=+1/2$) experience stimulated emission and transfer to the ground

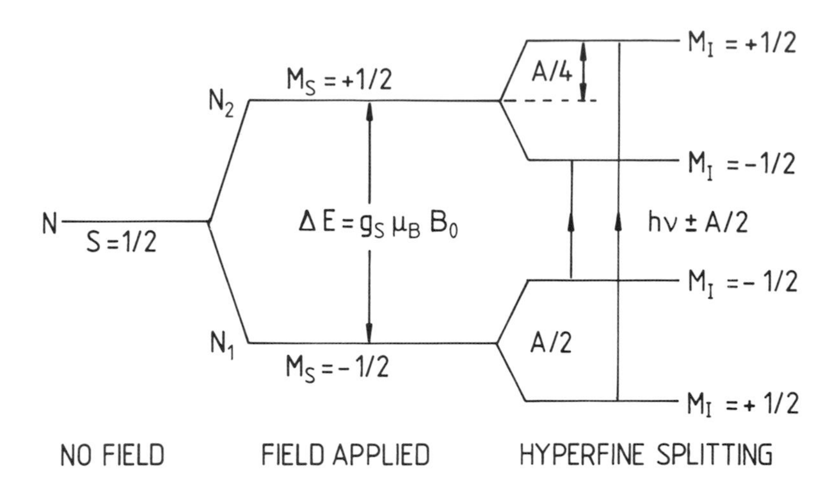

Figure 1

Energy levels of impurities each with an unpaired spin $s=1/2$. On the application of an external magnetic field B_o the two electron states $M_s=\pm1/2$ are split by an energy ΔE, with N_1 and N_2 centres in the lower and upper states respectively. If the impurity has a nuclear spin $I=1/2$, there is a further hyperfine splitting leading to the existence of two resonance conditions. For simplicity the hyperfine splitting has been treated only to first order, while the zero field interaction together with the direct nuclear Zeeman interaction has been omitted from the analysis and the diagram.

state. There is net absorption of radiation provided $N_2 < N_1$. It is clear that the effect of the radiation alone would tend to equalise N_1 and N_2 as the probabilities for the two types of transitions are equal. In the limit when $N_1=N_2$ the resonance could no longer be detected and is said to be saturated. Although there is a negligibly small rate of spontaneous emission at such low frequencies, there are mechanisms whereby the equilibrium distribution can be regained, involving a transfer of heat from the spins to the lattice. An electron spin interacts with an orbit (it could be that of another electron) by means of a magnetic coupling (spin-orbit coupling) and it has already been pointed out that the electron orbits relate to the bonding which in turn controls the vibrational modes of the crystal. The overall process of energy transfer occurs exponentially with time and is known as spin-lattice relaxation: the rate increases as the temperature of the sample is increased. It follows that saturation of a resonance from an electron in an s-state ($l=o$) occurs more readily than from an electron in a state with $l\neq o$. However details of the relaxation process are extremely complex (Feher and Gere 1959). Low temperatures are advantageous in increasing the value of (N_1-N_2), but a compromise is often necessary to prevent saturation of the resonance. In any case, the strength of the incident microwave radiation must be suitably limited.

4. EXPERIMENTAL DETAILS

Microwave absorption occurs as a result of magnetic dipole transitions since there is no change in parity between the ground spin state and the excited state. The incident radiation can induce such transitions only if the sample is located in a position where the microwave magnetic field $B(\mu\omega)$ is high. This field has to be perpendicular, or have a significant perpendicular component, to the magnet field B_o. A term of the form $H(t)=g_s\mu_B B_z(\mu\omega)s_z e^{i\omega t}$ will not induce transitions because it leads only to diagonal matrix elements according to the quantisation adopted. Shift operators $s\pm$ given by:

$$s\pm|s, m_s> = +\left[s(s+1)-m_s(m_s\pm1)\right]^{1/2}|s, m_s\pm1> \quad (7)$$

have the required effect of changing the value of m_s by either ±1 (Condon and Shortley 1939). By definition, $s\pm=(s_x\pm is_y)$ and so the operator $s_x=1/2(s_+ + s_-)$ can induce either absorption or stimulated emission: the time dependent interaction should therefore be of the form $H(t)=g_s\mu_B B_x(\mu\omega)s_x e^{i\omega t}$. A non-quantum mechanical argument has been given by Watkins (1975).

A simple experimental arrangement which satisfies the requirements is shown in Fig.2. A full-wave rectangular microwave cavity is used in which there are two lobes of the microwave magnetic field. The sample is located one quarter of the way along the cavity on the narrow side and the direction of B_o is perpendicular to the broad face of the cavity. Microwaves can be produced from a coupled Gunn diode (solid state generator), or more usually from a highly stabilised klystron. For

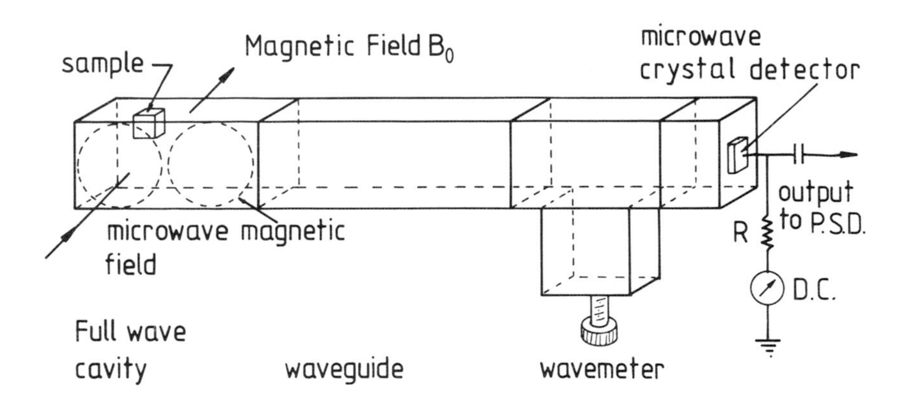

Figure 2

Simple equipment for the observation of EPR. The sample is located in a microwave cavity placed in an external field B_o. For simplicity the coupling of the microwave source to the cavity is not shown. Microwaves emerge from the cavity, pass a wavemeter and are then rectified to produce a DC signal which appears across the resistor R. When a 100kHz modulation field is added to B_o, an AC signal is produced and forms the input of a P.S.D.

absorption measurements the frequency is fixed by the dimensions of the sample cavity which is coupled by a waveguide to a crystal diode detector. The frequency can be measured by a calibrated wavemeter, a cavity which can be varied in length by means of a micrometer adjustment, or by a digital frequency counter. Hence the value of ν is determined. The microwave electric field incident on the crystal detector produces a DC signal which appears as a voltage across an external resistor R. When the magnet field B_o is increased slowly, the electron spins will pass through the resonance condition and there will be a reduction in the DC signal (Fig. 3) because power is absorbed. In practice the change in the output would be too small to detect directly. To overcome the problem phase-sensitive detection is used. An AC current at a frequency of 100 kHz is passed through a loop of wire around the sample to provide a magnetic field parallel to B_o. The amplitude of this field should be less than the linewidth of the resonance. As the system passes through resonance there is an AC output at the detector at 100 kHz corresponding to the first derivative of the absorption (Fig.4). The DC component is

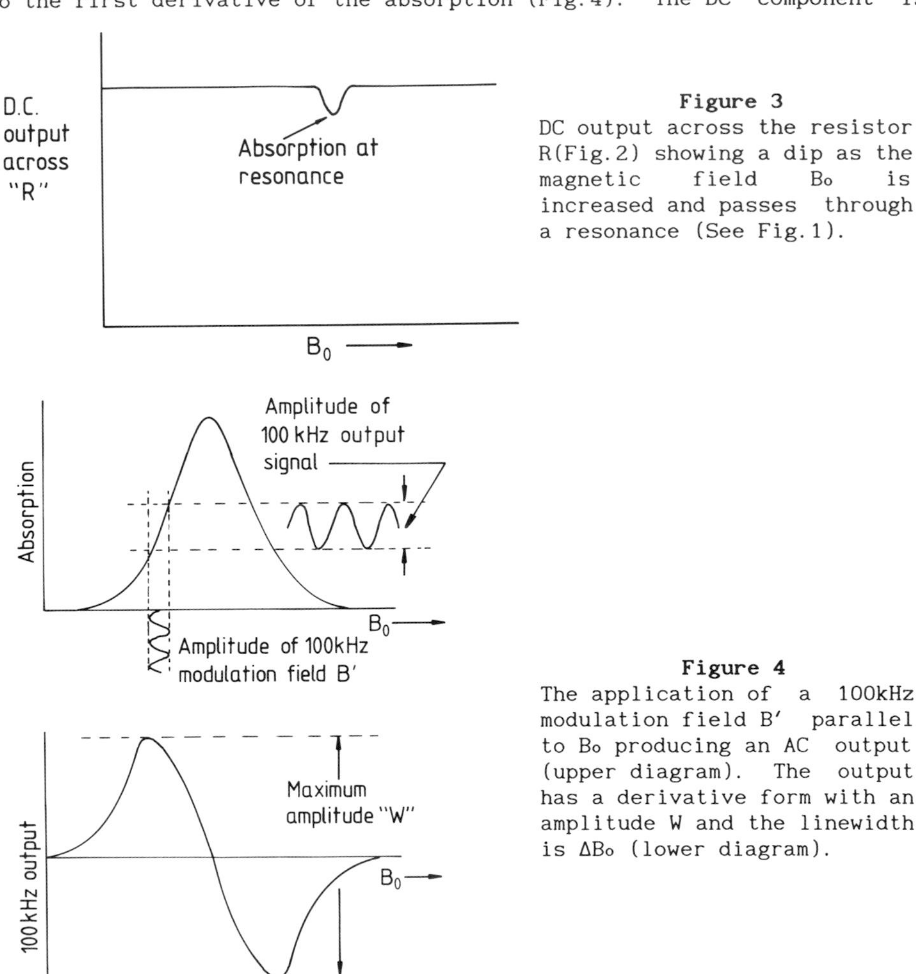

Figure 3
DC output across the resistor R(Fig.2) showing a dip as the magnetic field B_o is increased and passes through a resonance (See Fig.1).

Figure 4
The application of a 100kHz modulation field B′ parallel to B_o producing an AC output (upper diagram). The output has a derivative form with an amplitude W and the linewidth is ΔB_o (lower diagram).

blocked by a capacitor, the AC signal is amplified and then reconverted to a DC signal using a phase sensitive detector (P.S.D.) with the phase of the 100 kHz supply used as a reference signal. The output is fed to a pen recorder or digitised and stored using a data station. The latter process allows signal averaging to improve the signal/noise ratio, and spectral subtraction can be used to remove spurious signals from a contaminated cavity or to detect changes in a sample after a sequence of treatments.

The separation of the positive and negative peaks of the derivative output gives the linewidth ΔB_o of the resonance line (Fig. 4). The amplitude of the derivative trace (W) is clearly proportional to the strength of the original absorption. The integrated absorption, proportional to the number of spins in the sample, is therefore given by the product $W(\Delta B_o)^2$, multiplied by an unknown constant which depends on the experimental conditions of the measurement. A similar measurement made under identical conditions on a standard sample containing a known number of spins allows an absolute spin concentration to be determined (provided the spin quantum numbers are also known), usually to within a factor of two.

The field B_o is controlled by means of a signal derived from a calibrated Hall probe, using indium arsenide or a related semiconductor compound. For precise measurements of B_o, proton resonance is used. Protons also have a magnetic dipole moment due to their nuclear spin I=1/2. When placed in the magnet field two nuclear Zeeman levels are produced with an energy separation $\Delta E' = g_N \mu_N B_o$, analogous to the result given in Equation 5 for electrons. μ_N is the nuclear magneton $e\hbar/2M_p$, where M_p is the mass of the proton and g_N is the proton g-value, given by $g_N = 5.5854$: different g_N values are appropriate for different nuclei. It is important to note that $\mu_N \simeq 10^{-3} \mu_B$, so that the resonant frequency ν' given by $h\nu' = \Delta E'$ is close to 10 MHz. The radio frequency ν' is easily varied and measured for the same field as that at which the electron resonance occurs and since $g_N \mu_N$ is known the value of B_o can be determined with precision. the proton probe consists of some material containing a high concentration of hydrogen atoms such as water (in a container) or rubber on which a small coil is wound to couple the RF magnetic field that induces the transitions. At high magnet fields, nuclear magnetic resonance of deuterons can be used instead of protons.

If proton resonance is not available "g-markers" can be used to determine the field. An example is diphenyl picryl hydrazyl (D.P.P.H.) which has a g-value of 2.0036 (Assenheim 1966). Clearly a second marker is required with a different g-value to define the scale of scans of the magnet field, but it then has to be assumed that the equipment produces linear scans.

Details not discussed here relate to methods of coupling the microwave cavity to the waveguide, various types of cavity, microwave bridges, the use of higher microwave frequencies, the requirements for rotating the sample in a cavity and/or rotating the field B_o around the cavity etc.

5. HYPERFINE INTERACTIONS AND SHALLOW IMPURITIES

It was assumed in section 3 that the unpaired electron spin interacts only with the external field B_o. However, it has also been stated that nuclei have magnetic moments and they will therefore give rise to dipolar fields B_{INT}. The unpaired electron will interact with the combined fields rather than just B_o. Initially it may be supposed that there is just one nuclear dipole located at a vector distance **r** from the electron spin **s**. The field from the nucleus is calculated at the position of the electron and the energy of interaction is $H_{HF}=-\mu_s.B_{INT}$, where

$$H_{HF} = g_s g_N \mu_B \mu_N \left[\frac{\mathbf{s}.\mathbf{I}}{r^3} - \frac{3(\mathbf{s}.\mathbf{r})(\mathbf{I}.\mathbf{r})}{r^5}\right], \quad (8)$$

and is known as the hyperfine interaction. A detailed discussion of Equation (8) is given by Slichter (1963, p.46). If the only quantum mechanical operators that are retained are s_z and I_z (first order terms giving diagonal matrix elements), a simplified result follows

$$H_{HF} = \frac{g_s g_N \mu_B \mu_N \, s_z I_z}{r^3} (1-3\cos^2\theta). \quad (9)$$

where θ is the angle between **r** and the z-axis. The field produced by the nucleus is $B_{INT} \simeq g_N \mu_N / r^3$ which is about 25 Gauss for a proton separated from the electron by a distance of 1Å. Thus the energy described by Equation (9) is small compared with the direct Zeeman energy of the electron (Equations 3 and 4) since $B_o \simeq 3500$ Gauss. H_{HF} may then be treated as a perturbation of the Zeeman splitting.

An inherent difficulty has been ignored which is crucial to a hydrogenic defect where the electron is in an s-state and the central atom has a non-zero nuclear spin (e.g. ^{31}P with I=1/2 and 100% abundance in a silicon host). The electron density is given by $|\psi(r)|^2$, where $\psi(r)$ is the donor wavefunction. The value of H_{HF} averaged over the angular function in Equation 9 is zero. On the other hand, $\psi(o) \neq o$ for an s-state and there appears to be a logarithmic singularity in H_{HF} as $r \to o$. The apparent dilema is resolved by noting that there is a finite probability $|\psi(o)|^2$ dv of finding the electron in the nucleus which has a small volume dv. This interaction, called the Fermi contact term, is finite and independent of the assumed nuclear volume (Slichter 1963). The analysis yields an extra term to be added to Equation 9, given by

$$H_{HF}\,(\text{Fermi}) = g_s g_N \mu_B \mu_N \,(8\pi/3)|\psi(o)|^2 \mathbf{s}.\mathbf{I}. \quad (10)$$

Taking only first order terms and substituting the eigenvalues m_s and m_I for s_z and I_z leads to a simplification of Equation 10 to give,

$$H_{HF} = A\, m_s m_I, \quad (11)$$

where A is the central hyperfine parameter. The magnitude of this term is again usually smaller than the electron Zeeman term, and corresponds to an additional interaction comparable with that given by Equation 9.

The unpaired s- electron will experience a total field equal to the vector sum of B_o, the fields from neighbouring nuclei with $I \neq o$, together with that arising from the central nucleus. If the interactions with the neighbours (superhyperfine terms) are first ignored, the resulting energy levels for s=1/2 and I=1/2 (central atom) are shown in Fig.1. Two resonances will be observed at external fields B_1 and B_2 ($B_1 < B_2$). The internal field adds to B_1 to produce the resonant field B_0 while the internal nuclear field opposes the applied field B_2 to produce the resonant field B_0 for a second time. Thus the difference (B_2-B_1) is a measure of the internal field. If the nucleus has a spin quantum number I, there will be (2I+1) resonance signals corresponding to the possible quantisation directions of the nuclear dipole (m_I values). It follows that a resonance from phosphorus donors (I=1/2) in silicon is easily distinguished from that of ^{75}As donors (I=3/2, 100% abundant), where four signals are produced. To obtain the total spin concentration of the donors from a calibration measurement it is necessary to add the contributions to the integrated absorption from each of the (2I+1) signals.

Because there is a direct interaction of the external field with the nuclear moment there are further corrections to the energy levels of the system not shown in Fig. 1. These effects are usually small and in practice may be smaller than second order quantum mechanical corrections to the central hyperfine interaction. For many systems, contributions to the energy from the nearest neighbour dipolar interactions are also small and lead only to a broadening of the resonance lines as for phosphorus in silicon which contains 4.7% ^{29}Si(I=1/2)(Feher 1959).

The conduction bands of materials with indirect band gaps have a more complicated structure than that implied here, and the donor ground state may not be a singlet level (spin not included). As a consequence not all donors give EPR signals. Thus substitutional group V donors in silicon do show resonances, but interstitial shallow lithium donors do not. Likewise in GaP, the group VI donors S, Se and Te show resonances but not group IV donors such as silicon located on gallium lattice sites. A related problem exists for shallow acceptors in semiconductors where the ground state is like an atomic state with j=3/2, corresponding to an orbital state with l=1. Random strains in the crystal (Feher et al 1960, Goldstein 1966) split the quartet level into two spin doublets but with a range of energy separations. Consequently, there are differences in the energies of the Zeeman levels after the field B_0 is applied and the resonance is smeared over a range of the field values so that it is not detected. In very homogeneous samples of silicon a resonance has been found for boron acceptors (Neubrand 1978 a,b), providing a method for assessing the crystal quality.

6. ANTISITE DEFECTS IN GaAs AND GaP

Group V atoms occupying group III lattice sites or vice versa in compound III-V semiconductors are called antisite defects. In its neutral charge state an As^{o}(Ga) should be a double donor as there are two excess electrons. If other defects which act as deep electron traps are present the As-antisite defects will become ionised. As^{+}(Ga) defects would then resemble hydrogenic atoms except that the ionisation energy is greater and the wavefunction of the single excess electron in its ground state would be more concentrated around the impurity than that of a simple shallow donor. In fact, As^{+}(Ga) centres show a resonance spectrum consisting of four lines due to the isotropic central hyperfine interaction of the unpaired s- electron with the ^{75}As nucleus (I=3/2). Many GaAs samples exhibit an additional resonance in the middle of the pattern due to an unrelated defect. The two spectra can be separated experimentally if the incident microwave power is increased. The As^{+}(Ga) spectrum saturates, whereas the second spectrum does not (Fig.5) suggesting that there is an orbital contribution to the magnetic moment for this latter defect (Goltzené et al 1985).

The strength of the antisite spectrum can be increased by subjecting GaAs crystals to plastic deformation, electron irradiation, neutron irradiation, but not all the observed defects are equivalent (Beall et al 1984). Some centres show saturation whereas others do not. An explanation is that the second class of defects are really complexes with a second defect in a lattice site adjacent to the As(Ga) atom. The latter defects are not detected directly because of the large linewidth of the As^{+}(Ga) resonance of 400 Gauss, but their presence has been revealed in Electron Nuclear Double Resonance (ENDOR) spectra, which will not be discussed here (Spaeth et al 1985).

Similar measurements made on gallium phosphide crystals reveal the presence of P^{+}(Ga) antisite defects which produce relatively sharp lines (Fig.6) in two groups separated by the isotropic hyperfine interaction of

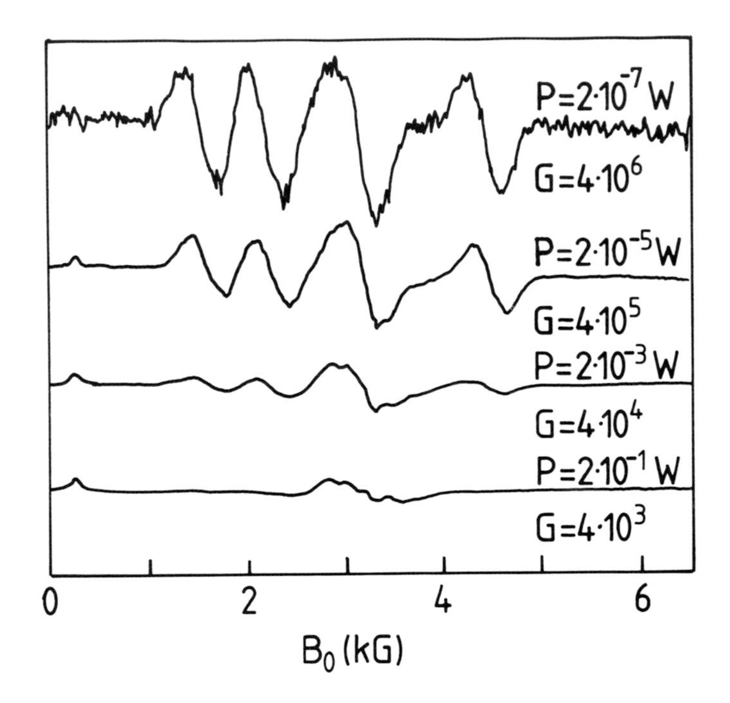

Figure 5

The EPR spectrum of the As^{+}(Ga) antisite defect in GaAs showing the derivative form illustrated in Fig.4. Four main peaks are evident (upper trace) due to the hyperfine interaction of the ^{75}As atom with I=3/2. As the incident microwave power P is increased the signal/noise ratio is improved but the resonance becomes saturated: to accommodate the increased output the amplifier gain G has to be reduced. At high values of P residual structure due to a second resonance that does not saturate is observed with $B_0 \simeq 3$ kG. The two spectra are therefore separated by the method of selective saturation (After Goltzené et al 1985).

the electron with the phosphorus nucleus (I=1/2) (Kaufmann et al 1976). Further superhyperfine interactions (Equation 8) with the four neighbouring phosphorus nuclei are resolved. A simple pattern of lines is obtained if the applied magnetic field is aligned along a crystallographic [100] direction. The field then makes equal angles with the <111> bond axes and there is superposition of the anisotropic hyperfine interactions with each neighbour. Each group of lines should have relative strengths of 1,4,6,4,1 corresponding to the various combinations of "spin up" and "spin down" of the four neighbours, each with I=1/2: there is excellent agreement with the observations.

Other defects labelled P-P3 give rise to a similar anti-site spectrum but with a somewhat modified central hyperfine splitting which still appears to be isotropic. However the fine splitting has a different structure indicating only three nearest neighbour phosphorus atoms to the P(Ga) antisite defect. There could be a fourth neighbour with a nucleus that does not have a nuclear spin, but other evidence suggests that there may be a phosphorus vacancy in that location (Beall et al 1984). This possibility implies that the complex is a rearrangement of a gallium vacancy in which one neighbour has made a single atomic jump. In illuminated p-type material there is a different spectrum which has been attributed to the gallium vacancy (Kennedy and Wilsey 1981). Thus the local atomic arrangement appears to depend on the charge state of the

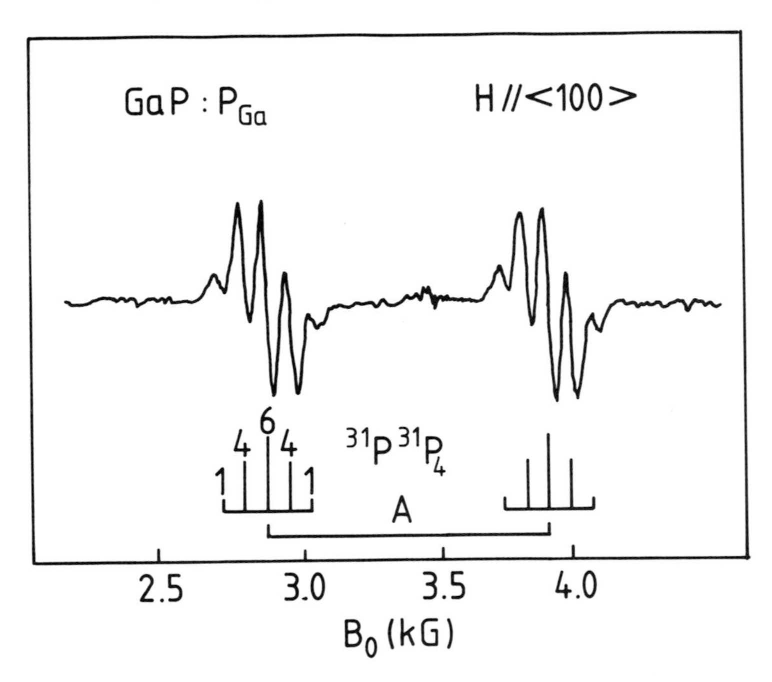

Figure 6

The EPR spectrum of the P^+(Ga) antisite defect in GaP showing a splitting A due to the central hyperfine interaction with the ^{31}P nucleus with I=1/2 (Fig.1). Each line is then further split into components due to superhyperfine interactions with the four phosphorus nearest neighbours. Similar additional splitting would be expected in Fig.5 but cannot be resolved because of the large widths of the four lines in the spectrum (see Kaufmann et al. 1976).

defect. Such observations are important as they give clues to the mechanisms of bi-stability shown by certain classes of defects (Chantre 1988).

7. ORBITAL ANGULAR MOMENTUM

The concept of quenched orbital angular momentum will be discussed further by considering a defect with orthorhombic C_{2v} symmetry involving an unpaired spin and a p-type wave-function. For a free atom the eigenstates would be given by $|1,1\rangle$, $|1,0\rangle$ and $|1,-1\rangle$, using the notation $|1, m_l\rangle$. Orbitals suitable for the reduced crystal symmetry consist of linear combinations of the atomic states and are designated p_x, p_y and p_z. If normalising constants are ignored, the angular parts of the states $|1,1\rangle$ and $|1,-1\rangle$ are given by $\sin\theta e^{i\phi}$ and $\sin\theta e^{-i\phi}$ respectively: l_x will then involve the combination $r\sin\theta(e^{i\phi}+e^{-i\phi})$ and l_y involves $ir\sin\theta(e^{i\phi}-e^{-i\phi})$. As the symmetry is low the energies of the three states will in general have different values because of interactions with the neighbouring atoms. It is assumed that the state l_x has the lowest energy and is occupied by the electron when the EPR measurement is made. Clearly the "current circulation" given by the $|1,1\rangle$ part of the orbital state is equal and opposite to that from the $|1,-1\rangle$ part. There is therefore no net angular momentum or magnetic moment, apart from that of the unpaired electron spin. Similar comments apply to the excited states.

This analysis is incomplete since the spin-orbit interaction, mentioned earlier in connection with the spin-lattice relaxation time, has been ignored. This term, $H_{s.o.}=\xi\mathbf{l.s}$, leads to an alignment of the spin relative to the orbit in a free atom. When the magnetic moment of the electron in the crystal is aligned along an arbitrary direction by the externally applied field there will be a torque acting on the crystal orbit, causing it to be somewhat modified. If the combined Zeeman and spin orbit terms are analysed using second order perturbation theory, taking account of excited states, there are shifts in the energy levels and mixing of the orbital wave functions (the modification referred to above). The latter effect can then lead to an imbalance between the clockwise and anticlockwise circulating currents and there will be a net contribution of the orbital motion to the total magnetic moment. Measured g-values will depend on the direction of the applied field and will show deviations from the "pure spin" value of $g_s=2.0023$. The deviations may be positive or negative depending upon the sign of the spin-orbit coupling parameter ξ. The magnitude of the shifts will be of the order of ξ/Δ, where Δ is the energy separation of the ground state from the excited orbital states. It has been assumed here that ξ/Δ is small, but it should be remembered that ξ increases rapidly with increasing atomic number. For the ^2F ground state of Yb^{3+} with an electron configuration $4f^{13}$, the spin-orbit splitting is close to 1eV. Discussion of rare earth impurities is omitted, although they might be of importance as electroluminescent defects in semiconductor devices.

If a defect with s=1/2 has axial symmetry (trigonal or tetragonal), rather than orthorhombic symmetry, we obtain

$$H_{ez} = g_{\parallel}\mu_B B_z s_z + g_{\perp}\mu_B(B_x s_x + B_y s_y) \tag{12}$$

where B_x, B_y and B_z are the components of the applied field related to the crystal axes. $g_{\parallel}$ is the value of "g" determined experimentally (c.f. Equation 4), when the field is parallel to the symmetry axis of the defect which is designated the z-axis. $g_{\perp}$ is the value of "g" when the field is perpendicular to the z-axis (in the x-y plane). For intermediate angles θ between the field direction and the z-axis the measured g-value is given by

$$g_{exp}^2 = g_{\parallel}^2 \cos^2\theta + g_{\perp}^2 \sin^2\theta. \qquad (13)$$

By rotating the field relative to the crystal axes (usually about the [110] axis) and plotting the values of g_{exp}, it is possible to determine the symmetry type of the defect. More than one line will be observed because of defects occurring with equivalent but different crystallographic axes. For example, a trigonal defect will have a [111] crystal axis (z-direction) and the application of the external field parallel to this axis will lead to the determination of $g_{\parallel}$. However, other defects lying parallel to $[\bar{1}11]$, $[1\bar{1}1]$ and $[11\bar{1}]$ all have z-axes making an angle of $\cos^{-1}\theta=1/3$ to the [111] axis. Thus a second resonance, three times as strong as the first, will occur at a value of g_{exp} intermediate between $g_{\parallel}$ and $g_{\perp}$ according to Equation 13. For an arbitrary direction of the applied field, resonances would be observed at four distinct values of B_o.

8. CENTRES WITH S>1/2

Many defects in semiconductors can exist in several charge states depending upon the position of the Fermi level. The charge states As^o(Ga), As^+(Ga) and As^{2+}(Ga) of the antisite defect in GaAs have already been implied. If a defect involves two additional electrons (say As^o(Ga) in GaAs), it may have no resultant spin (S=0), but there is a possibility that the two spins are aligned to give S=1 (for examples see Watkins (1965)). The application of a magnetic field would then split the spin triplet into three Zeeman levels: the two separations would be equal and only one resonance feature would be observed experimentally. However, higher lying orbitals should again be taken into account because there will be interactions with each spin via the spin-orbit coupling. The first spin will modify the orbit leading to a change in its coupling to the second electron. The spins will therefore interact with each other to give energy terms proportional to ξ^2. In addition, the proximity of the two spin magnetic moments will give a dipole-dipole interaction, similar in form to that given in Equation 8. As a consequence there is a splitting of the original triplet state into three levels for a defect with orthorhombic symmetry, prior to the application of the magnet field. These levels then shift as the field is increased and there is again a dependence on the direction of the field relative to the defect axes.

For such a defect, extra terms have to be added to the energy

$$D\left[S_z^2 - S(S+1)/3\right] + E(S_x^2-S_y^2), \qquad (14)$$

where D and E are constants. For centres with axial symmetry E=0 while D is also zero if the centre has tetrahedral symmetry. The terms in expression (14) lead to additional splittings as explained above and in general there are 2S so-called fine structure lines corresponding to the number of separations between the Zeeman levels. There is anisotropy which has to be compounded with the anisotropy in the g-values already discussed. In addition any hyperfine splittings will also be anisotropic. The terminology of fine structure and hyperfine structure can be misleading as the latter splittings may be larger than the former.

Some impurities have $S>1$. For example Fe^{3+} and Mn^{2+} (Goldstein 1966) have half-filled $3d^5$ sub-shells giving $^6S_{5/2}$ ground states. The $S=5/2$ state splits into a doublet and a quartet levels in a tetrahedral environment as a result of interactions with the neighbours. When the magnetic field is applied six Zeeman levels are produced with unequal separations for a general direction of the field. As a consequence five fine-structure lines are found in the resonance spectrum for Fe^{3+} (Goldstein 1966) (Fig.7). To distinguish fine structure effects from hyperfine splittings it is useful to analyse EPR spectra obtained with two different microwave frequencies. The hyperfine splittings depend only on the internal fields generated by nuclei and are not dependent on the frequency. However, to obtain the resonance condition $h\nu=\Delta E$, it is necessary to increase B_o as ν increases. In that case, the ratio of the Zeeman splitting to the fine structure (crystal) splitting will change and the relative dispositions of the fine structure lines will change if ν is increased.

DISCUSSION

An attempt has been made to illustrate physical principles rather than present detailed mathematical results which are given elsewhere. It is clear that the analysis of spectra is complicated and it should be pointed out that some interactions such as that of nuclear quadrupole moments with electric field gradients have not been mentioned. In general, a spectrum has to be unravelled to find the defect symmetry, the principal g-values, the hyperfine structure and the fine structure. The

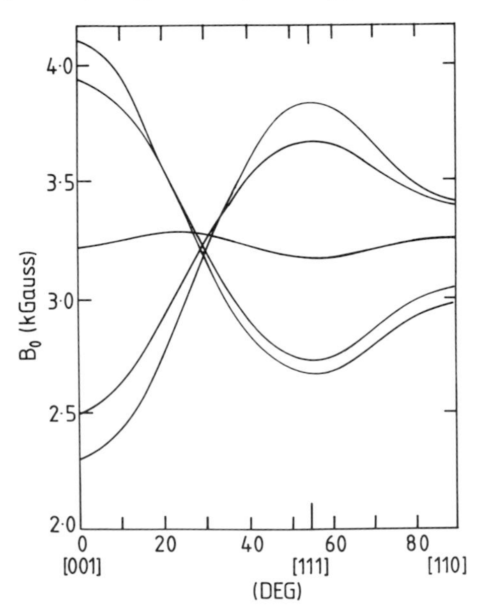

Figure 7
The spectrum of GaAs containing substitutional Fe^{3+}(Ga) impurities with $S=5/2$ shows five fine structure lines. Their positions in terms of the values of B_o change as the direction of the field is changed relative to the crystal axes. The measured angular variation is characteristic of the impurity site symmetry (see Goldstein 1966).

technique works only when the defect is paramagnetic and parallel measurements using other diagnostic methods are essential in many cases. The example of oxygen-vacancy centres in silicon has already been cited in section I. It is worth commenting on a second example. As(Ga) antisite defects in GaAs can be generated by high energy (2MeV) electron irradiation and they then show an EPR spectrum. On annealing, the strength of the spectrum eventually decreases, but the question then has to be asked whether the centres have been removed by diffusion-controlled defect reactions, or whether there has only been a shift in the position of the Fermi level leading to a change in the charge state of the antisite defect. This is a complex topic which will not be pursued here but it illustrates an obvious difficulty of interpretation.

The sensitivity of the EPR technique varies according to the defect being examined. The detection limit for $As^{+}(Ga)$ in GaAs using a conventional spectrometer is about 10^{16} defects cm^{-3}, due to the large linewidth of 400 Gauss. If the linewidth is lower by a factor of 10^{2}, as found for other centres the limit of detection is improved by up to a factor of 10^{4}. These numbers refer to the number of spins being examined in bulk crystals. It is clear that the technique is less appropriate for thin MBE or MOCVD epitaxial films. Nevertheless, the defects in such layers detected by other methods are often the same as those which are present in bulk crystals and examination of the latter material is therefore vital to characterise those defects.

REFERENCES

Assenheim, H.M., 1966 Introduction to E.S.R. (Hilger-London).

Beall, R.B., Newman, R.C., Whitehouse, J.E. and Woodhead, J., 1984, J.Phys.C:Solid St.Phys. 17, L963-8:1985, J.Phys.C.Solid St.Phys. 18, 3273-83.

Cavenett, B.C., 1981, Adv.in Phys. 30, 475.

Chantre, A., 1988, Mater.Res.Soc.Symp.Proc. 104, 37-46.

Condon, E.U. and Shortley, G.H., 1939, The Theory of Atomic Spectra (University Press:Cambridge).

Feher, G., 1959, Phys.Rev. 114, 1219-44.

Feher, G. and Gere, E.A., 1959, Phys.Rev. 114, 1245-56.

Feher, G., Hensel, J.C. and Gere, E.A., 1960, Phys.Rev.Lett. 5, 309-11.

Goldstein, B., 1966, Semiconductors and Semimetals, Edited by R.K. Willardson and A.C. Beer (Academic Press:New York), 2, 189-201.

Goltzené, A., Meyer, B., Schwab, C., Beall, R.B., Newman, R.C., Whitehouse, J.E. and Woodhead, J., 1985, J.Appl.Phys. 57, 5796-8.

Heine, V., 1960, Group Theory in Quantum Mechanics, (Pergamon:Oxford).

Kaufmann, U., Schneider, J. and Räuber, A., 1976, Appl.Phys.Lett. 29, 312-3.

Kennedy, T.A. and Wilsey N.D., 1981, Phys.Rev.B. 23, 6585-91.

Low, W., 1960, Paramagnetic Resonance in Solid State Physics: Advances in Research and Applications, Edited by F. Seitz and D. Turnball (Academic Press:New York), Supplement 2.

Neubrand, H., 1978a Phys.Stat.Solidi b86, 269-75: 1978b Phys.Stat.Solidi b90, 301-8.

Slichter, C.P., 1963, Principles of Magnetic Resonance, Edited by F. Seitz, (Harper and Row:New York).

Spaeth, J.-M., Hofmann, D.M. and Meyer, B.K., 1985, Mater.Res.Soc.Symp. Proc., 46, 185-94.

Watkins, G.D. 1965, Radiation Damage in Semiconductors (Dunod: Paris) p.97: 1975, Point Defects in Solids, Edited by J.H. Crawford Jr. and L.M. Slifkin. Volume 2, Semiconductors and Molecular Crystals (Plenum: New York) pp333-392.

Watkins, G.D. and Corbett, J.W., 1961, Phys.Rev. 121, 1001-14.

Photoluminescence Characterisation

E C LIGHTOWLERS

1. INTRODUCTION

Luminescence spectroscopy is a very sensitive tool for investigating both intrinsic electronic transitions and electronic transitions at impurities and defects, generally of atomic or molecular dimensions, in semiconductors and insulators. The most common technique is photoluminescence (PL), meaning luminescence excited by photons (usually a laser) as distinct from accelerated electrons (cathodoluminescence) or carrier injection (electroluminescence) or by other means. Impurity and defect related luminescence is generally dominant, particularly at low temperatures, except in the very purest material or in thin layer structures which exhibit quantum confinement. Much of the vast amount of work reported in the literature, predominantly during the last three decades, would be classed as fundamental research. However, because of the need for high quality material and controlled incorporation of specific impurities and defects to carry out these investigations, fundamental research and material characterisation are not really separable, particularly in the case of semiconductors.

Photoluminescence detects an optical transition from an excited electronic state to a lower electronic state, usually the ground state. If there is a multiplicity of excited states, only transitions from the lowest excited state can generally be observed at low temperatures because of rapid thermalisation. In the simple situation depicted in Figure 1, optical absorption can detect transitions from the ground state to all the excited states. In addition, optical absorption is an absolute measurement; the absorption coefficient α determined from the change in intensity of the transmitted light is proportional to the concentration of absorbing centres (Figure 2). However, absorption is often more difficulty to measure than the corresponding luminescence, particularly in thin layers or when the defect concentration is low. A useful technique for detecting weak absorption, or absorption in a thin layer on a heavily absorbing substrate, is photoluminescence excitation spectroscopy (PLE) which is illustrated in Figure 3. If the absorption process creates an excited state which can

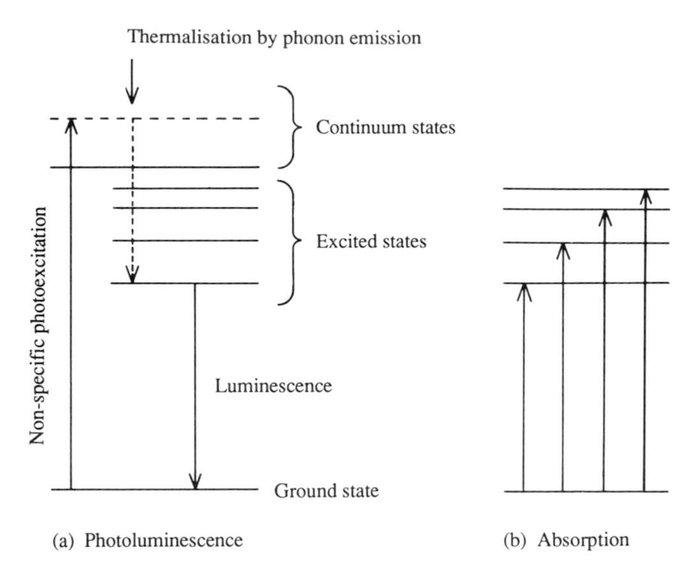

Figure 1. Schematic representation of low temperature photoluminescence and optical absorption

decay by luminescence emission at a different (lower) energy, measurement of the intensity of the luminescence emitted as a function of the energy of the exciting radiation can be much more sensitive than measuring a very small change in transmission, particularly if the excitation source is a tunable laser rather than the output of a monochromator from a broad band radiation source. If the absorption process results directly or indirectly in the creation of free carriers, photoconductivity can be employed with high sensitivity, the sample acting effectively as the radiation detector. However, electrical contacts have to be made in this case which is not always straight forward. In addition to photoluminescence spectroscopy, the main

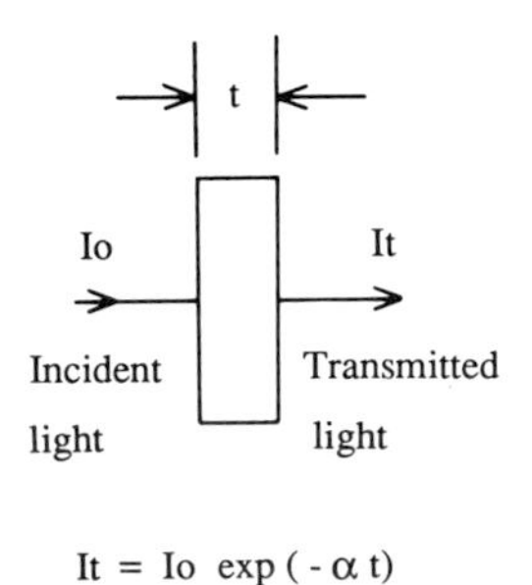

Figure 2. Optical absorption. The absorption coefficient α is proportional to the concentration of absorbing centres

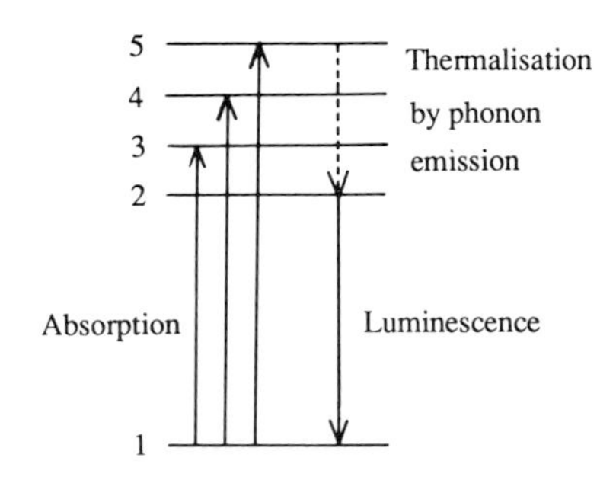

Figure 3. Schematic representation of luminescence excitation. Weak absorption into excited states 3, 4 and 5 is detected by luminescence emission from state 2

theme of this chapter, some attention will also be given to absorption and luminescence excitation. Luminescence decay time will also be briefly considered since this relates to the absorption coefficient to the defect concentration.

Only a limited range of topics can be addressed, and not in very great detail, and these are arranged as follows. Section 2 is concerned with experimental methods and equipment employed in absorption, photoluminescence and luminescence excitation measurements, in the near infrared spectral region appropriate for investigating both intrinsic and extrinsic electronic transitions in the indirect gap group IV semiconductor Si, and the direct gap group III V semiconductors GaAs and InP and heterostructures using these materials as substrates. Section 3 is concerned with free and bound-exciton luminescence from Si, and its employment for measuring donor and acceptor concentrations, and with the problem of identifying luminescence features with specific defect complexes. Section 4 provides a very brief account of near-band-edge luminescence from GaAs and InP and Section 5 an even briefer account of the spectroscopic characterisation of quantum well structures.

2. EXPERIMENTAL METHODS AND EQUIPMENT

The energy gaps Eg at low temperatures of the materials of interest are 1.16eV for Si, 1.42eV for InP and 1.52eV for GaAs. In order to incorporate heterostructures involving AlGaAs and InGaAs and defect luminescence from Si, mainly at $h\nu >$ Eg/2, we need to cover the spectral energy range $\sim$ 0.5 to 1.9eV. In terms of the units often employed in spectroscopy, the wavelength range is $\lambda \sim$ 650nm to 2.5μm and the wavenumber range $\nu \sim$ 4000 to 15,000 cm^{-1}. (Note: $h\nu(eV) = 1.239852 \times 10^3/\lambda vacuum$ (nm) $= 1.239852 \times 10^{-4}$ ν vacuum (cm^{-1}), and that $\lambda vacuum/\lambda air =$ 1.00028).

The simplest possible experimental arrangement is shown in Figure 4. For absorption measurements, light from a tungsten strip lamp *S* is focused by a lens *L*1 onto the sample mounted in a cryostat, and the transmitted light is focused on the entrance slit of the monochromator by the lens *L*2. The sample surface should have a suitable finish for efficient optical transmission, though accurately parallel faces in a thin sample can lead to the generation of interference fringes. The *f*-number of *L*2 at the monochromator should match the *f*-number of the monochromator for maximum efficiency and minimum scatter. The dispersed light falls on the detector *D* and additional optical components will be needed here if the detector area is small. An order-sorting filter *F*1 will generally be necessary with a grating instrument. If AC amplification and phase-sensitive detection (lock-in amplifier) are employed, the light will need to be modulated with a mechanical chopper *C*1. The output of the

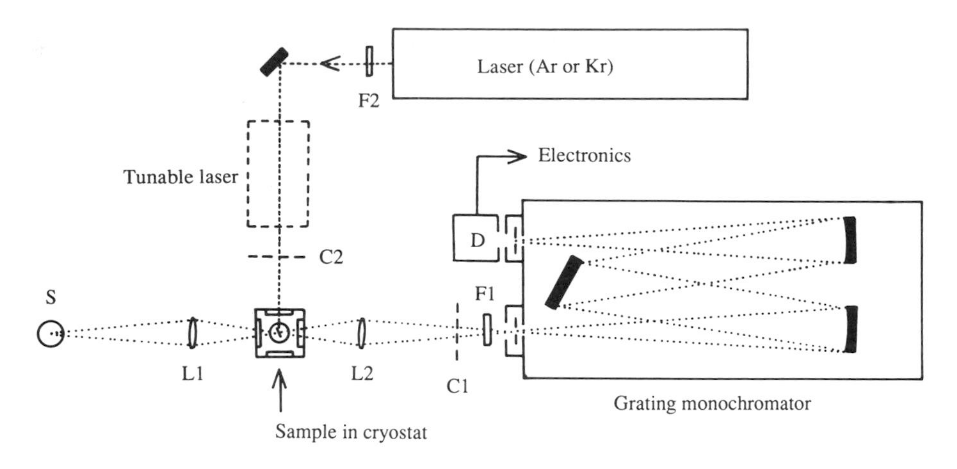

Figure 4. An experimental arrangement for absorption, photoluminescence and photoluminescence excitation measurements

detector is the product of the lamp spectrum, the transfer function of the monochromator and external optical system and the response function of the detector, as well as the transmissivity of the sample; some computing power at the end of the electronics will be an asset in deriving absorption spectra. For high precision measurements mirror optics would be superior to lenses to overcome the problem of chromatic aberation (the variation in focal length with wavelength). Optical quality glass has a transmission > 90% over the spectral region of interest and can be used for the lenses and cryostat windows. Note however that there are problems with atmospheric water vapour absorption between ~875 and 918 meV, which means that the total optical path has to be flushed with dry nitrogen when spectral features of interest occur in this region.

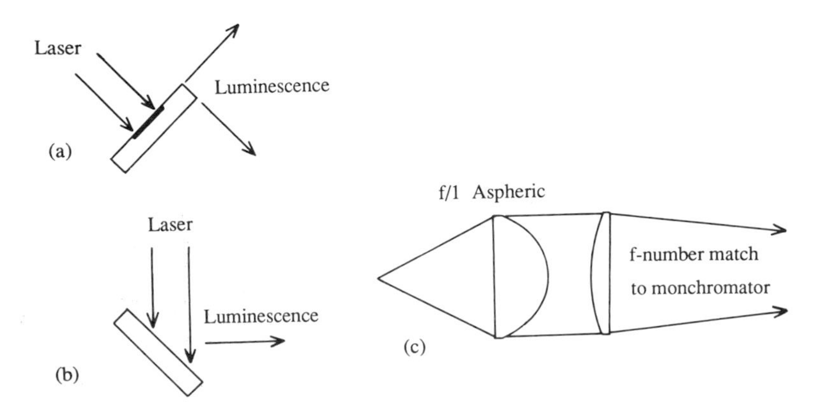

Figure 5. (a) Optimum geometry for a material with a large refractive index. (b) Conventional front surface geometry (c) Optimum light collection with minimum spherical aberration.

For luminescence measurements, the most convenient excitation source is a laser with $h\nu > Eg$, for example the 514.5nm line of an Ar laser or the 647.1 nm line of a Kr laser. For selective excitation, for example into an absorption band below the energy gap, a tunable source such as a dye laser would be employed. For optimum light collection from a thick sample with a high refractive index (n=3.5 for Si) and for minimum image distortion on the entrance slit of the monochromator, the arrangement shown in Figures 5(a) and 5(c) have been found to be most appropriate for luminescence measurements. However, if the light originates in the surface layer or would be strongly attenuated in passing through a long path in the sample or substrate, a front-surface geometry (Figure 5(b)) is necessary with about a factor of two to three less collection efficiency. The filter $F1$ (Figure 4) should be capable of preventing the exciting laser light from entering the monochromator, and a second filter $F2$ might be needed, particularly if the luminescence is weak or a front surface geometry is employed at 45°, to eliminate interference from non-lasing lines in the laser plasma which fall within the spectral region of interest. When using AC detection, it is more appropriate to chop the laser at $C2$ rather than the luminescence at $C1$.

Basically the same system can be used for luminescence excitation measurements with a tunable laser and with the monochromator set to detect the luminescence emitted as the energy of the exciting radiation is varied. Dye lasers are available only above about 1.24eV. Lower energies can be accessed by pulsed lasers based on an optical parametric oscillator or a Raman shifter, and in part by a CW colour centre laser. A broad band source and monochromator can be used instead of the tunable laser but only when the luminescence efficiency is very high and/or the resolution requirement is relatively low.

Spectroscopy based on monochromators is termed dispersive spectroscopy. Much higher sensitivity and data acquisition rates can be achieved by means of Fourier transform spectroscopy. This has been used for far-infrared measurements for several decades but has moved to the near infrared only recently and has only very recently been applied to luminescence measurements. The arrangement we have employed at King's College since 1985 for photoluminescence measurements on Si is shown in Figure 6 (Colley and Lightowlers 1987). The major component is a Michelson interferometer (I) which produces a composite interferogram associated with all the wavelength components in the incident light. The interferrogram is then Fourier transformed to yield the spectrum. This needs considerable computing power since an interferogram which will yield a high resolution low noise near infrared spectrum contains $\sim 10^5$ 20 bit data points. The gain in sensitivity occurs mainly because the whole of the spectrum is being detected the whole of the time rather than the small part which is incident on the exit slit of the monochromator at

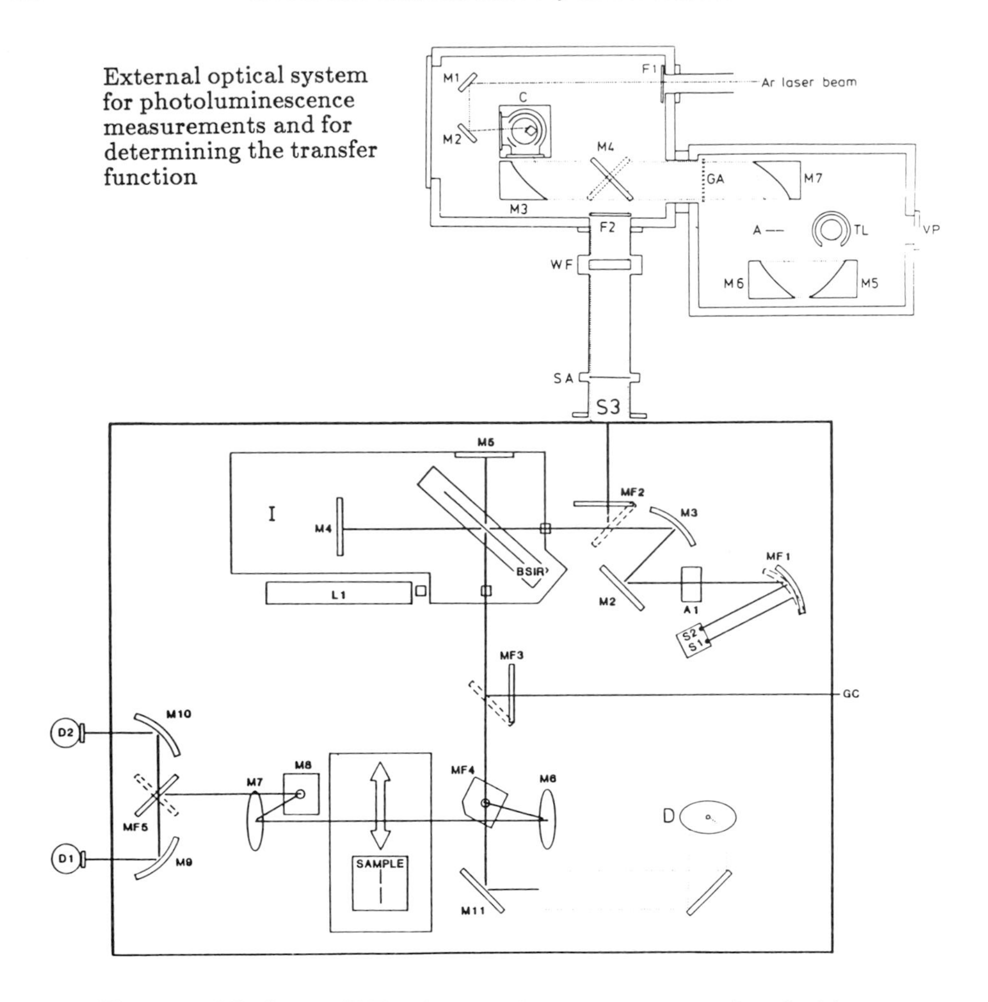

Figure 6. Nicolet 605X Fourier transform spectrometer fitted with a Si on CaF_2 beam splitter and modified for near infrared photoluminescence measurements (Colley and Lightowlers 1987)

any one time in dispersive spectroscopy. Fourier transform photoluminescence (FTPL) spectrometers are now commercially available.

Most spectroscopic measurements are carried out at low temperatures to prevent thermal ionisation of the optically active centres and to minimise the broadening of sharp spectral features by lattice vibrations. Generally the most convenient approach is to immerse the sample in liquid helium. Fluctuations in the light intensity caused by bubbling of the liquid boiling at atmospheric pressure at 4.2K can be eliminated by pumping the liquid helium below the λ-point at 2.16K (51mb pressure). Variable temperatures above 4.2K can most conveniently be achieved using a helium flow cryostat, but the temperature of the luminescence emitting region of a sample can be up to several degrees above the ambient helium gas temperature, measured by a temperature sensor, other than at very low excitation densities. Heat transfer is also a problem in luminescence measurements with the sample mounted on a cold finger in vacuum.

Finally in this section we must briefly consider the radiation detectors which can be employed. Above ~1.4eV a GaAs photocathode photomultiplier tube is most appropriate and above ~1.1ev an *S*1 photocathode tube. Both require cooling for optimum performance. The most sensitive detector in the range 0.7 to 1.1 eV is a cooled Ge diode detector with about a factor of 10 lower sensitivity than the cooled *S*1 photomultiplier at its maximum near-infrared response at ~ 1.55eV. Below 0.7eV PbS photoresistors at 196K or InAs photovaltaic detectors at 77K have been most generally employed. The signal/noise ratio of these detectors is ~10^3 less than the best Ge diode, in part due to the limit set by room temperature blackbody emission and in part to the low level of technological development of these materials. However, cold windowed InSb photodiode detectors have recently become available which are superior to PbS and InAs and are limited only by room temperature blackbody emission.

3. IMPURITIES AND DEFECTS IN SILICON

When silicon is excited at low temperatures with, for example, an Ar or Kr laser with photon energy above the band gap, electron hole pairs are produced. These can recombine in a number of ways, some of which give rise to luminescence (Figure 7). In general, the luminescence intensity associated with a particular path will be dependent on the relative and absolute concentrations of various impurities and defects, the excitation density and the temperature, and will involve capture cross-sections and branching ratios for radiative and non-radiative decay. Consequently luminescence spectroscopy is not generally a quantitative technique for measuring the concentrations of impurities and defects.

However, in material which is relatively defect free and in which the luminescence is dominated by free-exciton and donor and acceptor bound-exciton luminescence, it is possible to use the luminescence for the quantitative determination of donor and acceptors concentrations. At very low concentrations this is superior to any other approach and has the advantages of being impurity specific and capable of measuring both the compensated and uncompensated donor and acceptor concentrations (Electrical measurements and far-infrared absorption can only measure $N_A - N_D$ in *p*-type material and $N_D - N_A$ in *n*-type material). The quantitative determination of donor and acceptor concentrations in Si by photoluminescence measurements will be

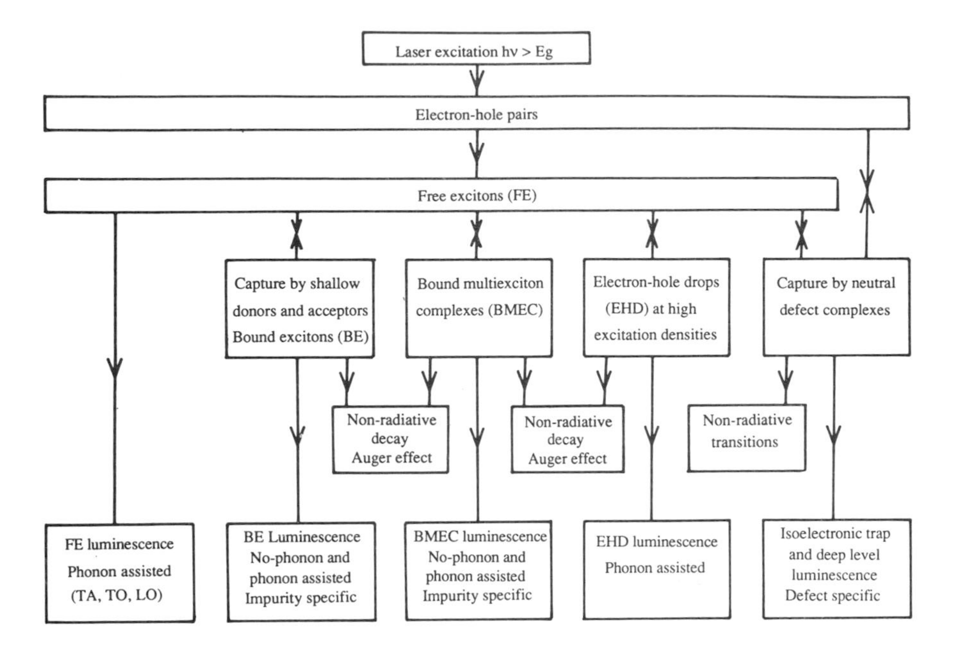

Figure 7. Luminescence decay processes stimulated by above band gap photoexcitation

considered in some detail in Section 3.2. The extreme right hand path in Figure 7 is responsible for well over 100 reported luminescence systems in silicon, in material which has been radiation damaged and/or heat treated or contaminated with transition elements (Davies 1989). Only one example will be dealt with here, in Section 3.3, with the emphasis on structural identification of defect complexes using perturbation methods.

3.1 Free and bound exciton luminescence

In high purity material, at low temperatures and relatively low excitation densities, the excess electrons and holes predominantly form excitons which subsequently decay giving rise to free-exciton luminescence (the extreme left-hand path in Figure 7). Because of the indirect band gap of silicon, the decay must be accompanied by the emission of a phonon which conserves momentum or wavevector (Figure 8). Three luminescence bands are observed associated with the TO, LO and TA phonons with the same wavevector **kc** as the conduction band minimum. The threshold energy Eth is given by

$$Eth = E_g - E_x - \hbar\omega \qquad (1)$$

where E_g is the energy gap, E_x is the exciton binding energy and $\hbar\omega$ is the energy of

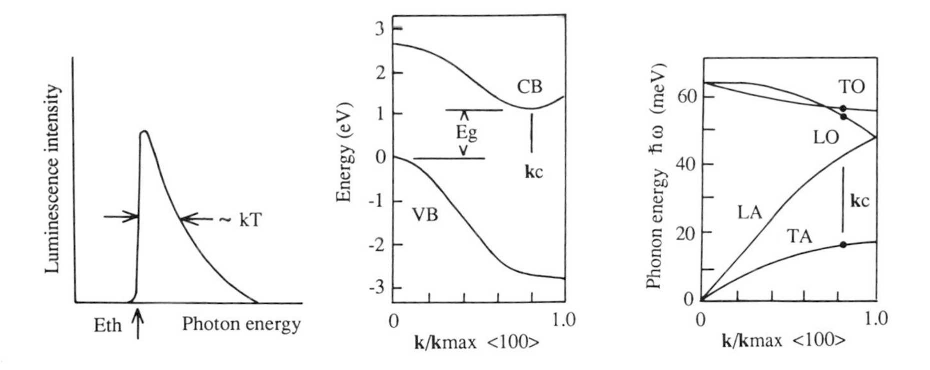

Figure 8. Free exciton luminescence in silicon, an indirect gap semiconductor. The threshold energy $E_{th} = E_g - E_x - \hbar\omega$ (phonon) where $E_x =$ exciton binding energy

the momentum-conserving phonon. A luminescence spectrum obtained from near-intrinsic silicon is shown in Figure 9. The TO, LO and TA phonon-assisted free-exciton bands are the dominant features. The sharper lines are due to the presence of B, P, Al and As at quite low concentrations.

If the material contains donors and acceptors with concentrations $> 10^{15}\text{cm}^{-3}$ then at 4.2K virtually all of the free excitons are captured giving rise to impurity-specific bound-exciton luminescence (the second path in Figure 7). Bound-exciton luminescence lines are sharp with photon energy given by

$$h\nu = E_g - E_x - E_B\,(-\hbar\omega) \tag{2}$$

where E_B is the binding energy of the exciton to the neutral donor or acceptor, which is ~1/10 of the donor or acceptor ionisation energy. Bound exciton decay may proceed with or without the emission of a momentum-conserving phonon. The branching ratio (no-phonon/phonon-assisted) increases with increasing binding energy. For example, in terms of integrated intensity as distinct from peak height, the ratio of the TO phonon-assisted to no-phonon process is ~110 for *B*, ~4 for *Al* and *P*, and ~0.3 for *In*.

At high excitation densities the donors and acceptors bind more than one exciton, and additional satellite lines are seen in the spectrum associated with the cascade decay of bound-multiexciton complexes. The decay schemes for phosphorus and aluminium are shown in Figure 10. The splitting of the bound-exciton states for boron is much smaller than for aluminium and can only just be resolved. The transitions drawn with heavy lines give rise to the dominant-luminescence features

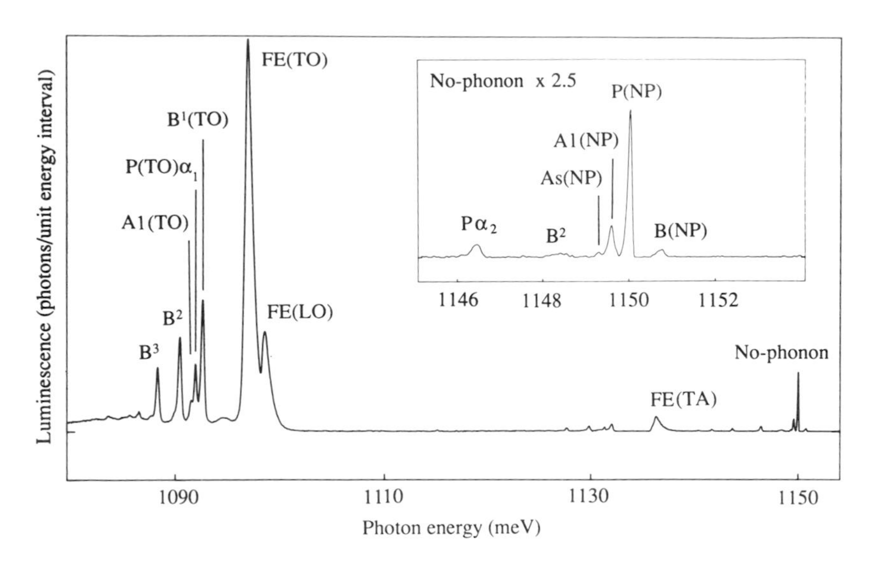

Figure 9. Photoluminescence spectrum of near intrinsic silicon ($\rho >$ 20 kΩcm)

at 4.2K. The bound-multiexciton transitions drawn with broken lines are relatively weak and dependent on the excitation density. The transitions drawn with full thin lines only become significant as the temperature increases above 4.2K. A detailed review of bound-multiexciton complexes is given by Thewalt (1982). At very high excitation densities the electrons and holes condense into a plasma, generally referred to as an electron-hole liquid or electron-hole drops (EHD). This gives rise to very

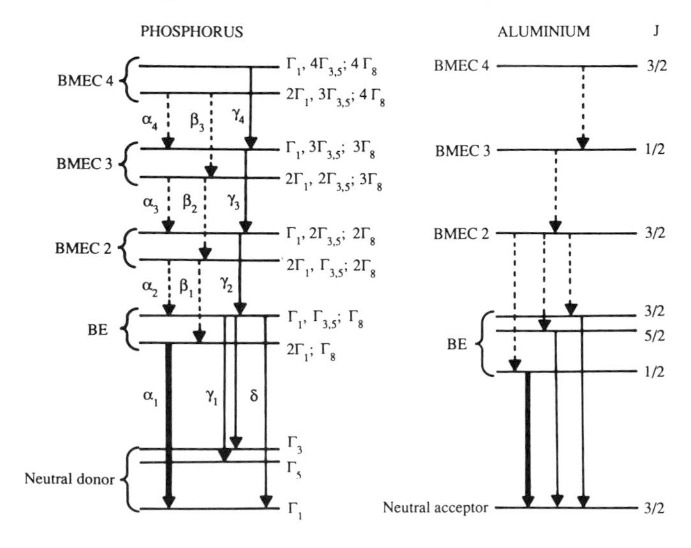

Figure 10. Decay of bound multiexciton complexes giving rise to luminescence features

broad luminescence bands associated with electron–hole recombination within the drops.

The luminescence features associated with the transitions discussed above are illustrated in Figures 11, 12 and 13. The spectra were all obtained with the samples immersed in liquid helium at 4.2K and were excited with 250mW of the 514.5nm line of an Ar laser. They were all obtained in about 10 minutes using Fourier transform spectroscopy and have all been corrected for the system response to photons/unit energy interval. It is clear from the spectra shown in Figures 9,11,12, and 13 that photoluminescence is a very sensitive tool for detecting the presence of various donors and acceptors. Further, both compensated and uncompensated species can be seen because at the excitation densities employed the electron and hole quasi–Fermi levels are in the conduction and valence bands, respectively, and all the electronically active species are neutral; excitons only bind to neutral donors and acceptors in silicon. For example, for the near–intrinsic material in Figure 9, the electrically active impurity concentrations are: $[B] = 1.36 \times 10^{12} cm^{-3}$, $[P] = 1.69 \times 10^{12}$ cm^{-3}, $[Al] = 0.61 \times 10^{12} cm^{-3}$ and $[As] = 0.14 \times 10^{12} cm^{-3}$ giving $N_A - N_D = 1.4 \times 10^{11} cm^{-3}$. The determination of this quantitative information from photoluminescence measurements is considered in Section 3.2.

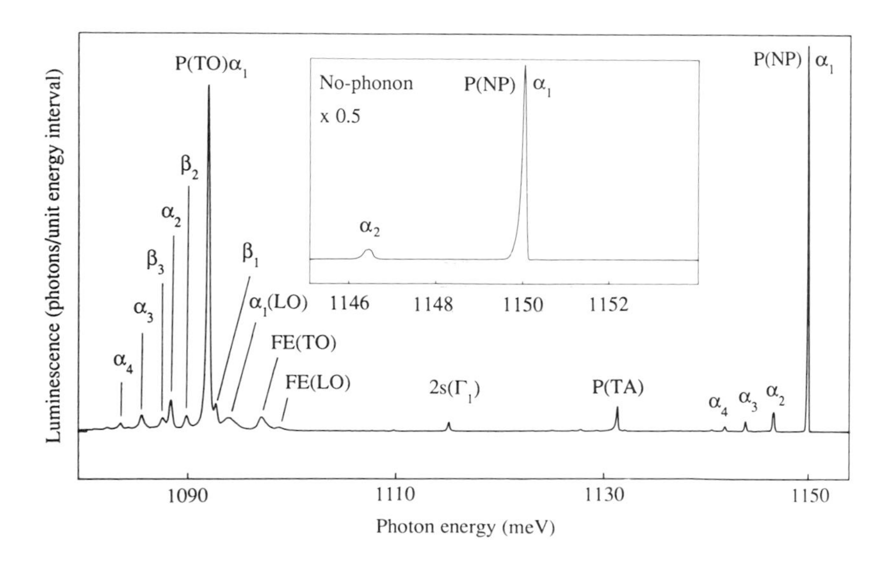

Figure 11. Photoluminescence spectrum from a Si sample doped with $3 \times 10^{14} cm^{-3}$ P with negligible B, Al and As.

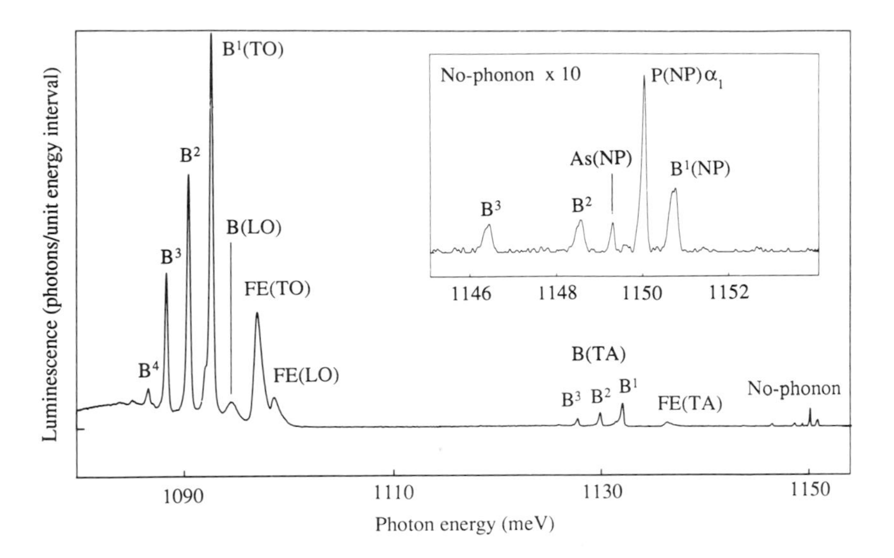

Figure 12. Photoluminescence spectrum from a Si sample doped with $1.3 \times 10^{13} cm^{-3}$ B, contaminated with $1.8 \times 10^{12} cm^{-3}$ P and $3 \times 10^{11} cm^{-3}$ As

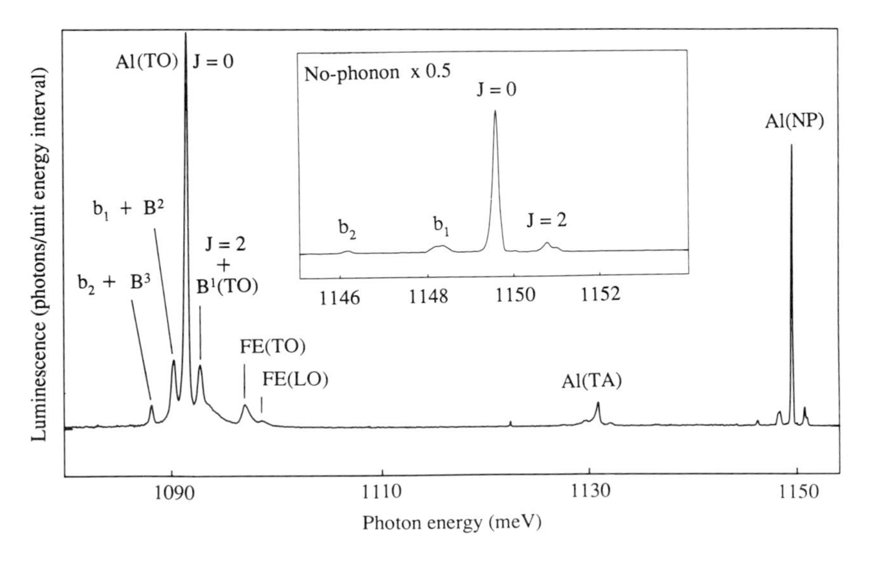

Figure 13. Photoluminescence spectrum from a Si sample doped with $2.7 \times 10^{14} cm^{-3}$ Al. Boron contamination can be seen in the TO phonon–assisted region

3.2 Determination of donor and acceptor concentrations from photoluminescence measurements

The original work in the mid to late 1970's on bound-exciton luminescence in Si and Ge and other materials was principally concerned with the fundamental physics and led, for example, to the energy level schemes seen in Figure 10 and an understanding of electron-hole drops. Tajima (1978) was the first to show that quantitative information on donor and acceptor concentrations could be derived from the ratios of bound-exciton to free-exciton luminescence features. Calibration of the technique is critically dependent on the excitation density and the temperature, which can be interelated if thermal contact with the refrigerant is inadequate, and serious consideration has to be given to the spectral resolution and the spectral response function of the measuring system if the calibration is to be used in different laboratories using different equipment. A detailed account is given by Colley and Lightowlers (1987)

Consider the spectrum from the sample doped with 3.2×10^{14}cm^{-3} phosphorus shown in Figure 11. The quantitative technique employed by Colley and Lightowlers (1987) is based on the height or area of $P(NP)$ compared with the height of $FE(TO)$, at an excitation density which as near as possible, saturates the multiexciton luminescence but produces no significant heating of the luminescence emitting region above 4.2K with the sample immersed in liquid helium boiling at atmospheric pressure. The use of area is preferable to peak height since this eliminates some of the problems of defining the spectral resolution and the effects of strain broadening. Clearly the spectrum has to be corrected to absolute units, here photons/unit energy interval. This can be is achieved by measuring the spectrum of a tungsten lamp with a known filament temperature, and calculating the theoretical spectrum from the Planck radiation formula, corrected for the emissivity of tungsten as a function of wavelength. Earlier calibrations are based on the spectral features in the TO phonon-assisted region. However, the phonon-assisted lines are broad and not clearly separated (see Figure 9), and there are weak underlying features due to LO-phonon assisted transitions and multiexciton transitions which are not allowed in the no-phonon region. These factors make peak height and area determination more difficult in the TO-phonon region especially for minor components. In the no-phonon region all of the spectral features can be clearly resolved except for antimony which is almost coincident with phosphorus.

The spectrum shown in Figure 12 was obtained with a sample doped with 1.3×10^{13} cm^{-3} boron. The no-phonon region, expanded in the inset, shows that the sample is contaminated with phosphorus and arsenic. The no-phonon transitions are very weak for boron but can be used for analysis down to 2 or 3×10^{12}cm^{-3}. At lower concentrations $B(TO)$ can be employed, down to $\sim 10^{10}\text{cm}^{-3}$ in the absence of interference from much higher concentrations of other impurities. An aluminium spectrum is shown in Figure 13, obtained from a sample containing $2.7 \times 10^{14}\text{cm}^{-3}$. The weak high energy doublet (J=2) coincides with the principal boron bound exciton line and care has to be taken with boron analysis based on no-phonon transitions when aluminium is also present. The ratio of J=2 to J=0 is a very precise indicator of the temperature of the luminescence emitting region of the sample, here 4.5 ± 0.1K. Comparison of the relative intensities of the weaker features in the no-phonon and TO phonon-assisted regions indicates that there is some boron contamination.

The calibration for *P*, *B* and *Al*, based on the ratio of the no-phonon bound-exciton peak area and the strength of the free exciton band is shown in Figure 14. The slopes are within 3% of unity on the log-log plots over three decades showing that the relationships are effectively linear. The concentrations were determined from electrical measurements and an iterative approach had to be employed to allow for co-doping at lower concentrations. The low concentration limit in this calibration was set by the availability of material without serious co-doping; the upper limit is set by the difficulty in measuring the height of the free-exciton band at concentrations $>10^{15}\text{cm}^{-3}$. Better low concentration material has since been obtained and the calibration has been extended to $\sim 10^{17}\text{cm}^{-3}$ by making

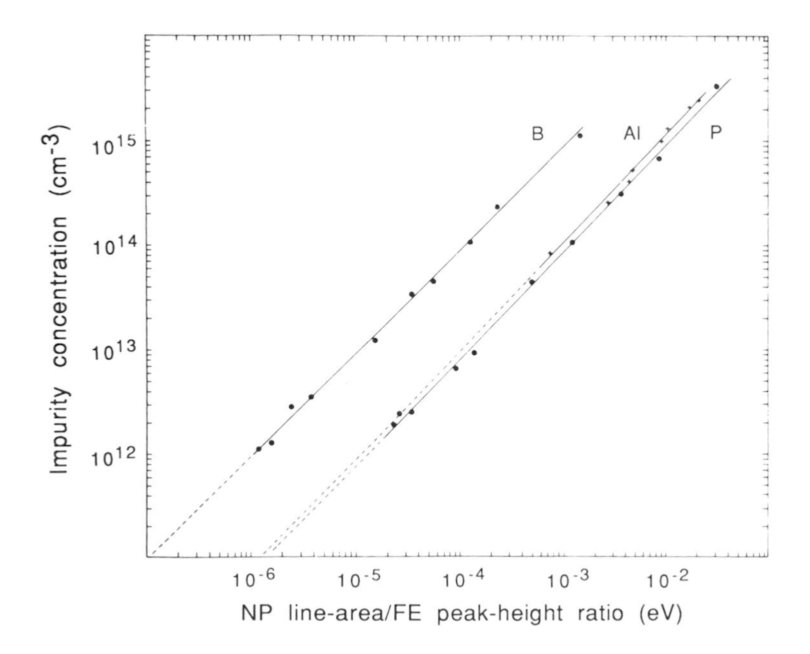

Figure 14. Calibration of the photoluminescence technique for measuring *B*, *P* and *Al* concentrations in Si

measurements at 20K where the bound excitons are substantially ionised. More recent work in this area has been principally concerned with the additional problems of characterising thin epitaxial layers as distinct from bulk material.

3.3 Identification of defect complexes

An essential precursor to materials characterisation, either qualitative or quantitative, is identifying specific spectral features with the defect responsible. This is relatively straight forward for donor and acceptor bound–exciton luminescence since deliberate doping with various donor and acceptors is reasonably well under control. For other impurities and particularly for complexes, either formed during growth or by contamination or by radiation damage and/or thermal treatment, this is not the case. Luminescence spectroscopy has yielded a large amount of data and much speculation about the defects responsible, but only in a very limited number of cases do we have a detailed structural model of the defect responsible for a particular luminescence system (Davies 1989). To arrive at a detailed structural model for a defect generally needs a combination of several techniques, for example mid–infrared local–mode vibrational spectroscopy and electron spin resonance as well as electronic absorption and luminescence spectroscopy. One particular case will be considered, namely a radiation–induced defect in Czochralski silicon, which illustrates the problem and relates to the characterisation techniques discussed in another chapter.

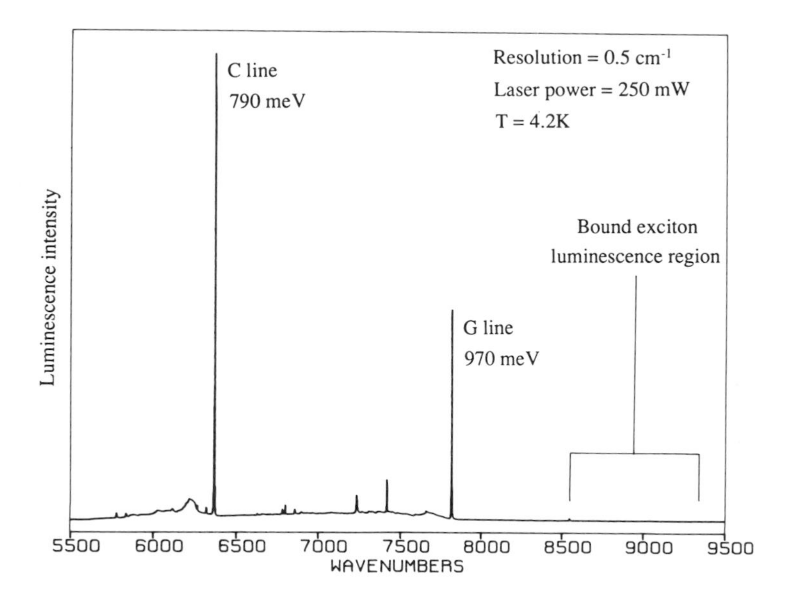

Figure 15. Photoluminescence spectrum from CZ Si containing $7 \times 10^{17}cm^{-3}$ oxygen and $9 \times 10^{16}cm^{-3}$ carbon, irradiated with $2 \times 10^{17}cm^{-2}$ 2 MeV electrons

Czochralski (CZ) silicon, that is silicon pulled from the melt in a silica crucible heated by a graphite susceptor, the way most device material is produced, generally contains oxygen in the range 5×10^{17} to $10^{18}cm^{-3}$ and carbon in the range 10^{16}–$10^{17}cm^{-3}$. The oxygen is mainly present as a bonded interestial in a supersaturated solid solution and the carbon is predominantly substitutional. Oxygen is mobile at high temperatures and heat treatment leads to complex formation; the formation of oxygen precipitates which punch out dislocations is the basis of intrinsic gettering. The carbon is benign and basically immobile on a substitutional site, but any process which releases Si self intestitials, in particular radiation damage including ion implantation, plasma etching and neutron transmutation doping, or simply the growth of a thermal oxide, results in the transfer of carbon to interstitial sites by the Watkins replacement mechanism. Interstitial carbon is mobile at room temperature and appears to form complexes with virtually every other impurity and defect in the material (Davies et al 1987).

The two major luminescence systems produced by electron radiation damage are the *C*–line system and the *G*–line system shown in Figure 15. These consist of very sharp no–phonon lines at 790 and 970 meV, respectively, with vibronic side bands at lower energies which represent electronic transitions accompanied by the emission of phonons with a wide range of energies. The same electronic transitions can be seen in absorption but the vibronic side bands are on the high energy side of the no–phonon line in this case. The spectrum in Figure 15 has not been corrected for the system response and the low energy limit is set by the cut–off of the germanium diode detector employed. The high energy limit is set by the band gap of silicon but it can be seen that the near–edge bound exciton luminescence has been completely

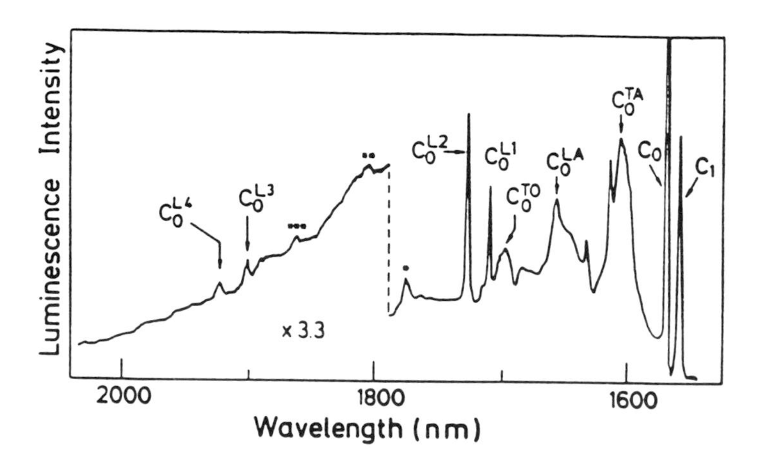

Figure 16. Vibronic side band of the *C*–line luminescence system (Wagner et al 1984)

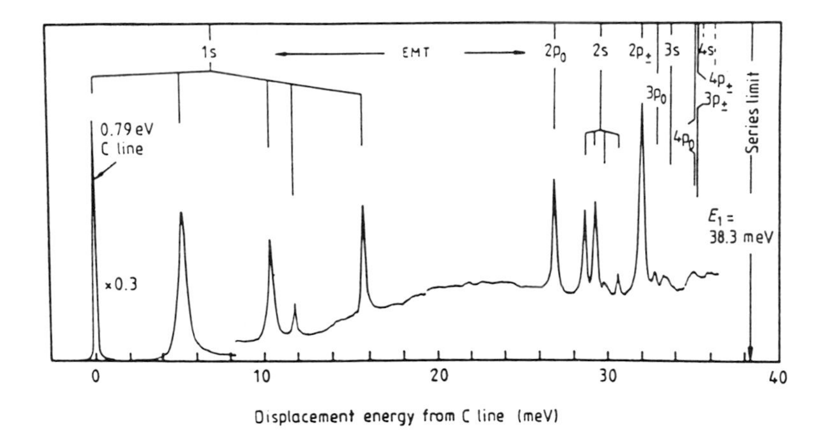

Figure 17. Luminescence excitation spectrum of the C–line defect showing the electronic excited states (Thonke et al 1985)

suppressed by the radiation damage. The *G*–line defect is known to be a carbon–carbon complex which has been investigated in great detail (Davies et al 1987). We will concentrate here on the *C*–line.

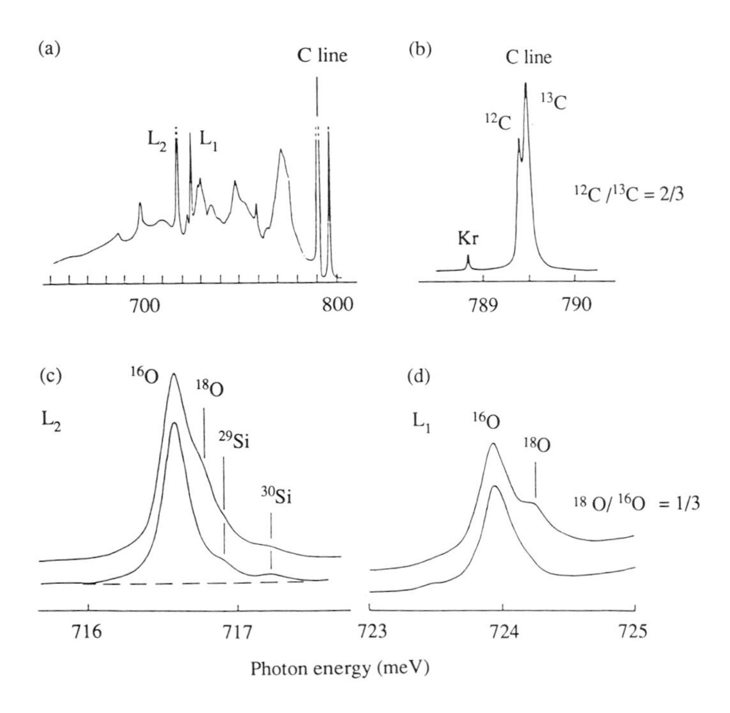

Figure 18. Isotopes splitting of the no–phonon line and local mode lines of the C–line luminescence system (Davies et al 1985)

The vibronic–side band of the *C*–line in luminescence is shown in detail in Figure 16 (the *C*–line is labelled Co) and it can be seen that the electronic transition couples to a wide range of lattice modes as well as local vibrational modes identified as L1 to L4. The luminescence excitation spectrum in Figure 17 shows that there is a large number of electronic excited states, which can be modelled in terms of a donor–like bound–exciton. We cannot go into details on either the phonon coupling or the excited state structure here, and we will concentrate on identification of the defect.

We would hazard a guess from the circumstantial evidence that the defect is a complex involving both carbon and oxygen, but this needs to be shown unambiguously. This can be achieved by isotope–substitution. Figure 18(b) shows the no–phonon line in an irradiation damaged CZ sample doped with both ^{12}C and ^{13}C. The splitting of the no–phonon line into two components in roughly the same ratio as the carbon isotopes, measured independently by mid–infrared vibrational spectroscopy, suggest that the defect contains one carbon atom. An oxygen isotope shift is not detected in the no–phonon line but is clearly seen in the local mode features *L*1 and *L*2 in a sample doped with both ^{16}O and ^{18}O (Figures 18(*c*), (*d*)). Structure associated with the natural isotopic composition of silicon is also apparent in *L*2. It is clear that there is one oxygen atom in the defect.

Perturbation methods can provide information about the symmetry of a defect which has to be consistent with any proposed structural model. The most powerful perturbation technique for studying defect complexes in silicon has proved to be uniaxial stress spectroscopy. Zeeman measurements and the application of large electric fields can provide symmetry information, but have not proved to be very

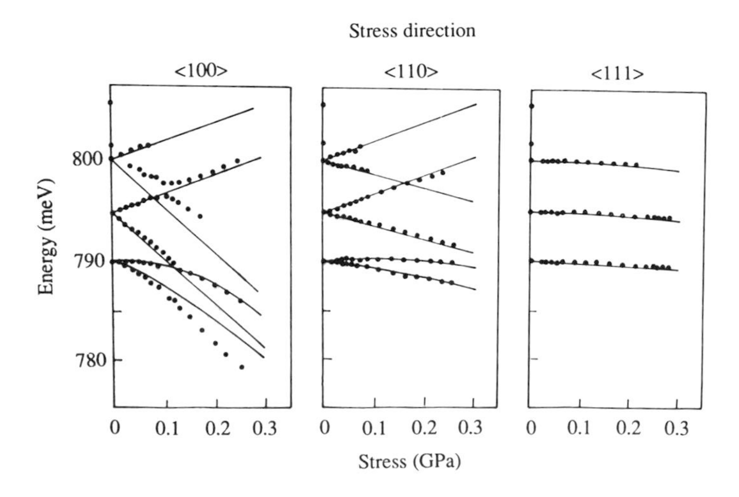

Figure 19. Uniaxial stress splitting of the three lowest energy lines of the C–line defect, measured in absorption

useful for defect complexes in silicon. The application of uniaxial stress along the major crystal axes, usually <100>, <110> and <111>, lowers the symmetry of the environment about the defect and produces a shift and splitting of the electronic states and consequently in the sharp no–phonon lines produced by electronic transitions between these states. A recent review is given by Davies (1988).

Figure 19 shows the stress splitting of the three lowest energy electronic transitions at the *C*–line defect measured in absorption; the experimental points are represented by the closed circles and the theoretical model ignoring interactions between electronic states is represented by the lines. This analysis gives the symmetry as monoclinic I (C_{1h}) but close to tetragonal (D_{2d}), i.e., there is a major symmetry axis of the defect close to the <100> direction but distorted into a (110) plane. These and more detailed stress results preceeded the model of the electronic states derived from the data in Figure 16 but are consistent with this model. However, none of this tells us the exact locations of the oxygen and carbon atoms but it does limit the options.

Detailed independent and collaborative work in various laboratories has shown that the defect responsible for the *C*–line is also responsible for the *C*3 mid–infrared vibrational bands at 1115, 865, 742, 550 and 529 cm^{-1} in its neutral charge state, for the *G*–15 spin resonance signal and a DLTS level at $E_v + 0.32$ eV in a positive charge state, and the *G*–16 spin resonance signal and a DLTS level at E_c–0.43 eV in a negative charge state. A correlation between the strength of the *C*–line absorption and the *C*3 absorption is shown in Figure 20.

The point indicated in Figure 20 was used to check the expressions often employed for determining defect concentrations from optical absorption data. For vibrational

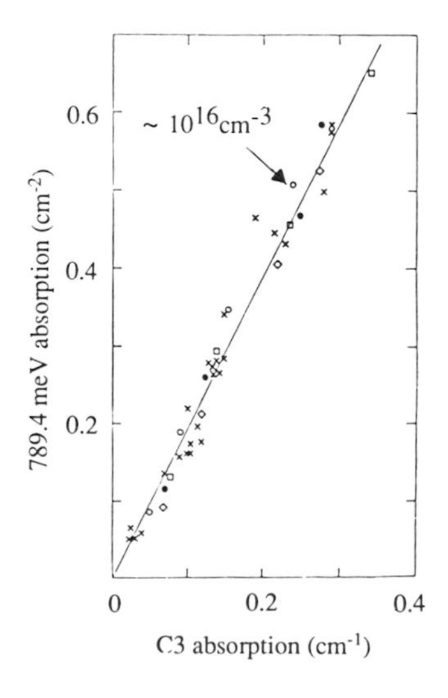

Figure 20. Correlation between the 790 meV *C*–line absorption and the *C*3 local vibrational mode absorption. The different symbols represent different radiation doses and annealing stages for a series of samples with different oxygen and carbon concentrations (Davies et al 1986)

absorption, the defect concentration and total integrated absorption are related by

$$[C3\text{ defect}] = \frac{4\,ncm\,\epsilon_0}{h\,\eta^2} \int \mu(E)\,dE \tag{3}$$

where μ is the absorption coefficient, n is the refractive index, m is the mass of the vibrating atom and η is the effective charge. The other symbols have their usual meanings. Using the mass of the carbon atom for m and the same effective charge $\eta = 2.5e$ known to apply to the substitutional carbon vibrational absorption, on the basis that the $C3$ lines exhibit a large carbon isotope shift, gives $[C3\text{ defect}] = 2 \times 10^{16}\text{cm}^{-3}$.

The total electronic absorption, no–phonon lines and vibronic sideband, is related to the defect concentration by

$$[C\text{–line defect}] = \frac{9E^2\tau\, n^2}{\pi^2\hbar^3 c^2(n^2+2)}\,\frac{g_f}{g_i} \int \mu(E)\,dE \tag{4}$$

where g_i and g_f are the degeneracies of the initial and final states of the electronic transition, τ is the luminescence decay time and E is the mean transition energy (Dexter 1958). Using $g_f/g_i = 1$ and a measured luminescence decay time of 95μs (at 20K), $[C\text{–line defect}] = 8 \times 10^{15}\text{cm}^{-3}$. The agreement is remarkably good considering the assumptions made.

The detailed atomic structure of the defect has recently been determined from uniaxial stress measurements on the G–15 spin resonance spectrum by Trombetta and Watkins (1987). This is shown in Figure 21. The model is consistent with the

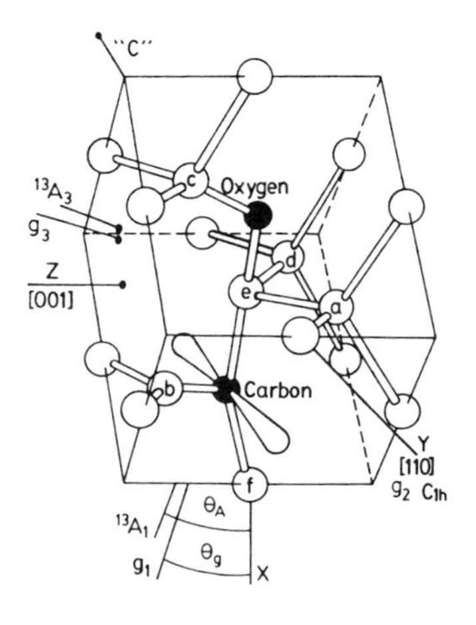

Figure 21. Atomic model for the C–line and $C3$ defect derived from electron spin resonance measurements (Trombetta and Watkins 1987)

data derived from spectroscopic measurements. In essence, the defect consists of a bonded interstitial carbon atom at approximately a second neighbour distance from a bonded interstitial oxygen atom, and is produced by the capture of a mobile interstitial carbon by an interstitial oxygen atom.

4. NEAR BAND–EDGE LUMINESCENCE FROM GaAs AND InP

GaAs and InP are the major indirect gap III V compound semiconductors with current and potential device applications. There is a very large literature on these materials dealing with both near–band–edge luminescence and with deep level luminescence, particularly resulting from the incorporation of various transition elements. Only a fairly superficial account of free and donor and acceptor bound–exciton luminescence can be given here. A recent detailed review is given by Skolnick (1987).

A near–band–edge luminescence spectrum obtained from high purity InP grown by vapour phase epitaxy is shown in Figure 22. High purity GaAs and InP generally contain > 10^{13}cm^{-3} electrically active impurities and impurity related luminescence dominates. The broad weak component at high energy, centred about 1.4185 eV, is associated with the decay of free excitons. In direct gap semiconductors there is a strong coupling of the free excitons with photons of the same energy which results in a multi–mode excitation called an exciton–polariton. The shape and intensity of the exciton–polariton luminescence band is a complicated function of the excitation density, the impurity concentration and surface preparation and we cannot enter into a discussion of this here. A detailed account is given by Steiner *et al.* (1986).

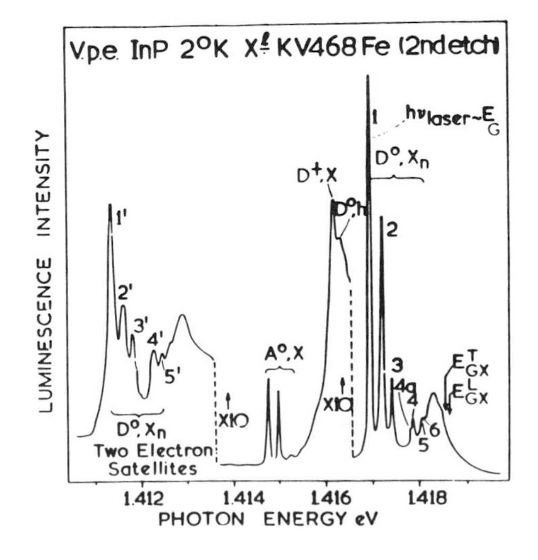

Figure 22. Photoluminescence spectrum obtained from high purity InP sample grown by vapour phase epitaxy(Skolnick 1987))

The sharp lines bracketed as D^0, X_n between ~1.417 and 1.418 eV are associated with the recombination of excitons bound to neutral donors. The various lines are not due to different donor species, which have virtually indistinguishable exciton binding energies. They are associated with different electronic states of the neutral bound exciton complex, composed of a donor ion, a hole and two electrons, arising from electron–hole exchange interaction, crystal field splitting and different angular momentum states of the hole. The features labelled D^0,*h* and D^+,*X* are ascribed to free hole recombinations at a neutral donor and the decay of an exciton bound to an ionised donor, respectively. These two processes are not observed in silicon. The doublet labeled A^0, *X* is associated with the decay of an exciton bound to a neutral acceptor, and the doublet structure is due to different electronic levels produced predominantly by hole–hole interaction, as in the case of acceptor bound excitons in silicon. The lowest energy features in Figure 22 are identified as two–electron satellites of the donor bound exciton transitions. In a so–called two–electron transition, the donor is left in an excited state rather than the 1s ground state following exciton recombination. This is illustrated in Figure 23. Photoluminescence spectra for three high purity GaAs samples are shown in Figure 24 and exhibit similar features to those described for InP. We will continue the discussion in terms of InP.

The spread in donor binding energies in InP is ~0.3 meV at ~7 meV, and the corresponding spread in donor bound–exciton binding energies is less than the D^0,*X* transition line widths which are ~0.1 meV in very high quality material. However, since most of the difference in donor binding energies appears between the chemically shifted 1s levels and the effective–mass–like 2p and 2s levels, the two–electron $(D^0,X)'$ transitions for the different donors differ in energy by up to ~0.3 meV. Hence different donor species can just about be distinugished in the

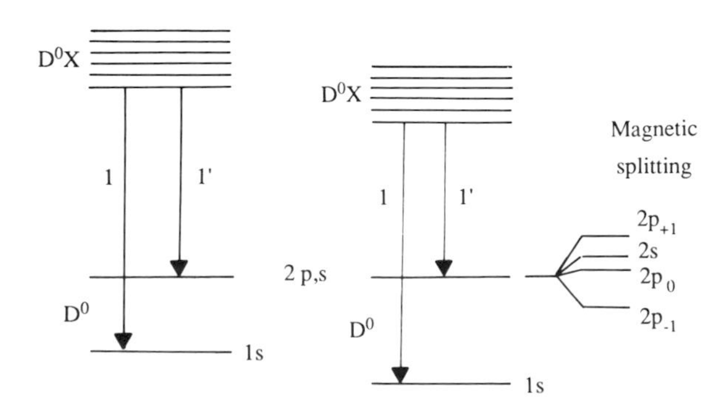

Figure 23. The D^0,*x* and $(D^0,x)'$ transitions for two different donors with different binding energies. Most of the difference in binding energy (the chemical shift or central cell correction) appears between the 1s and 2p and 2s levels

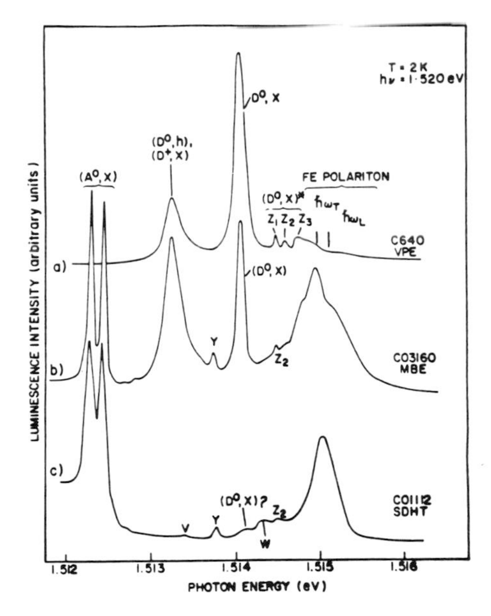

Figure 24. Photoluminescence from high purity GaAs. (a) Sample grown by vapour phase epitaxy. (b) Sample grown by molecular beam epitaxy. (c) Modulation doped GaAs/AlGaAs heterostructures grown by molecular beam epitaxy; the signal originates in the GaAs buffer layer (Skolnick 1987)

two-electron satellites. This is even more pronounced for the two hole satellites for the shallow acceptors for which the binding energies vary between 40 and 60 meV.

The situation is further improved by the application of high magnetic fields (Figure 25). This produces a shrinkage of the wavefunctions of the electronic states which reduces external energy broadening effects and increases the central cell corrections of the 1s states. This gives rise to a narrowing and increased separation of the luminescence lines, particularly in the two-electron spectrum as shown in Figure 25. Further line narrowing can be achieved by selective excitation and luminescence

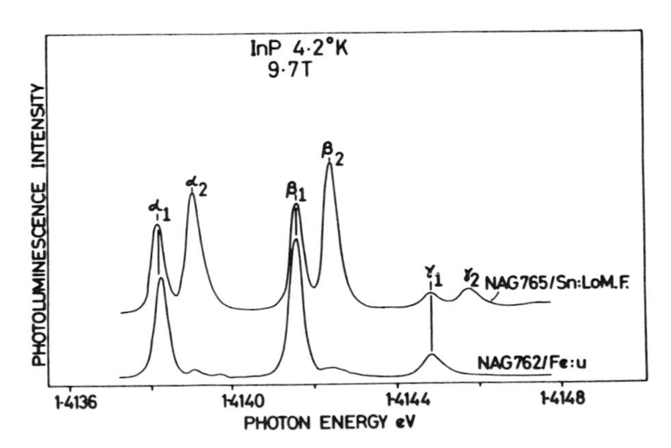

Figure 25. Luminescence spectra in the $(D^0,x)'$ satellite region for InP samples containing both sulphur and silicon donors (upper trace) and only sulphur donors (lower trace). The magnetic flux density is 9.7T (Skolnick 1987)

excitation spectroscopy, i.e., by using a very narrow line from a tunable laser to selectively excite the transition under investigation. Details of these techniques are given in the review article by Skolnick already cited (1987).

It is clear that qualitative analysis of donors and acceptors in InP and GaAs requires more sophisticated techniques than those which can be used for silicon. To derive a quantitative technique based on photoluminescence and related measurements will be very difficult. On the other hand, the spectroscopy is within the range of photomultiplier detectors and the luminescence efficiency of these direct gap semiconductors in orders of magnitude greater than for silicon.

5. LUMINESCENCE FROM QUANTUM WELL STRUCTURES

Since the pioneering work of Dingle (1975), a very large literature has developed on the spectroscopic assessment of quantum well structures, primarily GaAs/AlGaAs but also InP/In GaAs and others. Luminescence and luminescence excitation spectroscopy are particularly important characterisiation tools. Only a very brief and superficial account can be given here; the theory of quantum well structures and their characterisation are covered in detail in another chapter.

GaAs and the $Al_xGa_{1-x}As$ with x in the range 0.2 to 0.3 have very closely matched lattice constants, and layer structures made from these materials can be grown with negligible built in strain. The energy gap of AlGaAs is larger than that of GaAs and the conduction and valence band discontinuities in a structure made up of a thin layer of GaAs sandwiched between two layers of AlGaAs have the appearance shown in Figure 26. The relative size of the discontinuities or band offsets ΔC and ΔV in the conduction and valence bands have been the subject of much controversy and we will

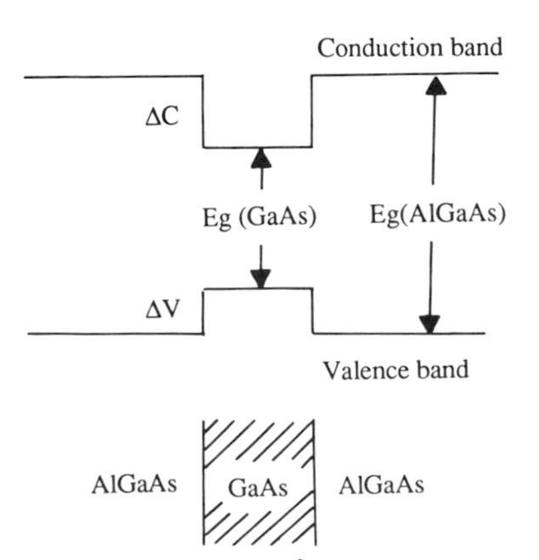

Figure 26. Energy discontinuities in the conduction and valence bands produced by sandwiching a thin layer of GaAs between two relatively thick layers of AlGaAs

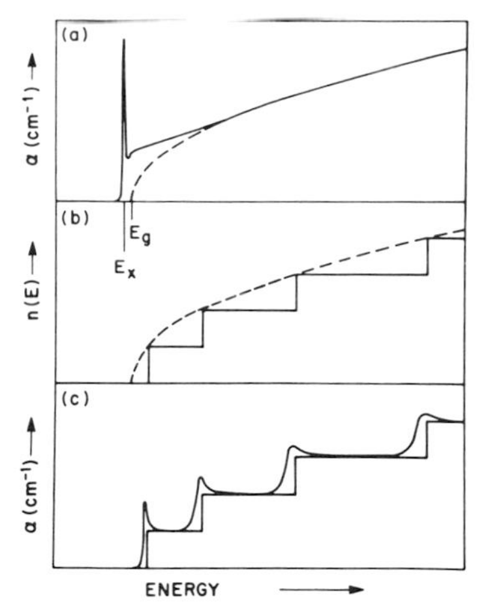

Figure 27. Absorption edge spectrum of GaAs including exciton effects and the bulk conduction band density of states. (b) Comparison of bulk and quantum limit (2 dimensional) density of states of a simple energy band. (c) Schematic absorption edge spectrum with and without exciton effects at each 2–dimensional band edge (Dingle 1975)

not enter into this here. Quantum size effects begin to appear when L becomes less than ~50nm. An electron or hole in the GaAs layer can be regarded as confined in a three dimensional potential well. If the well width is of the order of the de Broglie wavelength, the particle motion is quantised in one dimension and the energy levels are given by relatively straight forward solutions of the Schrodinger equation.
The eigenvalues for an infinite well are

$$E = \frac{\hbar^2}{2m}\left(kx^2 + ky^2 + kz^2\right) \tag{5}$$

where $k = \pi n/L$ and L is the length of the crystal in the appropriate direction. For $Lz << Lx$ and Ly we will have quantisation in the z direction and

$$E = E_n + \frac{\hbar^2}{2m}\left(kx^2 + ky^2\right) \tag{6}$$

where

$$E_n = \frac{\hbar^2}{2m}\left(\frac{n\pi}{Lz}\right)^2 \tag{7}$$

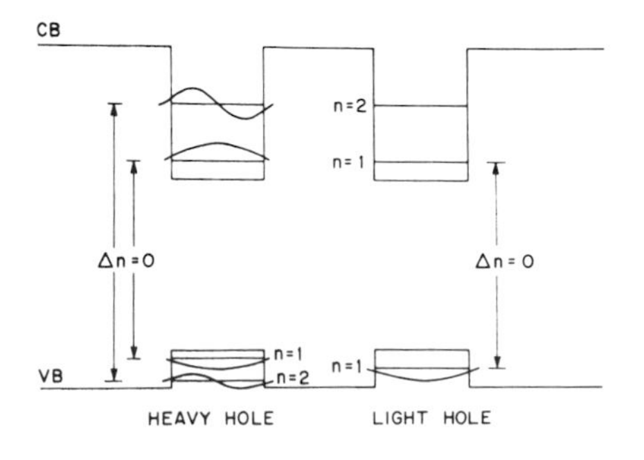

Figure 28. Bound states and wave functions of quantised valence and conduction bands demonstrating the $\Delta n = 0$ selection rule for interband transitions (Dingle 1975)

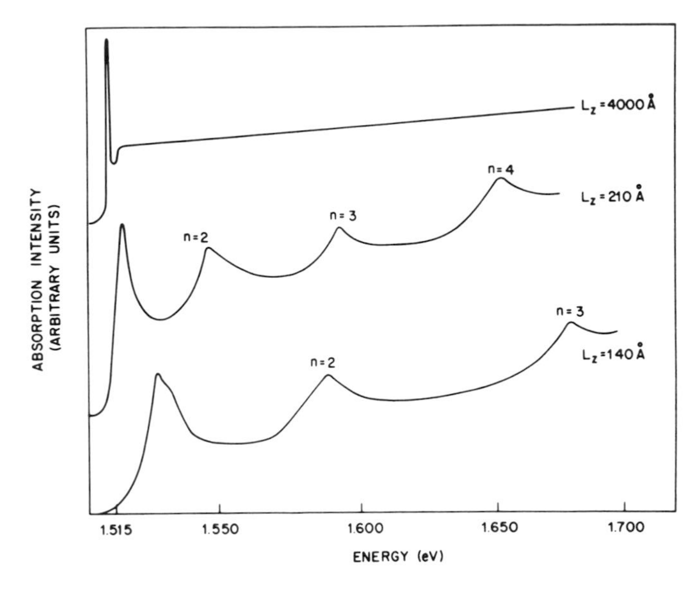

Figure 29. (a) Absorption spectra (2K) of 400, 21 and 14 nm thick GaAs quantum wells (Dingle 1975)

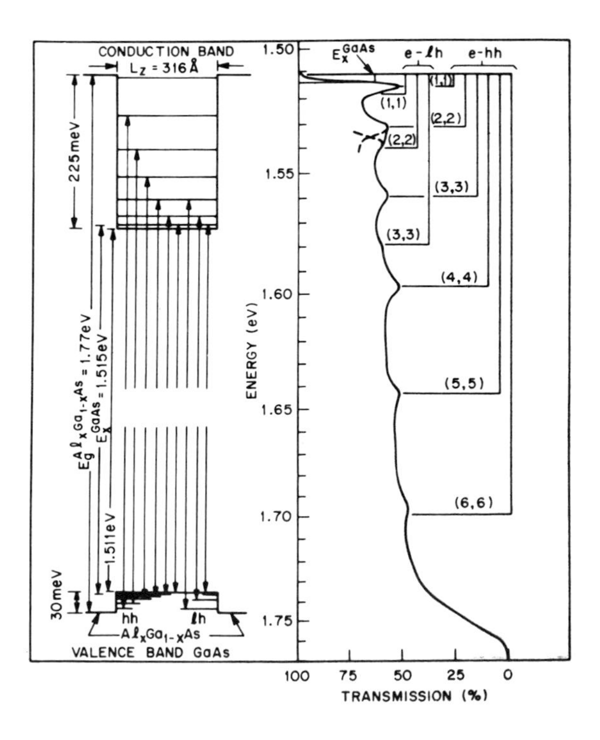

Fig.30. Transmission spectrum from a GaAs quantum well with the transitions assigned according to the $\Delta n = 0$ selection rule (Dingle 1975)

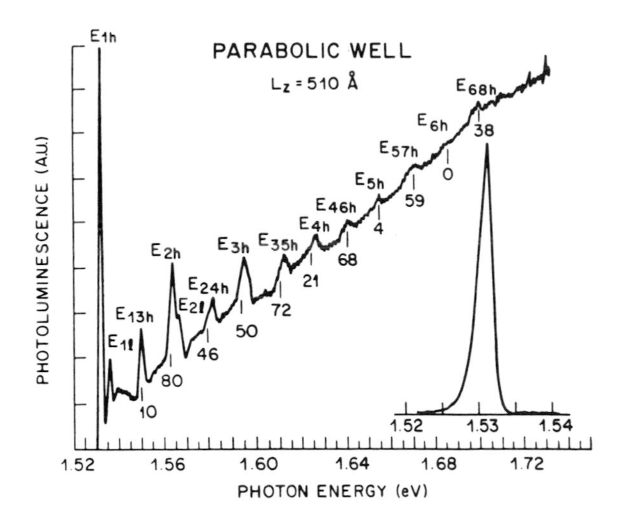

Fig.31. Luminescence excitation spectrum of a parabolic quantum well showing transitions which violate the $\Delta n = 0$ selection rule (Miller et al 1984)

For real wells of finite depth the situation is more complicated but this is a reasonably close approximation to reality. The energy levels in the kx and ky directions are very closely spaced and for each value of E_n there is a two-dimensional energy band in the kxy plane. A two dimensional energy band has a density of states which is independent of energy and at each value of E_n there is a sharp step in the cumulative density of states (Figure 27(b)). The bound electron and hole states in the quantum well are shown in Figure 28.

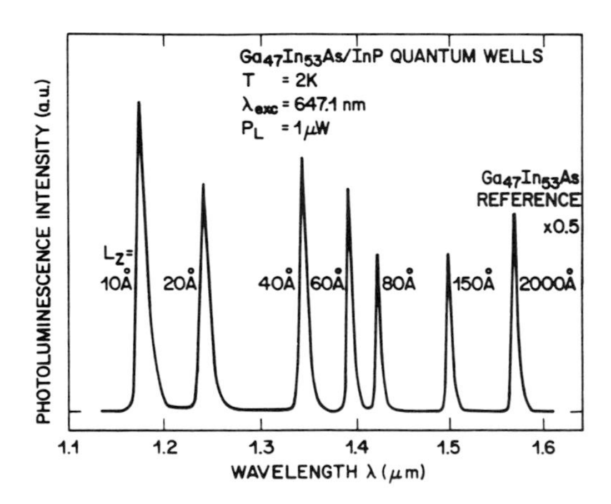

Fig.32. A photoluminescence spectrum from a stack of quantum wells with different thicknesses obtained at $T = 2$K (Tsang and Schubert 1986).

The absorption edge spectrum predicted for a GaAs quantum well with a step function density of states is compared with that for bulk GaAs, including the contribution for exciton creation, in Figure 27. Actual absorption spectra for 4000nm, 21nm and 14nm GaAs layers are shown in Figure 29. Dingle proposed that the transitions were between bound hole and bound electron states with equal quantum number n, the $\Delta n = 0$ selection rule (Figure 28). A more extensive spectrum with the assigned transitions is shown in Figure 30. More recent work has shown that transitions can occur which violate the $\Delta n = 0$ selection rule and that the band offsets ΔC and ΔV may be substantially different from those shown in Figure 30. A luminescence excitation spectrum obtained from a parabolic quantum well which shows the absorption transitions in more detail is shown in Figure 31 (Miller et al 1984).

Low temperature photoluminescence measurements primarily detect $n = 1$ exciton recombination. A photoluminescence spectrum obtained from a sample containing several quantum wells with different thicknesses is shown in Figure 32. Photoluminescence clearly provides a rapid assessment of well width, and the width of the luminescence peak a good measure of the quality of the interface since fluctuations in well width produce considerable broadening, particularly in narrow wells. Bound exciton luminescence can also occur from donor and acceptors in the quantum wells. The binding energy depends both on the well width and the position of the impurity within the well. It appears that impurities within wells are mainly segregated at the interface. A spectrum showing bound exciton transitions in a 40nm

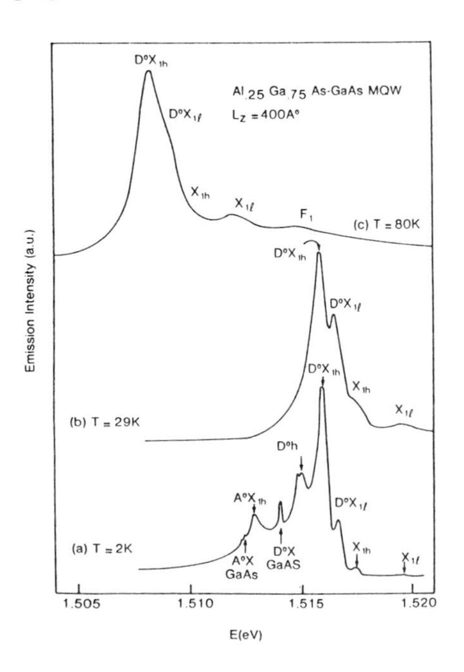

Fig.33. Photoluminescence spectrum of exciton transitions in a quantum well showing bound-exciton recombination (Yu et al 1985)

GaAs quantum well is shown in Figure 33 (Yu *et al*, 1985). A detailed account of the spectroscopic investigation of InGaAs/InP quantum wells is given by Skolnick et al (1986).

REFERENCES

Colley P McL and Lightowlers E C 1987 Semicond. Sci. Technol. **2** 157

Davies G 1988 Mat. Res. Soc. Symp. Proc. **104** 65

Davies G 1989 Physics Reports **176** 83

Davies G, Lightowlers E C, Woolley R, Newman R C and Oates A S 1985. *Proc. Thirteenth Int. Conf. on Defects in Semiconductors* ed. L C Kimerling and J M Parsey (New York: AIME) pp 725–31

Davies G, Lightowlers E C, Newman R C and Oates A S 1987 Semicond. Sci. Technol **2** 524

Davies G, Oates A S, Newman R C, Woolley R, Lightowlers E C, Binns M J and Wilkes J G 1986. J. Phys. C **19** 841

Dexter D L 1958 *Solid State Physics* Vol 6 ed. F Seitz and D Turnbull (New York; Academic Press)

Dingle R 1975 Festkorperprobleme **15** 21

Miller R C, Gossard A C, Kleinman D A and Munteanu O 1984 Phys. Rev. B **29** 3740

Skolnick M S 1986 Semicond. Sci. Technol. **1** 29

Skolnick M S 1987 *Brazilian School of Semiconductor Physics* (World Scientific Publications)

Steiner T, Thewalt MLW, Koteles E S and Salerno J P 1986 Phys. Rev. B **34** 1006

Tajima M 1978 Appl. Phys. Lett. **32** 719

Thewalt M L W 1982 *Excitons* ed. E I Rashba and M D Sturge (Amsterdam: North Holland) pp 393–485

Thonke K, Hangleiter A, Wagner J and Sauer R 1985 J. Phys. C **18** L795

Trombetta J M and Watkins G D 1987 Appl. Phys. Lett. **51** 1103

Tsang W T and Schubert E F 1986 Appl. Phys. Lett. **49** 220

Wagner J, Thonke K and Sauer R 1984 Phys. Rev. B. **29** 7051

Yu P W, Chaudhuri S, Reynolds D C, Bajaj K K, Litton C W, Masselink W T, Fischer R and Morkoc H 1985 Solid St. Commun. **54** 159

Hall, Magnetoresistance and Infrared Conductivity Measurements

R.A. STRADLING

1. INTRODUCTION

The most widely used test of electrical quality of semiconductor materials is the Hall measurement. Yet there is frequently a lack of appreciation of the very substantial systematic errors and pitfalls in interpretation that can occur in simple electrical experiments. This failure to appreciate the difficulties in obtaining accurate data perhaps arises from an over-familiarity brought about by introducing the Hall effect early on in undergraduate courses. At this stage the force produced by the Hall electric field is assumed to balance the Lorenz force exactly Conventionally the current density J, and the drift velocity v are taken to be in the x-direction, the Hall field E_y in the y-direction and the magnetic field B in the z-direction. The Lorenz balance condition is then $E_y = vB$. The Hall coefficient R and the mobility μ are given by $R = E_y/JB = 1/nq$ and $E_y/E_x = \mu B$ (if μ and B are in SI units) $= \Omega_c\tau$ where n is the free carrier concentration, q is the charge of the carriers, Ω_c is the cyclotron frequency, τ is the collision time and E_x is the electric field in the x-direction.

If the balancing of the Hall force by the Lorenz force were achieved for all carriers (i.e. if all carriers had the same drift velocity v), then, once the Hall field had been established, all carriers would move through the sample undeflected by the magnetic field. This would then imply that there would be no change in the sample resistance on applying a magnetic field. In the case of a few simple metals (eg Na, K), this prediction is obeyed to very high order and the magnetoresistance can be very small (e.g. 1 part in 10^5 or less even in high quality samples at very high magnetic fields). In most semiconductors the magneto-resistance is quite large (e.g. >10% for good quality n-GaAs at 1T), demonstrating the Lorenz balance condition is not achieved for most of the carriers.

2. THE RESISTIVITY TENSOR AND THE INFLUENCE OF SAMPLE GEOMETRY ON HALL MEASUREMENTS

In a Hall measurement the current direction and the electric field direction are not colinear although it is assumed that the two are proportional for small values of electric field (Ohm's Law) at all points in the sample. In practise the experimentalist will need to ensure that the electric field is sufficiently small that the current is indeed Ohmic unless it is intended to study the properties of 'hot' electrons. As a rule of thumb a bias field of 1V/cm is reasonable for most semiconductors at room temperature but with high mobility samples the bias field may have to be reduced below 100mV/cm at low temperatures. It should be noted that the electric fields applied by multimeters on the 'resistance' ranges are usually considerably in excess of these values. Furthermore such meters operate in a 'two-contact' mode and and consequently the contact resistances are included in such measurements. Thus multimeters can only be used to give a very rough indication of the electrical properties of semiconductors.

In high purity metals and small semiconductor structures the assumption of a point relationship between current and field may not be valid and Monte Carlo or other theoretical techniques have to be used to determine the probable particle trajectories. However, for the remainder of this article, it will be assumed that $\underline{E}$ and $\underline{J}$ are related by the conductivity or resistivity tensors given by

$$\underline{J} = \underline{\sigma}\, \underline{E} \quad \text{or} \quad \underline{E} = \underline{\rho}\, \underline{J} \tag{1}$$

In most experiments, there is no closed current path for the carriers in the y-direction. The carriers therefore accumulate at the surface. The time-dependent solution of Poisson's equation describes the build-up of free charge. If the plasma oscillations at the angular frequency given by $\Omega_p^2 = ne^2/\epsilon\epsilon_o m^*$ are strongly damped (where ϵ is the dielectric constant or relative permittivity of the material) , the Hall field becomes established in a time of the order of the dielectric relaxation time ($\tau = \epsilon\epsilon_o/\sigma = 1/\Omega_p^2\tau$). After this dynamic equilibrium is established, the boundary conditions at the surface of the sample require that there is no net current in the y-direction (i.e. $J_y = 0$). The current density and electric field in the x-direction are determined by the external bias conditions and the Hall electric field varies with the applied magnetic field. Two elements of the resistivity tensor are then determined directly in an experiment (provided that the energy bands and the scattering of the carriers are isotropic).

$$E_x = \rho_{xx} \cdot J_x \quad \text{and} \quad E_y = \rho_{yx} \cdot J_x \tag{2}$$

where the electric fields and current density are imposed externally by the voltages appearing across the sample, the bias current and the sample dimensions. On the other hand, it is generally easier to calculate elements of the conductivity tensor which is then inverted to provide a comparison with experiment. In the simplest case of a single type of carrier with concentration n, an isotropic effective mass (m^*) and a single or energy independent collision time (τ), the elements of the conductivity tensor are

$$\sigma_{xx} = \frac{\sigma_o}{1 + \Omega_c^2\tau^2} \qquad \sigma_{xy} = \frac{\pm\sigma_o\, \Omega_c\tau}{1 + \Omega_c^2\tau^2} \tag{3}$$

where the ±signs refer to whether the ratio q/m^* is positive or negative, i.e. whether one is discussing electrons or holes. These expressions can be generalised to include ac effects as discussed in sections 7 and 8. In the limit of low magnetic fields, $\sigma_{xx} \rightarrow \sigma_0$ and $\sigma_{xy} \rightarrow \pm\sigma_0\Omega_c\tau$. However, in the high field limit, $\sigma_{xy} \gg \sigma_{xx}$ and $\sigma_{xx} \propto 1/\tau$; i.e. provided that there is no electric field in the y-direction, the current in the x-direction becomes very small and is proportional to the scattering rate $(1/\tau)$ rather than the collision time (τ). If the collision time became very large, i.e. were the electrons to be moving in empty space and starting from rest under the influence of an electric field in the x-direction and a magnetic field in the z-direction, their paths would be cycloidal with no net velocity in the x-direction, the result expressed mathematically by $\sigma_{xx} \rightarrow 0$ (Fig. one). Without collisions, the electron motion in the x-direction is oscillatory with angular frequency Ω_c and amplitude $r = (E/\Omega_c B)$. Each collision therefore moves the electron forward a distance between 0 and $2r$, depending at what point in the cycloidal path the motion is interrupted by the scattering event. The current in the x-direction, which is zero without scattering, becomes proportional to the scattering rate.

If an electric field is suddenly applied in the x-direction with the magnetic field already present, the carriers will execute cycloidal paths transiently until the polarisation and the Hall field build up in the y-direction and dynamic equilibrium is established.

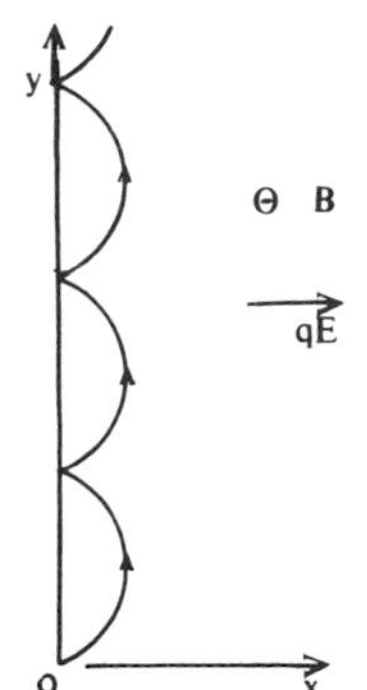

Figure one shows the path of a carrier released with zero velocity from the origin and allowed to move in free space under the influence of crossed electric and magnetic fields if there is no scattering and no polarisation charge or space charge. Note that the mean velocity along the electric field is zero.

Occasionally a current path is provided in the direction of the Hall field. If the external current leads can be considered to be of zero resistance, the Hall field is zero and the carriers continue to execute interrupted cycloidal paths. The elements of the conductivity tensor are then measured directly in the experiment from J_y , J_x and E_x. Any net current flow in the y-direction resulting in a drop in the Hall field will cause an increase in the magnetoresistance.

Shorting current paths can sometimes occur within the sample. When a NiSb:InSb eutectic is grown, highly-perfect and strongly-conducting cylindrical inclusions of NiSb are formed within an InSb matrix [Weiss,1967]. These structures are used as magnetic-field sensors on account of their large magneto -resistance. Epitaxial samples are particularly prone to layered conductivity because of accumulation regions at the surface or interface with the substrate or because of non-uniform dopant incorporation. In heterostructure or quantum well structures, these parallel conductance effects can be strong.

The success or failure of any Hall experiment often rests on the ability to make metallic contacts which are both electrically and mechanically satisfactory. Low resistance contacts can be formed by a variety of methods including spot-welding, alloying, soldering, plating and pressing suitable probes onto the surface. The lowest resistance abrupt contact between a metal and semiconductor is formed by the ideal Schottky barrier where the resistance is determined by the depletion length in the semiconductor. This can be reduced by increasing the local doping (e.g. by alloying or ion-implanting suitable dopants).

Large area metallic contacts will short out any electric field parallel to the plane of the contact. Careful consideration must therefore be given to the current configuration within the sample. The simplest sample shape is that of a rectangular bar with current contacts placed at the ends and point probes placed down its length to measure voltages proportional to the resistance and Hall field. The resistance must never be found by measuring a voltage across a contact carrying the bias current and the voltmeter employed must always have an impedance which is much greater than the sample resistance. Provided that the length to width ratio is greater than five and the voltage probes are kept away from the ends, accuracies approaching a few percent can be achieved in determining the Hall coefficient and mobility [Beer, 1960; Weiss, 1967] defined as the product $R\sigma$.

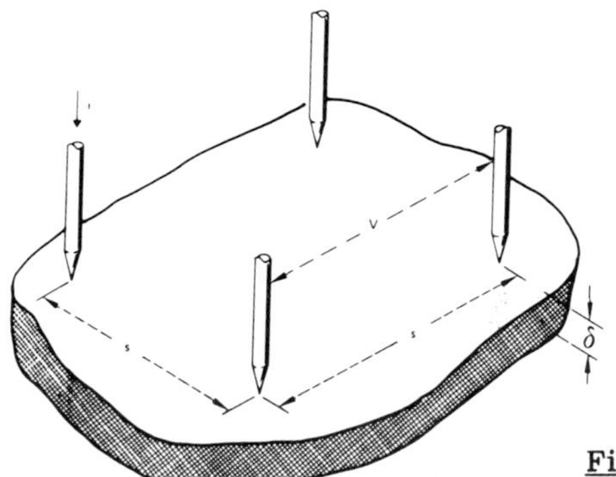

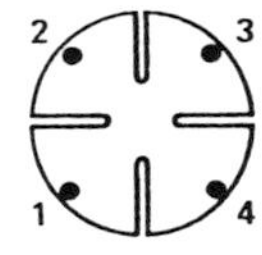

Figure two shows sample configurations commonly employed for van der Pauw measurements.

In some sample configurations, the current lines are allowed to diverge deliberately and the resultant resistance is calculated by potential mapping techniques. A number of common arrangements of probes are shown in figure two. With thin epitaxial films having high mobilities, the Hall angle, $\tan^{-1}(E_y/E_x = \mu B)$, can be very large and shorting effects in the region of the contacts can be particularly troublesome. The van der Pauw four-contact technique [van der Pauw, 1958] is usually used for epitaxial films. In this case the 'clover-leaf' shape shown in figure two has the advantage of keeping the current lines away from the Hall contacts if current is passed through diametrically opposite pairs of contacts with the Hall voltage being measured between the second opposite pair. The resistivity is found by passing the current through an adjacent pair of contacts and measuring the voltage between the other adjacent pair. As with other sample geometries, the effects of contact misalignment, sample inhomogeneity, spurious thermoelectric effects or other irregularities can be much reduced by taking the appropriate averages of all voltages obtained by commuting all possible contact combinations and by reversing the magnetic field. The resistivity is then determined from the relation

$$\rho = (\pi d/\ln 2)R \qquad (4)$$

where d is the thickness of the film and R is the resistance calculated after averaging the results for adjacent pairs of contacts and reversing the current directions.

3. MULTIPLE CARRIER EFFECTS

The previous section has outlined the experimental geometries commonly employed in Hall measurements. Even if the systematic errors that can arise from faulty measurement techniques or inhomogeneous samples are avoided, perfectly good experimental data can be misinterpreted through a failure to appreciate the assumptions implicit in the simple classical treatment of conduction in a magnetic field expressed in equations (2) and (3). The most basic of these is the assumption that all carriers have the same drift velocity. If this were correct, there would be no magnetoresistance whereas all semiconductors show a magnetoresistance to a greater or lesser extent.

The most common reason why a single carrier system with an isotropic effective mass shows a magnetoresistance is that the scattering is energy dependent. Appropriate averages have then to be taken for the collision times over the range of carrier energies as outlined in appendix A. The Hall coefficient R is no longer independent of magnetic field but decreases from r/nq in the limit of low magnetic fields ($\mu B \ll 1$) to 1/nq at high fields ($\mu B \gg 1$) where n is the carrier concentration and $q = \pm e$ depending on carrier type. The Hall factor or constant (r) is given by

$$r = \langle\tau^2\rangle/\langle\tau\rangle^2 \qquad (5)$$

In the case of ionised impurity scattering, using a simple Rutherford model, the numerical value for r is found to be $315\pi/512$. Consequently, although carrier concentrations deduced from the simplistic use of $R = 1/nq$ in the low field regime are frequently quoted to three places, it should be realised that the systematic error could approach 100% if ionised impurity scattering is limiting the mobility unless correction is made for r. Smaller but still very substantial errors are present with other scattering mechanisms if classical statistics apply to the carrier distribution.

If two or more distinct carrier species are present, then both the magnetoresistance and the difference between the low and high field Hall coefficients become larger. Multiple carrier groups can occur because of the presence of light and heavy holes, electrons and holes in semimetals or near intrinsic semiconductors or because of degenerate conduction band minima such as occur with silicon and germanium. Parallel conduction in a layered system is also an example of conduction involving multiple carrier types although the carriers do not occupy the same space. The case of electrons in silicon and germanium must be treated with particular care as the tensor effective masses concerned imply that the drift velocity may not always be colinear with the accelerating force. Unless the magnetic field is applied along a direction where all the constant energy ellipsoids are equivalently oriented, several groups of carriers, all with different masses, will result.

The Hall coefficient for a two carrier system is given by

$$R = \frac{\pm n_1\mu_1^2 \pm n_2\mu_2^2}{e(n_1\mu_1 + n_2\mu_2)^2} \qquad \text{at low fields} \qquad (6)$$

or by

$$R = \frac{1}{e(\pm n_1 \pm n_2)} \qquad \text{at high fields} \qquad (7)$$

The low-field mobility, as measured in a Hall experiment and defined as $R\sigma$, is then given by

$$\mu_H = \frac{|\pm n_1\mu_1^2 \pm n_2\mu_2^2|}{(n_1\mu_1 + n_2\mu_2)} = R\sigma \qquad (8)$$

where $q = \pm e$ and the positive and negative signs refer to hole and electron transport respectively. Equation (7) is derived by writing $\sigma_{xy} \gg \sigma_{xx}$. However, in the intrinsic case where the electron (n) and hole (p) concentrations are equal, $\sigma_{xy} \to 0$ and equation (7), which predicts an infinite Hall voltage when n=p, is not correct. Instead $R \to 0$ at high fields for intrinsic conduction.

The variation with temperature of the Hall coefficient can be used to measure the energy gap or the impurity activation energy involved whereas the Hall mobility ($R\sigma$) is generally used to determine the nature of the scattering mechanisms concerned. With a single carrier it would appear to be preferable to use the high magnetic field limit as it may not be possible to estimate the Hall factor r accurately, particularly when the averaging over the carrier distribution is complicated by the presence of two scattering process with different energy dependencies. However, in most cases, the magnetic fields required to reach the high field limit are unavailable. Even if it is possible to satisfy the condition $\mu B >> 1$, the onset of various quantum limit phenomena discussed in section [6] may complicate the interpretation still further. Consequently systematic errors can creep into the measurement of both energy or mobility.

When two or more carriers are present, the analysis is even more complicated as the individual carrier concentrations and mobilities must be estimated if activation energies are to be derived at all accurately.

Despite these difficulties, Hall effect and conductivity measurements are used widely and successfully to determine the electrical properties of semiconductor samples including carrier concentrations, mobility, band gap and impurity activation energies, density-of-states for the bands concerned and to determine the scattering processes concerned (see general reviews by Putley, 1960; Wieder 1979).

4. A CASE STUDY OF THE DIFFICULTY AND DANGERS INVOLVED IN THE INTERPRETATION OF HALL DATA.

Electrical measurements with epitaxial InAs films provide a good example of how Hall measurements can be misinterpreted. The InAs surface contains a high density of surface states which normally lead to a conducting skin which is formed by an electron accumulation layer.

Recently some very high mobility films of n-InAs were grown on GaAs substrates in the Imperial College MBE Facility [Holmes et al, 1989]. As can be seen in figure (3), the mobility derived from the product ($R\sigma$) decreases rapidly with decreasing film thickness. A similar variation had been noted by another group and attributed to the onset of scattering by misfit dislocations located close to the interface . Cyclotron resonance and magneto-impurity spectroscopy measurements (details of these techniques are discussed in sections 7 & 9) showed the presence of a high mobility region in the centre of the film with sharp donor states together with a low mobility two-dimensional electron gas (2DEG) caused by surface states. No evidence for dislocation scattering was found as the cyclotron resonance line width was independent of the film thickness. The mobilities in the two regions could be estimated from the magneto-optical line widths and the number of carriers in each of the two regions could be determined; in the one case from the Shubnikov-de Haas effect (section 6.1) observed by the 2DEG formed by the surface accumulation layer and, for the bulk electrons, from the absolute value of the absorption coefficient found from the intensity of the cyclotron resonance line (see section 7).

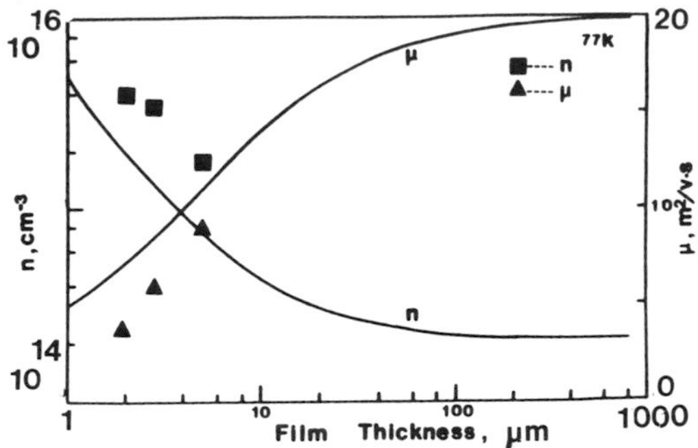

Figure three shows the predicted effects of parallel conduction as a function of thickness for a two-carrier layered system consisting of a low-mobility high-carrier concentration surface skin and a high-mobility low-carrier concentration bulk region. The curves show the predicted Hall mobility and apparent carrier concentration of such a sample as a function of thickness. The symbols are experimental Hall data for MBE InAs at 77K. The values of n and μ used to generate the solid curves are derived from magneto-optical experiments. For such a system Hall measurements at 77K will only give the true bulk values of n and μ for thicknesses ≥ 20μm and if fields ≤ 0.01T are employed so that the low-magnetic-field limit is attained.

The boundary condition for the Hall current generated with a magnetic field perpendicular to the film is identical to that for two carrier conduction in a homogeneous medium (ie the net current has to be zero). Hence equations (6) and (8), modified by replacing the volume carrier densities by the concentrations per unit area, can be used to model the field dependence of the Hall coefficient and the magnetoresistance, and also the thickness dependence of the apparent carrier concentration deduced from a simplistic interpretation of the Hall coefficient (figure 3). It can be seen that the two carrier modelling predicts the observed drop in Hall mobility as the sample thickness is decreased without the need to include dislocation scattering.

In addition it should be noted that, because of the very high mobility for the electrons in the bulk of the film, the magnetic field has to be reduced to the extremely low value of 0.01T before the low field limit applies and the Hall coefficient is independent of field. The absolute values for the magneto-resistance are predicted accurately, the values for the apparent mobility and n less so. The reason for this difference between experiment and model probably lies in the existence of a third conducting region close to the InAs/GaAs interface rather than in a significant error in the fitting parameters.

It should also be noted that, despite the carrier concentrations in both the surface and bulk regions being independent of temperature between 290 and 77K, the Hall coefficient increased by more than a factor of two in this temperature range. This temperature dependence arises because the mobility in the bulk region is increasing rapidly with decreasing temperature whereas that in the low mobility surface skin is independent of temperature. A freeze-out (see section 6.2) of electrons onto impurity sites would be deduced if a single carrier model were assumed although in fact no such deactivation was taking place.

5. LONGITUDINAL MAGNETORESISTANCE

With classical carriers in a single isotropic band, there will be no coupling between transverse and longitudinal motion and no longitudinal magnetoresistance is expected. In practise a longitudinal magnetoresistance is very often observed in many samples. This can be due to a number of effects e.g.

i) Anistropic band structure - e.g. the carriers may be moving in ellipsoidal or warped constant energy surfaces.
ii) Electrical inhomogeneities within the sample. These can distort the current lines locally so that there is a current component perpendicular to the magnetic field. This distortion therefore mixes in part of the transverse magnetoresistance.
iii) The presence of a scattering mechanism which is dependent upon the magnetic field, e.g. spin-dependent scattering which can give a negative magnetoresistance at low temperatures (the Kondo effect). In the 'quantum limit' ($\hbar\Omega_c >> k_BT$), the scattering generally increases because of the field-induced increase in the density-of-states.

6. QUANTUM TRANSPORT EFFECTS

The simple Drude-Sommerfeld approach used so far in this article, and other classical treatments such as the Boltzmann equation, break down in the quantum limit when the separation between the Landau levels becomes greater than the thermal energy ($\hbar\Omega_c > k_BT$). Under these conditions, a number of new phenomena become apparent in the electrical resistance of semiconductors.

6.1 THE SHUBNIKOV-DE HAAS EFFECT

The Shubnikov-de Haas effect represents an oscillatory variation of the electrical resistance which is periodic in reciprocal magnetic field. The scattering-rate peaks when each Landau level empties. In order that this variation be observable, the carrier distribution must be degenerate (ie the Fermi energy $E_F >> k_BT$), the sample must be of high purity and the magnetic field must be high ($\hbar\Omega_c >> k_BT$ and $\Omega_c\tau >> 1$). Under these conditions the resistance in a magnetic field should be described by the following expressions [see Roth & Argyres, 1966, Kubo et al, 1965]. The scattering rate and hence the resistance show peaks at values of magnetic field close to the fields where the Landau levels empty. This result arises from the form of the one -dimensional density-of-states function which varies as $E^{-\frac{1}{2}}$ where

E is the kinetic energy along the magnetic field. Provided that the peaks are not too sharp, the resistance is given by

$$\rho_{zz}/\sigma_0 = 1 + b\cos(2\pi F/B - \pi/4) \quad (9)$$

$$\rho_{xx}/\sigma_0 = 1 + 2.5\, b\cos(2\pi F/B - \pi/4) + R \quad (10)$$

where

$$b = -\left[\frac{\hbar\Omega_c}{2E_F}\right]^{\frac{1}{2}} \frac{\cos(\pi v)\ X\ \exp(-2\pi/\Omega_c\tau)}{\sinh X} \quad (11)$$

and

$$X = 2\pi^2 k_B T/\hbar\Omega_c \qquad F = E_F m^*/\hbar e \qquad v = g^* m^*/2m \quad (12)$$

In the equations above, E_F is the Fermi energy and g^* is the effective Landé g-value for the carriers. The cosine terms in equations (9) and (10) determine the period in inverse magnetic field (1/B) of the resistance oscillations. For a single isotropic band, $F = \hbar k_F^2/2e = \hbar(3\pi^2 n)^{2/3}/2e$ where k_F is the wavevector of the carriers at the Fermi surface. Hence the observed period of the oscillations gives the number of carriers directly without the uncertainty associated with the estimation of the Hall factor (r). With an anisotropic Fermi surface, the dependence of the period on the orientation of the magnetic field with respect to the crystal axes gives the anisotropy of the extremal cross-sectional areas ($F = hA/2\pi e$). Note that measurements of the size or shape of the Fermi surface are completely independent of the effective mass of the charge carriers. Equations (9) and (10) are in effect the first terms of Fourier expansions describing the periodic variation of the scattering rate as each Landau level empties with increasing magnetic field. The b term contains the temperature dependence of the oscillations through the factor X/sinhX which for large values of X is dominated by the term exp-X. The effective mass of the charge carriers can be determined from fitting the temperature dependence of the amplitude to this expression. Note that the Shubnikov-de Haas oscillations reach their maximum amplitude only when X/sinh X -> 1 or when $\hbar\Omega_c \geq 20k_BT$ which provides a demanding requirement on the experimental low temperature/high field facilities.

The R term in equation (10) represents a phase shift of the oscillations arising from the competition between inter and intra Landau level scattering processes. This phase factor changes as a function of field. The exp $(-2\pi/\mu B)$ term in equation (11) arises from the broadening of the Landau levels by scattering. By fitting the magnetic field dependence of the oscillations to this expression the mobility can be measured. The cos (πv) term is generated by the spin-splitting of the Landau levels. If $\Omega_c\tau \gg 1$, higher order Fourier terms must be included in eqtns. (9) to (11) and spin-splitting of the Landau levels becomes apparent enabling the effective g-value for the carriers to be determined directly.

Hence all the main parameters for the free carriers can be

measured directly from the Shubnikov-de Haas effect; i.e. the carrier density and the anisotropy of the bands from the positions of the peaks, the effective mass and the mobility of the carriers from the temperature and field-dependence of the amplitudes, and the effective g-value from the spin-splitting of the peaks.

The above analysis refers specifically to a normal bulk sample. However much of the treatment can be taken over for quasi-two dimensional samples which exhibit the Quantum Hall Effect in addition to Shubnikov-de Haas oscillations in the magneto-resistance. The density-of-states function for a two dimensional system lacks the $E^{-\frac{1}{2}}$ dependence on energy along the field and, as a result, the magnetic field separation of the spin-split peaks does not depend simply on the electronic g-value but the effective mass and the mobility of the charge carriers can be found from the temperature and field-dependence of the oscillations in the same way as for a bulk sample. The formula from which the electron concentration is deduced must be modified in the case of a 2DEG and spin effects are not treated so simply but otherwise the effect is qualitatively similar to the three dimensional case.

<u>Figure four</u> shows the Shubnikov-de Haas effect and Quantum Hall effect for the two-dimensional electron gas (2DEG) in a GaAs/GaAlAs hetero-structure measured at Imperial College.

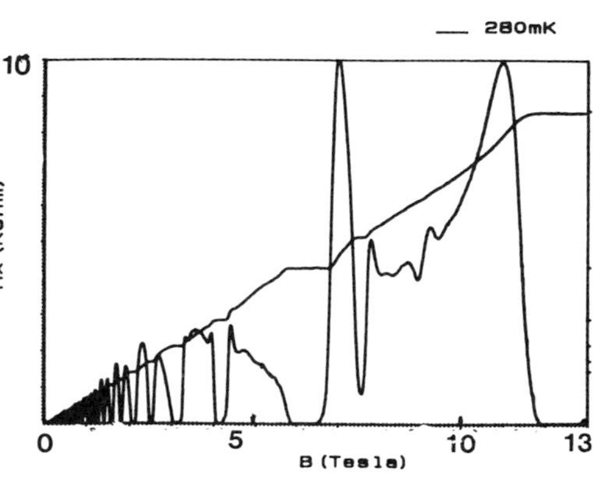

6.2 MAGNETIC FREEZE OUT

A further quantum effect, which can cause a large magneto-resistance in the high field limit and also difficulties in determining the Hall factor r from the saturation of the Hall coefficient at high fields, is the decrease in free carrier concentration which will occur as a consequence of the decrease in extent of the impurity wavefunctions and increase in binding energy with increasing magnetic field ("magnetic freeze-out"). The effect is very much accentuated if the shrinkage of the size of the impurities induces the "metal-insulator" phase transition. The carriers which give rise to the Shubnikov-de Haas effect normally originate from impurities at concentrations which are above the critical concentration for the metal-insulator transition where the binding energy to the impurities goes to zero. It is therefore far from obvious that freeze-out will necessarily occur at higher fields than the low index ($N \approx 1$) Shubnikov-de Haas peaks. The following analysis shows that the magnetic-field induced metal-insulator phase transition and magnetic freeze-out occur close to but at slightly higher magnetic field than the $N = 1$ peak. If spin-splitting is not observed, the $N = 1$ Shubnikov-de Haas peak occurs at a field given by

$$B_{SH} \approx \frac{\hbar(3\pi^2 n)^{2/3}}{3e} \propto n^{2/3} \qquad (13)$$

The Mott-Hubbard theory of the metal-insulator transition predicts that [Mott,1974]

$$n^{1/3}a^* \approx \tfrac{1}{4} \qquad (14)$$

where a* is the effective Bohr radius for the impurities. In a magnetic field the impurity shrinks in size and becomes ellipsoidal in shape. In the high field limit, the extent of the impurity wavefunctions perpendicular to the magnetic field approximates to the cyclotron radius (r) in the quantum limit which is given by

$$r^2 = \hbar/eB \qquad (15)$$

At first sight, the extent of the impurity along the magnetic field should be given by the effective Bohr radius a*. In this case, the critical magnetic field B_c may be found by substituting the mean radius of the impurity $(r^2a^*)^{1/3}$ into equation (14) to give

$$B_c \propto n \qquad (16)$$

On comparing equation (13) and (16) it is clear that there is an increasing difference between the critical field and the fields where the Shubnikov-de Haas effect is observed as n increases . However the assumption that the size of the impurity along the magnetic field is independent of field is incorrect. Yafet, Keyes and Adams (1956) showed that the extent of the impurity wavefunction along the field shrinks 'in sympathy' and the two fields diverge more slowly. As is shown below, the ratio of the two fields is independent of carrier concentration in the two dimensional case. For a 2DEG, the equivalent equation to (13) for the position of the N=1 Shubnikov-de Haas peak is:

$$B_{SH} \approx hn/3e \propto n \qquad (17)$$

The condition for magnetic freeze-out can be written as

$$n^{\frac{1}{2}}r \approx \tfrac{1}{4} \qquad (18)$$

giving

$$B_c \approx 16hn/2\pi e \qquad (19)$$

Comparing (17) and (19) it is seen that magnetic freeze-out for a 2DEG occurs close to but at rather higher fields than the lower order Shubnikov-de Haas peaks.

6.3 THE STATISTICAL TRANSITION

If magnetic freeze-out should not occur, for example if the carriers do not originate from the normal impurity shallow impurity states, then the carriers will become trapped in the lowest Landau level (N=0) in the extreme quantum limit. In this case, as the density-of-states within this level increases linearly with magnetic field, the Fermi level eventually drops beneath the bottom of the Landau level and classical rather than quantum statistics apply. A large longitudinal magnetoresistance then results above a critical field given by (Askenazy, 1973).

$$B \approx \frac{(3h^2n)}{8e}\left[\frac{1}{2m^*k_BT}\right]^{\frac{1}{2}} \qquad \propto \; n[m^*T]^{-\frac{1}{2}} \tag{20}$$

6.4 THE WIGNER CRYSTAL

If the sample contains very little disorder, the electron gas itself should take on an ordered arrangement at very low temperatures. With the alloy mercury-cadmium-telluride, it has been suggested that this could occur but this is a subject of continuing controversy.

6.5 THE MAGNETOPHONON AND MAGNETO-IMPURITY EFFECTS

Two further quantum effects should be mentioned in passing. Both of these effects involve inelastic scattering in contrast to the elastic scattering processes usually responsible for the Shubnikov-de Haas effect.

The first is the magnetophonon effect where scattering with high energy optic phonons gives rise to peaks in the electrical resistance when

$$\Omega_{ph} = N\Omega_c \qquad \text{(N is an integer)} \tag{21}$$

The effect will therefore give the effective mass of the charge carriers if the phonon frequency (Ω_{ph}) is known from Raman measurements or otherwise. Alternatively the phonon frequency can be derived if the effective mass is known.

At low temperatures resonant cooling of a hot electron distribution occurs when

$$\hbar\Omega_{ph} = N\hbar\Omega_c + E_I(B) \tag{22}$$

$$\text{or} \qquad \hbar\Omega_{ph} - E_I(B) = N\hbar\Omega_c$$

This effect, known as the impurity-shifted magnetophonon resonance, can be observed under non-ohmic conditions.

The carriers can be inelastically scattered off neutral impurities giving peaks in the resistance at fields given by

$$E_I = N\hbar\Omega_c \tag{23}$$

where E_I is the energy difference between two impurity levels. This effect is known as the magneto-impurity effect.
Both the impurity-shifted magnetophonon effect and the magneto-impurity effect can in principle identify the impurities present from the energies involved in equations (22) and (23). With the magneto-impurity effect, the resistance peaks can be sufficiently sharp that the ~1% differences between the ground state energies of the shallow donors in GaAs and InP can be measured (see figure 5) (Nicholas & Stradling, 1976, 1978; Eaves & Portal, 1979)

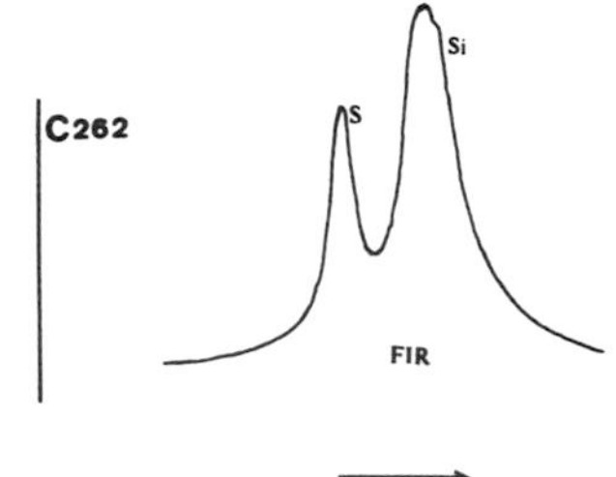

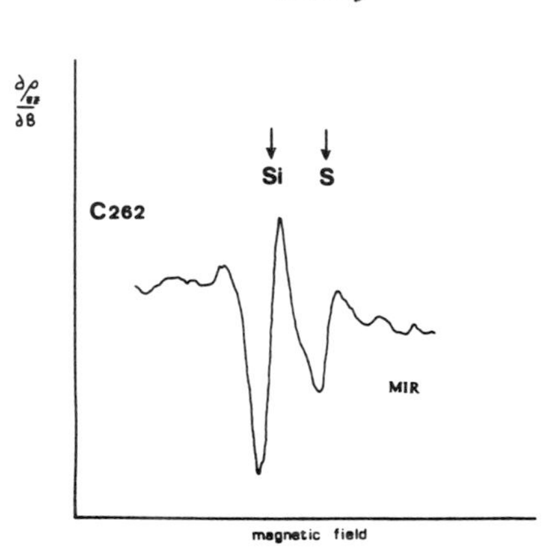

Figure 5 shows a recording of the derivative of the resistance against magnetic field for a high purity sample of InP where a magnetoimpurity peak involving a 1s to 2p- impurity transition is split into two by te presence of two donor species whose binding energies differ by 1%. Also shown is the far infrared magneto-optical spectrum of the same sample demonstrating that the magneto-impurity effect can achieve comparable resolution.

7 CYCLOTRON RESONANCE

The treatment of the conductivity tensor outlined in the earlier sections can be generalised to include AC phenomena in a magnetic field. At frequencies (Ω) much higher than the plasma frequency ($\Omega_p^2 = ne^2/\epsilon\epsilon_0 m^*$), depolarisation charges do not have time to become established and Hall fields are not present. It is helpful to resolve a linearly-polarised electric field applied in the x-direction into circularly-polarised components of the form

$$\hat{E} = E_0(1 \pm j)\exp(j\Omega t) \tag{24}$$

Provided no depolarisation charge has built-up, the resultant circulating currents in the x-y plane can be written as

$$\hat{J} = \hat{\sigma}\hat{E} \tag{25}$$

where a complex ac conductivity is given by

$$\sigma = \sigma_o/[1 + j(\Omega \pm \Omega_c)\tau], \quad \Omega_c = qB/m^* \tag{26}$$

where the ± symbol refers to the different senses of circular polarisation of the electromagnetic field.
The resultant power loss per unit volume arising from the circulating currents can be written as

$$P = J.E = \sigma_o E_o^2/[1 + (\Omega \pm \Omega_c)^2 \tau^2] \tag{27}$$

If the carrier concentration is so low that the skin depth is much greater than the thickness of the sample, the electric field at all points in the sample is independent of the magnetic field and the total power absorbed as measured by the transmission of radiation from the sample or the power loss in an external resonant circuit is given by equation (27). As can be seen from this equation, the absorption of power by the free carriers is sharply peaked when $\Omega=\Omega_c$ for the negative sign in the denominator which corresponds to the sense of circular polarisation circulating in the same sense as the carriers provided that $\Omega\tau>>1$. This resonant absorption of power is known as cyclotron resonance and is the most direct and accurate method of measuring the effective mass of the charge carriers (from the resonant condition $\Omega=\Omega_c$) provided that the necessary experimental conditions can be established.

The collision time can be deduced directly from the width of the cyclotron resonance absorption from the relation $B/\Delta B = \Omega\tau$. An absolute value for the carrier concentration can be found from the peak absorption and the values for τ and m^* deduced from the width and position of the cyclotron resonance line.

The condition that $\Omega\tau>1$, which represents the requirement that the width of the cyclotron resonance line be sufficiently narrow for the observation of a distinct peak, is rather demanding as, even with the purest samples available of the common semiconductors (apart from Si and Ge), the values of τ are typically 10^{-11}s. The experiment thus requires submillimetre wavelengths for the sources of the electromagnetic radiation and the use of very high magnetic fields to attain the resonance condition. Low temperatures are also desirable in order to reduce phonon scattering.

8. PLASMA EDGE MEASUREMENTS OF CARRIER CONCENTRATIONS AND OTHER HIGH-FREQUENCY EFFECTS

If the carrier concentrations are high, depolarisation (or 'plasma') effects can become important. Even if depolarisation charges do not become established at the surface of the sample, the cyclotron motion can become coupled to the plasma oscillations and the maximum absorption corresponds to the peak response of the system as a whole. At the same time the skin depth is likely to become less than the sample thickness and dependent on the magnetic field. In this case, the 'equation of telegraphy' arising from Maxwell's equations

$$\nabla^2 E = \frac{\mu\epsilon}{c^2}\frac{\delta^2 E}{\delta t^2} + \mu\mu_o\sigma\frac{\delta E}{\delta t} \qquad (28)$$

must be used to derive a complex refractive index ($\hat{n}$) which includes the free carrier term

$$\hat{n}^2 = \mu\epsilon - j(\sigma\mu/\Omega\epsilon_o) \qquad (29)$$

If the carrier density is very high so that the first term in equation (29) can be neglected and the frequency is much less than the scattering rate (i.e.$\Omega\tau << 1$), the classical skin effect is found and the radiation penetrates a depth (δ) into the sample given by (30)

$$\delta^2 = 1/(\pi\sigma f\mu\mu_o) \qquad \text{where } f = \Omega/2\pi. \qquad (30)$$

If alternatively the frequency is much greater than the scattering rate ($\Omega\tau >> 1$), the complex refractive index is given by (31)

$$\hat{n}^2 \approx \epsilon[1 - \Omega_p^2/\Omega^2] \text{ where } \Omega_p^2 = ne^2/m^*\epsilon\epsilon_o \qquad (31)$$

provided the material is not magnetic ($\mu = 1$). Equation (31) predicts that the sample is completely reflecting at low frequencies because the refractive index is purely imaginary. When $\Omega > \Omega_p$, the refractive index is real and the radiation starts to penetrate into the sample. When $\hat{n}=1$, the sample becomes matched to free space and the reflection drops to zero (see figure 6). The abrupt change in reflectivity from the sample being totally reflecting to having zero reflection is known as the 'plasma edge' in the plot of reflectivity against frequency. This phenomenon is responsible for the ease of transmission of long-wavelength radiowaves which are reflected from the ionosphere. However, when the frequency of the radio waves is greater than the plasma frequency of the ionised gas in the ionosphere (~ 10MHz), near line-of-sight transmission must be used. With metals the plasma frequency is in the ultraviolet and consequently most metals appear silvery (i.e. totally reflecting) at visible wavelengths.

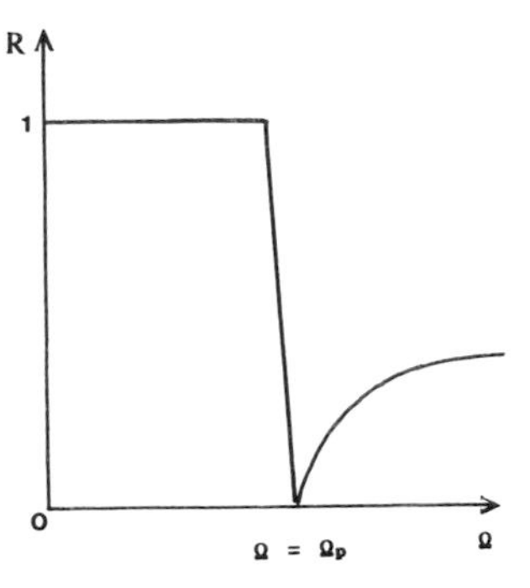

Figure 6 shows the reflectivity (R) against frequency (Ω) for a highly conducting sample in the limit of high frequencies ($\Omega\tau \gg 1$). The abrupt change from total reflection to transmission is known as the 'plasma edge'.

With semiconductors having carrier concentrations in the range $10^{14}$to 10^{20}cm^{-3}, the plasma frequency occurs in the mid-infrared region of the spectrum and the assumption that $\Omega\tau \gg 1$ applies. Consequently a well-defined 'plasma-edge' can be seen in the propagation constants and this can be used to provide an accurate and contactless method of determining the carrier concentration present in a sample provided that the effective mass is known. Such measurements can be used to determine the carrier concentrations and profiles in ion-implanted conducting skins in silicon extending only distances of the order of 5×10^{-5}cm from the surface.

9. INFRARED STUDIES OF ELECTRONIC ABSORPTION FROM IMPURITIES

Bound states of substitutional impurities are described by a 'hydrogenic' model in elementary texts. A single electron (or hole) is assumed to be bound by a single fixed charge of opposite sign on the impurity, the Coulomb potential is modified to include the dielectric constant ϵ and the inertial mass of the free electron is replaced by the effective mass of the free carrier to include the interaction with the periodic crystal potential. The energy levels for the impurities are then given by the well-known hydrogenic formula

$$W^*_n = -13.6[m^*/m]/\epsilon^2 n^2 \qquad \text{(eV)} \qquad (34)$$

Typically $m^*/m \approx 0.1$ and $\epsilon \approx 10$, so the ground state energy is reduced from the value for the hydrogen atom of -13.6 eV to about -10 meV. The wavelengths corresponding to typical binding energies lie in the range from the submillimetre to mid-infrared region of the spectrum. The Bohr radii of the bound states are very much extended in comparison with the hydrogen atom and are given by

$$a^*_n = 0.053 n^2 \epsilon/[m^*/m] \qquad \text{(nm)} \qquad (35)$$

This model provides an adequate first-order picture of the bound states of most shallow impurities. However, the theory breaks down to a greater or lesser extent for the following reasons:

i) Within the 'central cell' (i.e. the unit cell where the impurity is located), the potential is different from the $[-1/4\pi\epsilon\epsilon_o r]$ form assumed in the hydrogenic model. As the 1s ground state wavefunction has its maximum amplitude at the origin, substantial shifts in energy can occur for this state. The resultant chemical shifts are characteristic of the impurity concerned. It should also be noted that both the effective mass approximation and the assumption of a macroscopic dielectric constant break down close to the central cell region. Corrections arising from the failure of these assumptions will however be the same for all impurities to first order.

ii) The assumption of a single effective mass cannot take into account such complexities as degenerate valence bands, the ellipsoidal constant energy surfaces and tensor effective masses that occur for the conduction bands of Si and Ge, band-nonparabolicity, or more complex dispersion relations such as the 'camel's back' shape that occurs in the conduction band of GaP.

iii) Self-energy phenomena such as the polaron effect that arises through interaction with the phonons.

As electric dipole transitions are involved, the absorption is relatively strong and concentrations of shallow impurities of $10^{11}cm^{-3}$ can be detected with sample thicknesses of about 1mm. With thin films, even higher sensitivity can be achieved by detection of the photocurrent generated within the sample itself. The sharp bound-to-bound transitions can be observed through a two-stage 'photothermal' process where the carrier after photoexcitation into the higher bound state transfers into the conduction or valence band by the subsequent absorption of a phonon. By this means concentrations as low as $10^{7}cm^{-3}$ of minority impurities can be detected in ultrahigh purity germanium, and as few as 10^{4} neutral phosphorus donors have been detected at the edge of the depletion region of a silicon MOSFET structure [Nicholas et al, 1976].

Because the chemical shifts are characteristic of the impurities, infrared spectroscopy can be used to characterise the residual contaminants present in material grown without deliberate doping. Very high resolution can be achieved in the far-infrared by the use of laser techniques. Figure seven shows a recording of the photoconductivity obtained from the shallow donors by using a fixed frequency laser and sweeping the magnetic field applied to the shallow donors in a VPE sample of GaAs. This recording demonstrates the richness of the structure which can be observed. The fine structure which can be seen on the lines above 4T arises from the presence of several contaminating donors in this sample. Figure eight shows on an expanded scale the different donors present in a series of MOCVD and MOMBE samples of GaAs. [Armistead et al, 1989; Holmes et al, 1989b].

It should be noted that the difference in energy of the different donors is only about 50μeV i.e. approximately 1% of the effective mass binding energy (5.8 meV). Information about the nature of the contaminating shallow donors has been derived with comparable resolution with InP and InSb. In contrast, two activation energies cannot be distinguished in Hall measurements if they lie within a factor of two of each other.

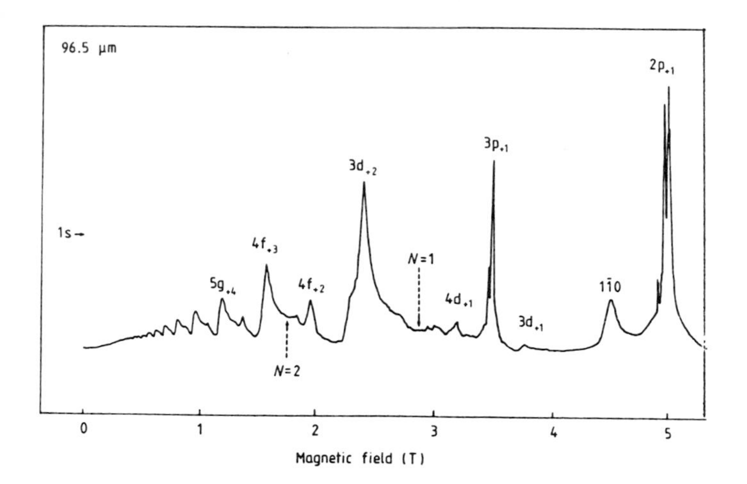

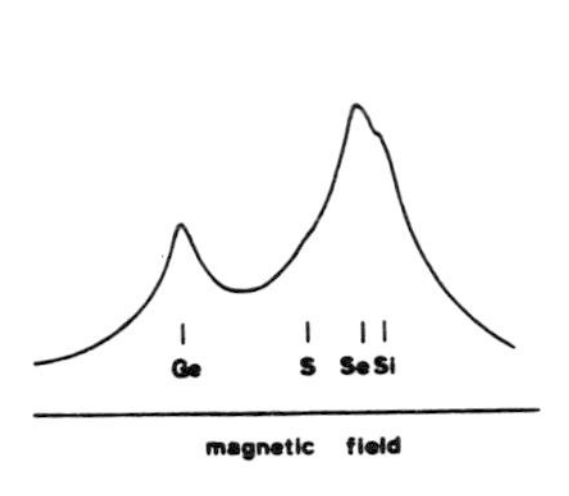

Figure 7 shows the photoconductive spectrum from the shallow donors in a high purity sample of VPE GaAs taken 4.2K with a far-infrared laser

Figure 8 shows an expanded field plot of the 1s to 2p+ line. Contaminating donors are resolved.

With materials having a higher effective mass, the binding energy itself and the chemical shifts as a proportion of the binding energy increase (eg for P in Si the binding energy is 40 meV and the chemical shift amounts to about 10meV). With some impurities the local potential in the central cell is so attractive that the chemical shift is much larger than the effective mass contribution to the ground state energy. In this case the impurity becomes 'deep' and the intensities of the transitions to the excited states drop dramatically. In this case, a rather broad edge is seen in both absorption and photoconductivity measurements near to the onset of transitions into the nearest energy band. This edge can also be used to characterise the nature of the impurities concerned as the high resolution necessary with the shallow effective-mass impurities is no longer required.

10. ACKNOWLEDGEMENTS

The author would like to thank Dr. K.W.J. Barnham, Dr. S. Holmes and Mr. N. Morawicz for the gift of certain of the figures used in this text.

11. APPENDIX - AVERAGES APPEARING IN DIFFERENT TRANSPORT COEFFICIENTS FOR A CLASSICAL DISTRIBUTION OF CARRIERS

As discussed in section 1, for carriers obeying classical Maxwell-Boltzmann statistics in a single isotropic band, the Hall coefficient in the low field limit ($\Omega_c\tau = \mu B << 1$) is given by $R = r/nq$ where the constant $r = \langle\tau^2\rangle/\langle\tau\rangle^2$. The symbol $\langle y \rangle$ denotes that the quantity y is averaged over the distribution function, i.e.

$$\langle y \rangle = \int \frac{4x^{3/2} y \exp(-x)}{3\sqrt{\pi}} dx \qquad \text{where } x = E/k_BT.$$

The values for r when the energy dependence of τ is $E^{3/2}$ (ionised impurity scattering) and $E^{-\frac{1}{2}}$ (acoustic phonon scattering) are given in the table at the end of the appendix.

The Hall mobility is defined as $\mu_H = R\sigma$ where σ is the zero field conductivity and R is measured in the low field limit (see eqtn. 8 for definition for a multicarrier system). The drift mobility for a single carrier is $\mu = (q/m^*)\langle\tau\rangle$.

The low field magnetoresistance found in the transverse or Hall configuration when there is no current path to short out the Hall voltage is given by

$$\frac{\Delta\rho}{\rho\mu^2B^2} = \frac{\langle\tau^3\rangle}{\langle\tau\rangle^3} - \frac{\langle\tau^2\rangle^2}{\langle\tau\rangle^4}$$

where the values for the appropriate averages for the two common scattering mechanisms are given in the table.

If the Hall voltage is completely shorted out (Corbino disc configuration), the low field magnetoresistance is greater as the last term in the previous equation is missing.

In the high field limit ($\Omega_c\tau = \mu B >> 1$), the Hall coefficient $R \to 1/nq$ (i.e. the r term is absent) and the transverse resistance will in the absence of quantum limit phenomena saturate at the values given at the foot of the table for the two different scattering processes. However, it should be remembered that, at very high magnetic fields, classical transport breaks down when one enters the 'quantum limit' ($\hbar\Omega_c >> k_BT$).

Table one shows the averages appearing in different transport coefficients for the two common scattering processes.

	$E^{3/2}$	$E^{-\frac{1}{2}}$
$\langle\tau^2\rangle/\langle\tau\rangle^2$	$315\pi/512$	$3\pi/8$
$\langle\tau^3\rangle/\langle\tau\rangle^3$	$15\pi/8$	$9\pi/16$
$\rho(\infty)/\rho(0)$	$32/3\pi$	$32/9\pi$

12. REFERENCES

Argyres, P.M. & Adams E.M. (1956) Phys. Rev. **104** 900.

Armistead C.J., Stradling R.A. & Wasilewski Z. (1989) Semicon. Sci. & Tech.

Askenazy, S., (1973) New Developments in Semiconductors p331 Ed. P.R. Wallace, pub: Noordhoff.

Beer, A.C. (1963) Supp. No. 4 to Solid State Phys. (Ed. Seitz & Turnbull)

Eaves, L. & Portal J.C. (1979) J. Phys. **C 12** 2809

Holmes, S.N. et al (1989) Semicon. Sci. & Tech.**4** 303.

Kubo R., Miyake S. & Hashitsume N. (1965) Solid State Physics **17** 269 (Ed. Seitz & Turnbull).

Mott, N.F. (1974) Metal-Insulator Transitions (pub. by Taylor & Francis).

Nicholas, R.J., Von Klitzing K. & Stradling R.A., (1976) Solid State Comm. **20** 77.

Nicholas R.J. & Stradling R.A. (1976) J.Phys. **C 9** 1253 & (1978) J.Phys. **C 11** L783

Putley E.H. (1960) The Hall Effect & Related Phenomena (Butterworth).

Roth L. & Argyres P.M. (1966) Semiconductors & Semimetals **1** 159.

van der Pauw L. Philips Research Reports (1958) **13** 1 & (1961) **20** 220

Wieder H.H. (1979) Laboratory Notes on Electrical and Galvanomagnetic Measurements (pub. Elsevier).

Weiss, H, (1967) Vol **1** Semiconductors & Semimetals & (1965) Solid State Electronics **8** 241

Yafet Y., Keyes R.W. & Adams E.N. (1956) J. Phys. Chem. Solids **1** 137

Characterisation of Semiconductors by Capacitance Methods

D W Palmer

1. INTRODUCTION

Measurements of the small–signal capacitance of a reverse–biased semiconductor p–n or Schottky junction, and of its steady–state and transient change with applied bias and temperature, allow considerable quantitative information to be obtained on the doping concentration in the semiconductor and its variation with distance, and on the presence, electronic energy levels and concentrations of deep–level electron and hole traps in the semiconductor material. This article outlines the physical principles on which these techniques of Capacitance–Voltage Profiling and Deep–Level Transient Spectroscopy are based, and describes typical practical procedures for their implementation. It is shown that, although experimental data using these techniques are fairly easy to obtain, great care has to be taken in the interpretation of the data. References to the original and other papers on the theory and practice of the various methods are given so as to allow further details to be obtained when required.

The use of measurements of electrical capacitance for studies and characterisation of semiconductors has developed over a number of years and relies mainly on the properties and capacitance of the depletion region of a reverse biased p–n or metal-semiconductor rectifying junction. Such measurements can give information on the concentration of donor or acceptor doping atoms as a function of distance from the junction (the doping/depth profile) and of the presence of electron and hole traps in the semiconductor and of the energy levels, electron or hole capture cross–sections and concentrations of those traps. By careful, well chosen experiments the concentration/depth profiles of the traps can also be determined, including whether there is an excess or deficiency of carrier traps in the vicinity of interfaces between semiconductor growth layers that differ in doping or composition. Information of such kinds is of crucial importance to the effective device use of semiconductor materials and in establishing the optimum conditions for epitaxial growth of semiconductor layers, for ion implantation and annealing, and for all the stages of semiconductor material processing.

At the present time, efficient apparatus for the characterisation of semiconductors by capacitance measurements can be readily assembled or purchased, and data can be easily obtained, sometimes in profusion, from such systems. This tends to give the impression that all is straightforward in characterising semiconductors by these means; but in fact although the main physical principles underlying the methods are mostly straightforward, accurate interpretation of what the data mean requires very careful consideration of the electronic processes occuring in the semiconductor during the measurements. Very often it is only under well chosen experimental conditions on some kinds of semiconductor sample that the data corresponds immediately to the quantitative information that is required.

The purpose of the present lectures and article is to give an introduction to the theory of, and experimental procedures for, capacitance-technique characterisation of semiconductors. The emphasis will be on describing the main principles for newcomers to this field, and references to the many published articles will be given to allow further details to be found and studied.

2. SEMICONDUCTOR P-N AND SCHOTTKY JUNCTIONS

2.1 Energy-Band Bending and Depletion Regions

As mentioned in Section 1, the capacitance methods for characterisation of semiconductors depend largely on the properties of semiconductor semiconductor or metal semiconductor rectifying junctions. Details of those properties can be found in semiconductor text-books (eg those by Bar-Lev, Carroll, McKelvey and Sze) and only an outline of the results is given here.

Figure 1a shows the electron energies of the edges of the valence and conduction bands of a semiconductor for separated p-type and n-type samples and for p-type and n-type layers in contact forming a p-n junction. The major effect of concern here is that for the p-n junction, diffusion of holes from the p region into the n region and of electrons in the opposite direction has resulted in the setting up of an equilibrium "built-in" potential difference V_{bi} across the junction (with the n-type material being positive with respect to the p-type) and of regions depleted of carriers in the p-type and n-type material; at this equilibrium condition, the Fermi level E_F is at a constant energy across the junction, and this condition determines the value of V_{bi}. The potential V_{bi} is the sum of the component potentials $(V_{bi})_p$ and $(V_{bi})_n$ across the depletion regions in the p and n materials respectively, and has a polarity equal to that of an applied reverse bias.

The depletion regions thus formed in the p-type and n-type material contain space-charges due to their individual, immobile doping ions, and this gives rise to electric fields in those regions. Use of Gauss's Law

$$dE(x)/dx = \rho(x) / (\epsilon_r \epsilon_0) \qquad (1)$$

relating the electric field E(x) at position x to the space charge density $\rho(x)$ (equal to $eN_a(x)$ and $eN_d(x)$ in the p-type and n-type materials respectively), where ϵ_r is the relative permittivity (dielectric constant) of the semiconductor and ϵ_0 is the permittivity of free space, and of the definition of potential difference as the integral of E(x) with respect to x, (in other words, using Poisson's Equation) allows determination of the relationship between $(V_{bi})_p$ and $(V_{bi})_n$ and the respective depletion region widths $(x_d)_p$ and $(x_d)_n$.

In the condition of uniform doping in the p-type and in the n-type material (this is called the "step junction" case), the result of such analysis is that

$$(x_d)_p = (2 \epsilon_r \epsilon_0 (V_{bi})_p / (e N_a))^{0.5} \qquad (2)$$

and

$$(x_d)_n = (2 \epsilon_r \epsilon_0 (V_{bi})_n / (e N_d))^{0.5} \qquad (3)$$

where N_a and N_d are the acceptor and donor dopings respectively of the two parts of the p-n junction.

From the expressions (2) and (3) above, it can be seen that if, for example, the p-doping is very much larger than the n-doping, then $(x_d)_p$ is much smaller than $(x_d)_n$; the effect is that almost all of the total depletion thickness and almost all of the total built-in potential occur on the n-side. Thus, for such a "one-sided" p^+-n junction (where p^+ indicates large p-type doping), the relationship between the total built-in depletion thickness $(x_d)_{bi}$ and the total built-in potential V_{bi} is given by

$$(x_d)_{bi} = (2 \epsilon_r \epsilon_0 V_{bi} / (e N_d))^{0.5} \qquad (4).$$

Because the data obtained in capacitance-technique measurements on a semiconductor depend on the properties, linear extent and volume of the depleted region, the use of a

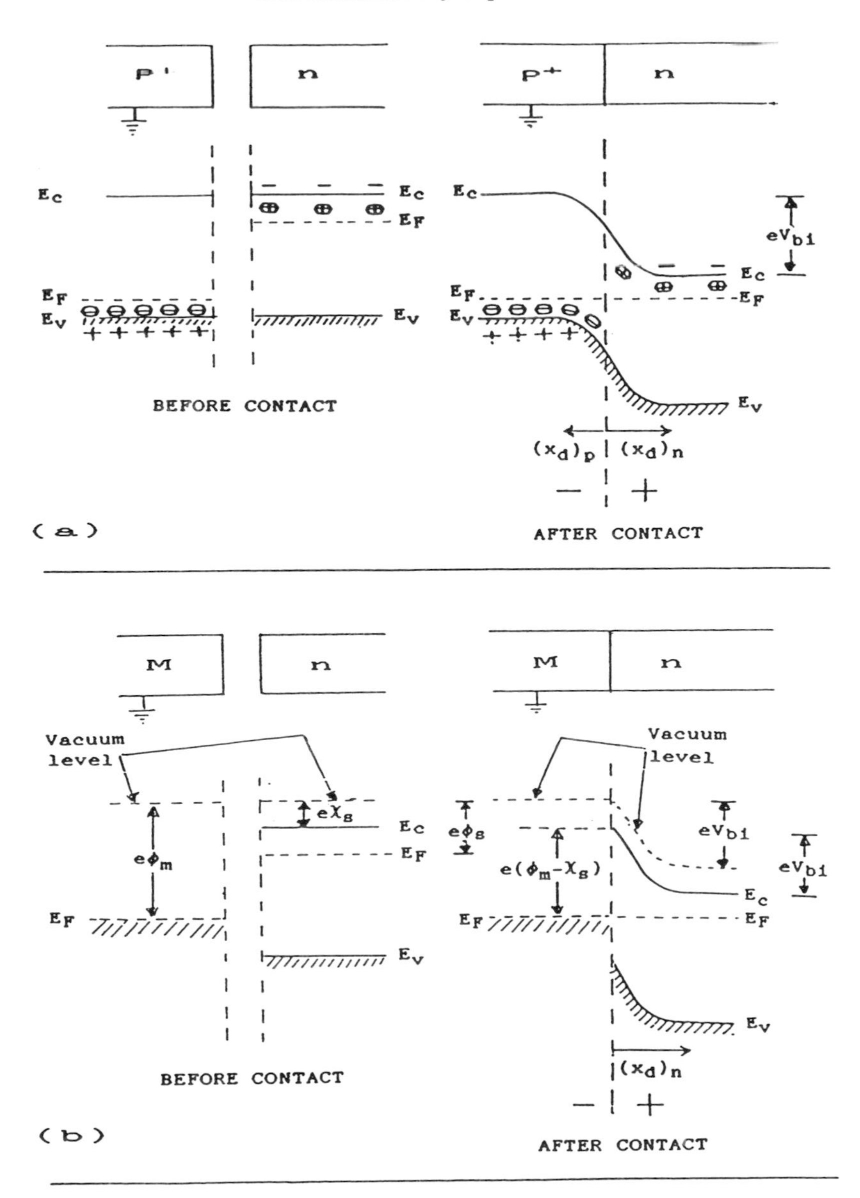

Figure 1. Electron energies as functions of distance
(a) for p^+ and n semiconductors before and after contact, and
(b) for a metal and an n-semiconductor before and after contact.

one-sided junction is of great benefit experimentally, since it is then clear on which side of the p-n junction the semiconductor is being investigated. Thus for investigation of a grown n-layer it can be advantageous to grow or fabricate a highly doped p-layer on top of the n-layer, so that the measurements can be made on the n-layer as a one-sided junction.

A very important alternative to the use of a one-sided p-n junction is the fabrication of a metal-semiconductor, Schottky rectifying diode, where the metal replaces either the p-type or n-type material. We shall see that this gives band-bending and depletion effects that are very similar to those of a one-sided p-n diode. It is conventional to consider the case of and derive the equations for the Schottky junction involving an n-type semiconductor, and to allow easy reference to other descriptions of capacitance-characterisation methods, that convention will be followed here also.

Figure 1b shows the valence band, conduction band and Fermi levels for a metal and an n-semiconductor without and with mutual contact. The energy differences $e\Phi_m$, $e\Phi_s$ and $e\chi_s$ are respectively the work functions of the metal and of the semiconductor, and the electron affinity of the semiconductor. If $e\Phi_m > e\chi_s$, then when the metal and semiconductor are in contact there is a potential barrier of magnitude $e\Phi_m - e\chi_s$ to electron flow from the metal into the semiconductor, the semiconductor side becomes positive with respect to the metal (the potential difference V_{bi} being equal to $\Phi_m - \Phi_s$), and a depletion region is formed in the semiconductor in the same way as in the ideal one-sided p^+-n junction. Thus the relationship between the built-in potential V_{bi} and the depletion width $(x_d)_{bi}$, equal to $(x_d)_n$, is given again by the expression (4) above.

So far in this section the energy diagrams and depletion regions in p-n and Schottky junctions have been considered only where the trans-junction potential is the built-in potential V_{bi}. When an external reverse bias Va is applied, the total reverse potential becomes $(V_{bi}+V_a)$, and then for the one-sided (p^+-n or metal-n) junction having a uniform donor-concentration N_d the width of the depletion layer is given by

$$x_d = (2 \epsilon_r \epsilon_0 (V_{bi}+V_a) / (e N_d))^{0.5} \qquad (5).$$

Thus an external reverse bias (positive on the semiconductor) increases the depletion width, whereas a forward bias decreases that width. Since the dielectric constants of silicon and gallium arsenide are each about 12, and the built-in potential energies eV_{bi} are usually approximately a half of the semiconductor valence-band to conduction-band energy gap, expression (5) shows that, for an applied reverse bias of zero to a few volts, the depletion widths are some fraction of a micron to several microns for typical doping concentrations of the order of 10^{14} to 10^{18} cm^{-3}.

If the donor concentration is not constant with distance into the semiconductor the expression (5) cannot be applied. If $N_d(x)$ is known, the potential difference across a depletion region of given thickness can be obtained by integration of Poisson's equation; but knowledge of the potential difference across a given depletion region does not of course allow the doping profile $N_d(x)$ to be deduced. This latter information can be obtained by measurement of the dependence of the electrical capacitance of the junction on applied reverse bias.

2.2 Depletion Capacitance and its Dependence on Applied Bias

As we shall now see, a p-n or metal-semiconductor reverse-biased junction has an electrical capacitance. That capacitance, called the "depletion capacitance", depends on the doping concentration, and measurement of that capacitance as a function of applied reverse bias gives the possibility of determining how the doping concentration varies with distance in the semiconductor. It is convenient, as discussed above, to consider a one-sided junction and to use a Schottky rectifier comprising a metal layer of large work function on an n-type semiconductor as a suitable example. The formulae that will be given for the depletion capacitance are equally applicable to the p^+-n junction.

Figure 2 shows the spatial variations of the energy bands and Fermi energy in a n-type semiconductor in the vicinity of a metal Schottky surface layer for two values V_{a1} and V_{a2} of applied reverse-bias voltage, for which $V_{a2} > V_{a1}$. The distance x is

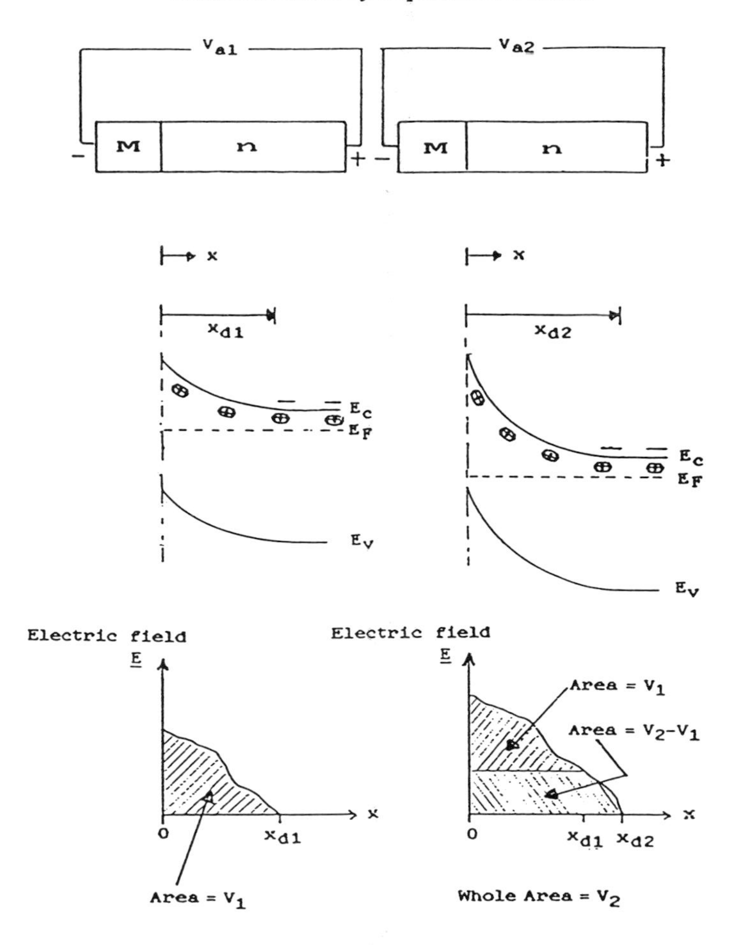

Figure 2. Electron energies and electric field values E as functions of distance in a metal/n-semiconductor Schottky junction for two applied reverse voltages V_{a1} and V_{a2}, for which $V_{a2} > V_{a1}$: x_{d1} and x_{d2} are the respective depletion distances. It is assumed here that the donor-doping concentration varies with distance x.

measured from the metal/semiconductor interface, and x_{d1} and x_{d2} are the depletion distances for V_a equal to V_{a1} and V_{a2} respectively. It is clear that increasing the applied voltage from V_{a1} to V_{a2} has caused movement of negative charge from the semiconductor into the positive terminal of the external circuit. The junction therefore acts as an electrical capacitor, of capacitance C equal to Q/V_a, where Q is the charge stored in the depletion region, and of small-signal (differential) capacitance C_s equal to dQ/dV_a.

If the doping concentration N_d in the semiconductor is independent of the distance x, then the charge Q stored in a depletion region of width x_d is $N_d x_d A$, where A is the junction area, and therefore from expression (5) the value of Q is

$$Q = A\,(2\,\epsilon_r\,\epsilon_0\,e\,N_d\,(V_{bi}+V_a))^{0.5} \qquad (6),$$

and C_s is given by the expression

$$C_s = dQ/dV_a = 0.5\ A\ (\ 2\ \epsilon_r\ \epsilon_0\ e\ N_d\ /\ (V_{bi}+V_a))^{0.5} \quad (7)$$

$$= \epsilon_r\ \epsilon_0\ A\ /\ x_d \quad (7a).$$

It is seen that this is exactly the formula for the capacitance of a parallel plate capacitor of inter-plate separation x_d.

Table 1 gives the depletion widths and depletion capacitances (for a junction area of 1mm^2) for one-sided diodes of gallium arsenide of three different doping concentrations. The depletion width and capacitance values would be very similar for silicon.

TABLE 1

Depletion Widths and Capacitances for one-sided Gallium Arsenide Diodes of Areas 1 mm^2 (for ϵ_r equal to 12.5 and V_{bi} equal to 0.7 Volt).

Doping Concentr. (cm^{-3})	1.0×10^{14}	1.0×10^{16}	1.0×10^{18}
Applied Reverse Bias (Volts)			
0	3.05µm, 36.3pF	0.31µm, 363pF	0.03µm, 3625pF
1	4.8µm, 23.0pF	0.48µm, 230pF	0.05µm, 2300pF
10	12.1µm, 9.1pF	1.2µm, 91pF	Elect. Breakdown

Convenient nomograms, that allow graphical finding of depletion region widths in silicon and gallium arsenide for wide ranges of applied reverse bias and doping concentrations or resistivities, have been given by Miller et al. 1977.

The more general doping situation is assumed in Figure 2, that the doping concentration $N_d(x)$ varies with distance x. The electric field variations then follow from Gauss's Law in the form of expression (1) with $\rho(x)$ equal to $eN_d(x)$. The areas under the respective E(x) curves are the voltage V_1, equal to $(V_{bi}+V_{a1})$, and the voltage V_2, equal to $(V_{bi}+V_{a2})$. If V_{a2} is only slightly larger than V_{a1}, such that we can write

$$V_{a2} - V_{a1} = \Delta V_a$$

and

$$x_{d2} - x_{d1} = \Delta x_d,$$

then Figure 2 demonstrates that

$$V_2 - V_1 = E(x_d).x_d, \qquad \text{where } V_2 - V_1 = V_{a2} - V_{a1},$$

and where

$$E(x_d) = (dE(x)/dx).\ \Delta x_d$$

$$= (e\ N_d(x)\ /\ (\epsilon_r\ \epsilon_0)\ )\ .\ \Delta x_d.$$

But $e\ N_d(x).\Delta x_d = \Delta Q(x_d)/A$

(where A is the junction area), and therefore we find that

$$V_{a2} - V_{a1'} = \Delta V_{a'} = \Delta Q(x_d)\ x_d\ /\ (A\ \epsilon_r\ \epsilon_0) \qquad (8).$$

Therefore $\Delta Q(x_d)/\Delta V_{a'} = C_{s'} = \epsilon_r\ e_0\ A\ /\ x_d$ (9).

This expression (9) for $C_{s'}$, valid for a non-uniform doping, is the same as (7) above obtained for uniform doping.

Thus the equivalence of the capacitance of the one-sided junction to that of a parallel-plate capacitor, giving the capacitance formula (9), is valid whether or not the semiconductor doping is uniform. This fact is crucial in allowing determination of the depletion width x_d from measurement of the junction capacitance for a semiconductor material of unknown doping profile. That value of x_d then gives the depth scale for the determination of the doping profile from capacitance-voltage measurements, as is now described.

From expression (9),

$$dC_s/dx_d = -\ \epsilon_r\ \epsilon_0\ A\ /\ x_d^2$$

$$= -\ C_s^2\ /\ (\epsilon_r\ \epsilon_0\ A) \qquad (10).$$

Then, using $dV_a/dx_d = (dC(x_d)/dx_d).(dV_a/dC(x_d))$,

and $C_s = dQ(x_d)/dV_a = e\ N_d(x)\ dx_d\ A\ /\ dV_a$ (11),

one can deduce that

$$C_s = -\ e\ N_d(x)\ \epsilon_r\ \epsilon_0\ A^2\ (dC_s/dV_a)\ /\ C_s^2.$$

Rearrangement of this expression gives

$$N_d(x_d) = -\ C_s^3\ (e\ \epsilon_r\ \epsilon_0\ A^2)^{-1}\ \ (dC_s/dV_a)^{-1} \qquad (12).$$

Thus from measurement of the small-signal capacitance C_s at applied reverse voltage V_a and of the rate of change of C_s with V_a, the doping concentration N_d at the depth x_d can be determined; the value of x_d is itself found from the measured capacitance from expression (9).

As has been explained above, the small-signal capacitance C_s of a reverse-biased junction arises formally because an increase of applied reverse bias causes movement of electric charge out of the depletion region. Kennedy et al (1968) suggested that therefore $N_d(x_d)$ on the LHS of expression (12) should be replaced by $n(x_d)$, where this latter is the free-electron concentration at the edge of the depletion region. In a semiconductor material in which the doping profile $N_d(x)$ is independent of x and there are no deep traps to capture free electrons, $N_d(x)$ and $n(x)$ are equal and so it does not matter which of these one has in mind. But in a material where $N_d(x)$ varies strongly with x, diffusion of the free electrons causes the $n(x)$ profile to be different from the $N_d(x)$ profile (Section 3.1.2), and certainly in such case it is right to consider the RHS of expression (12) as giving the value of $n(x_d)$, as Kennedy et al have stated.

However, when deep-level carrier traps are present (Section 3.4) the RHS of expression (12) is neither the thermal equilibrium value of $n(x_d)$ nor the doping concentration $N_d(x_d)$ at the edge of the depletion region, although it is related to both of these. Furthermore, in the presence of deep carrier traps, the RHS of (12) does not necessarily give a value valid for the depth x_d obtained from the measured differential capacitance. It is better therefore to consider the RHS of expression (12) as giving the value of a measured depth profile $N^+_{meas}(x)$, where interpretation of its value and of the x value obtained from the measured C_s need to be made. We note that the superscript "+" in $N^+_{meas}(x)$ is appropriate for measurements on n-type material and indicates that in the main the profile is that of the positive immobile space-charge which is dominated by the presence of the positive donor ions. We therefore rewrite (12) in the form

$$N^+_{meas}(x) = - C_s^3 \; (e \; \epsilon_r \; \epsilon_0 \; A^2)^{-1} \; (dC_s/dV_a)^{-1} \qquad (13).$$

and expression (9) as

$$C_s = \epsilon_r \; e_0 \; A \; / \; x$$

ie,

$$x = \epsilon_r \; e_0 \; A \; / \; C_s \qquad (14).$$

These expressions (13) and (14) constitute the basis of the C-V profiling method. Practical details of the method and consideration of what information it provides will be given in Section 3.

3. C-V PROFILING

3.1 General

As has been explained in Section 2.2 and summarized in expression (13), a capacitance-voltage profiling characterization of a semiconductor material determines a depth profile $N_{meas}(x)$ for the material by measurement of the rate of change dC_s/dV_a of the small-signal capacitance C_s of a rectifying junction with respect to applied reverse bias V_a. The present Section outlines the practical application of this technique and describes the information thereby obtainable. Further general discussion can be found in the papers by Kennedy et al 1968, Baxandall et al 1971, Bleicher and Lange 1973, Kimerling 1974, Wiley 1975, Wu et al 1975, Blood and Orton 1978 and Blood 1986.

3.2 Practical Procedures for C-V Profiling

3.2.1 Schottky and Ohmic Contacts

As seen above, the C-V profiling method requires use of the semiconductor material in the form of a p-n or Schottky rectifying junction. In fact, a major reason why capacitance characterisation of semiconductors has been found to be very convenient is that it is fairly easy to make metal-semiconductor Schottky diodes of good rectifying properties on the important semiconductors silicon and gallium arsenide; this avoids the necessity of having to fabricate a p-n or p^+-n junction for the C-V measurements, and the Schottky diode method is therefore very much used.

In C-V measurements, low reverse-bias current (less than a few nA/mm^2 if possible) is needed both to maintain the required electronic conditions in the depletion layer and to facilitate the actual measurement of the capacitance; this good reverse-bias behaviour for Schottky structures can be accomplished by use of vacuum evaporation to form the metal layer, provided that careful attention is paid to the cleanliness of the semiconductor surface and of the vacuum. Suitable metals include gold and molybdenum for n–type silicon, and aluminium, gold or silver for n–type GaAs. Less good but useful Schottky diodes can usually be made also on p–type material of these semiconductors. From Figure 1 it has been seen that the rectifying energy barrier against movement of electrons from the metal into the semiconductor should be equal to the difference $e\varphi_m - e\chi_s$ between the work function of the metal and the electron affinity of the semiconductor. However, in practice, it is found that the presence of a high density of electron states at the interface between the metal and the semiconductor causes the energy barrier to be rather independent of the actual metal being used.

Tabulated information on the energy barriers (and thus of the effectiveness of the Schottky diode rectification) can be found in the book by Sze. Other texts such as that by Rhoderick and the review paper by Rideout 1978 give more detailed information on the theory and fabrication of metal–semiconductor rectifiers.

A further method that can be very convenient for forming a Schottky diode for C–V measurements near room temperature on silicon and gallium arsenide is the use of a liquid-mercury contact (for silicon, eg Severin and Poodt 1972; for GaAs, eg Hughes 1972, and Binet 1975). In such a mercury-probe system, the semiconductor sample is usually arranged horizontally with the side of it that is to be examined facing downwards in contact with the top of a column of mercury held and slightly pressurized in a vertical tube. There is however often difficulty in knowing the exact area of the mercury semiconductor contact, and this can give much uncertainty in determining the capacitance per unit area. The best solution to this problem is to have available a standard semiconductor sample for which the capacitance per unit area in a C–V measurement has been found by use of an evaporated metal contact of known area; this is employed to determine the mercury-probe contact area under defined mercury pressure conditions. A well known commercial instrument in this field is the MDC Mercury Probe from the Materials Development Corporation.

It is found also that appropriate liquid electrolytes can also be used, instead of a metal, to produce a Schottky rectifying junctions on semiconductors. Features and applications of this technique will be outlined in Section 3.3.4.

The major qualification to be made concerning junctions for C–V profiling, whether they are p–n or metal semiconductor contacts is that effective rectifying junctions cannot be made on semiconductor material of high doping (doping concentrations of the order of 10^{18} cm^{-3} or greater), mainly because the thinness of the depletion region allows electron tunnelling to occur very easily. However the C–V profiling method is usually well applicable to semiconductors of reasonably large band-gaps of doping concentrations in the range 10^{14}–10^{17} cm^{-3}. As will be seen in Section 3.3.4, electro–chemical C–V profiling that incorporates etching of the semiconductor can allow assessment to somewhat higher doping concentrations.

Whether the structure on which the C–V measurements are to be made is a p–n diode or a Schottky diode, the connection of the semiconductor sample to the external circuit has to be via a non–rectifying ("ohmic") metal semiconductor contact. In a commonly used sample arrangement, the semiconductor material under investigation is grown on a highly doped semiconductor substrate of the same type (eg, n–Si on n^+–Si, n–GaAs on n^+–GaAs, n–$Al_xGa_{1-x}As$ on n^+–GaAs etc), and the ohmic back–contact is made to this substrate.

Information on the formation of low resistance, ohmic contacts can be found in semiconductor texts (eg, those by Sze, Rhoderick etc), and various research papers give recipes for particular semiconductors.

3.2.2 Equipment

The basic requirement for the C–V method is the measurement of the small–signal (differential) capacitance C_s of a reverse–biased semiconductor junction as a function of the applied bias. The measurement involves use of a voltage V_{in}, equal to the sum of the steady applied reverse bias V_a and of a small AC voltage $V_c(t)$, of amplitude $V_c(0)$ and of angular frequency ω_c, applied across the semiconductor junction being studied. The AC current I_{out} that passes through the junction, acting as a capacitor, is proportional to $C_s(V_{in})$, which is equal to the required $C_s(V_a)$ if the amplitude Vc(0) is small compared to V_a. Good sensitivity can be obtained by measuring I_{out} via a circuit tuned to the frequency ω_c, and the effects of any electrical resistance in the semiconductor junction can be reduced by use of a phase–sensitive detector set to a phase 90° ahead of that of $V_c(t)$. The well-known Boonton 72B capacitance meter works in this way, at a frequency $\omega_c/2\pi$ of 1MHz with $((V_c(t))_{rms}$ being 15mV. Its full–scale outputs cover 1pF–3000pF, it has both visual indication of C_s and a DC voltage output (with a response time of 1ms) proportional to that capacitance, and it has input terminals allowing the reverse bias V_a to be applied through the instrument to the semiconductor junction. The Boonton 72BD capacitance meter gives also a digital output of the measured capacitance but has a 2ms response time. The Hewlett–Packard HP–4271B LCR meter also works at 1MHz and has full–scale ranges of 10pF–10000pF, a minimum response time of 5ms and digital output; the HP–4297A 1MHz C–V digital–output meter has an incorporated programmable sweep–voltage upto 38 Volts for application as the reverse bias V_a on the semiconductor junction being investigated. The Keithley–590 capacitance meter gives the choice of operation at 100kHz or 1MHz, and an available sweep unit allows reverse biases upto 40V to be applied to the semiconductor junction. With use of any such instrument, all that is required is to determine C_s as a function of applied voltage V_a, and then to use expression (13) to find the effective doping–depth profile $N^+{}_{meas}(x)$, where x is in the simplest case equal to the depletion distance x_d deduced using expression (7a).

It is straightforward these days to use a measurement system comprising a capacitance meter linked to a computer, so as to measure $C_s(V_a)$ from the analogue output of the capacitance meter, and to use a software program to compute x_d, dC_s/dV_a, and thence $N^+{}_{meas}(x_d)$. For a capacitance meter giving an analogue voltage output proportional to C_s, an ADC linked to the computer via an IEEE488 interface or an ADC–card inside the computer is employed, and for a capacitance meter having a digital output, this latter may be able to be sent directly into the computer. The C–V Semiconductor Profiling facilities in the University of Sussex Physics Division include such a system using a Boonton 72B capacitance meter coupled to a BBC–B micro–computer, with software written so as to provide on–screen plots of $C_s(V_a)$ and $N^+{}_{meas}(x_d)$ (Irvine and Palmer 1988). The Keithley Package–82 C–V system is a commercial integrated instrument of this kind that includes the Keithley–590 capacitance meter linked to a Hewlett–Packard computer. The Bio–Rad DL–4600 equipment includes both a computerized C–V facility and a Capacitance–Transient Deep–Level Spectrometer (Section 5).

An automatic, non–digital method of finding both C_s and dC_s/dV_a, that has been much used, is based on the circuitry described by Baxandall et al 1971. It uses a low amplitude, high frequency $(\omega_c/2\pi)$ AC voltage applied across the semiconductor junction (as described above) to measure the small–signal capacitance C_s at the applied bias V_a, and that voltage at frequency ω_c is modulated at frequency $(\omega_{mod}/2\pi)$ to determine dC_s/dV_a. The current I_{out} that passes through the semiconductor junction is sent to a phase–sensitive detector PSD1 synchronised $(+90^o$ phase) to ω_c, and the rms output of PSD1 is proportional to $C_s(V_a)$; the output of PSD1 is sent also to a phase–sensitive detector PSD2 synchronised $(+90^o$ phase) to ω_{mod}, and the rms output of PSD2 is proportional to dC_s/dV_a. For precision measurements, the amplitudes $V_c(0)$ and $V_{mod}(0)$ of the ω_c and ω_{mod} voltages need of course to be much smaller than the DC applied bias V_a being employed. The capacitance–measurement and modulation frequencies used in the circuit are 100kHz and 1kHz respectively, with somewhat high rms amplitudes of 150mV. The unit includes a variable voltage supply for application as the reverse bias V_a to the semiconductor junction.

The commercial instrument using this measurement technique and circuitry was manufactured by JAC Electronics Ltd and called the IPP-366 Impurity Profile Plotter; a slightly modified version called the SS-367 was subsequently made by Shandon Southern Instruments Ltd. The system provides analogue output voltages proportional to C_s, x_d, V_a, dC_s/dV_a and $N^+_{meas}(x_d)$, all suitable as inputs to an X-Y chart recorder or to ADC/computer arrangements. The Bio-Rad PN4200 and PN4300 Electro-Chemical Profile Plotters employ the same principle (Ambridge and Faktor 1975, Blood 1986) for determination of N^+_{meas}, with operation at frequencies of 1kHz to 25.5kHz for the measurement of C_s, and of 10Hz to 2.55kHz for determining dC_s/dV_a; the special features of electro-chemical profiling are described in Section 3.3.4.

3.3 C-V Profiling in the Absence of Deep Traps

3.3.1 Capacitance-Voltage Data and Their Analysis

As explained in Section (2), the measured differential (small signal) capacitance C_s and its rate of change dC_s/dV_a with applied reverse bias V_a enable an experimental depth profile $N^+_{meas}(x)$ to be obtained for an n-type semiconductor being investigated, via the expression

$$N^+_{meas}(x) = -C_s^3 \, (e \, \epsilon_r \, \epsilon_0 \, A^2)^{-1} \, (dC_s/dV_a)^{-1} \qquad (13),$$

where

$$x = \epsilon_r \, e_0 \, A \, / \, C_s \qquad (14).$$

If the n-type semiconductor material contains only the intentional shallow-level donors at a concentration N_d that is uniform or changes only slowly with depth from the junction into the semiconductor, and if no dopant acceptor atoms or electron-trapping defects are present, then, in accordance with Figures 1 and 2, $N^+_{meas}(x)$ is equal to $N_d(x)$, and x is equal to the depletion distance x_d. That is, we return to expressions (12) and (9). If however the material contains dopant acceptor atoms of concentration N_a, having their electron energy levels just above the valence-band edge, then $N^+_{meas}(x_d)$ is equal to $N_d(x_d)-N_a(x_d)$. In either case, if there are no other deep-level carrier traps (impurities or lattice defects), the value of $N^+_{meas}(x_d)$ obtained is equal to the equilibrium value of the free electron concentration $n(x=x_d)$, provided that the measurement temperature is significantly above the carrier freeze-out temperature (eg above about 50K for silicon and above about 10K for GaAs). In such situation also, the depth profile $N^+_{meas}(x_d)$ is essentially independent of the frequency of the test voltage by which C_s is being measured or of the rate or frequency at which V_a is being changed in the measurement of dC_s/dV_a.

As an example of results of C-V profiling, Figure 3 gives $N^+_{meas}(x)$ data (from Rezazadeh 1983) for a commercial n-type GaAs layer grown by vapour-phase epitaxy. The measurements shown were made using a Shandon Southern SS-367 Impurity Profile Plotter (see Section 3.2.2) on a Schottky diode, fabricated by aluminium evaporation, held in an Oxford Instruments Ltd CF100 liquid-helium-cooled cryostat. Considering for now just the 295K data, one sees that the C-V profiling showed that the material had rather uniform doping at a concentration of about 2.3 x 10^{15} cm^{-3} over the measured depth region from near the surface to about 3.8μm. The starting depth for zero applied bias corresponds, in accordance with expression (4), to the depletion region thickness of about 0.7μm for the built-in potential V_{bi} of approximately 0.7V. The applied reverse bias to reach the depth of 3μm was about 15V (expression (5)), and the noise evident on the N^+_{meas} output signal for depths at and beyond 3μm is probably an indication of the start

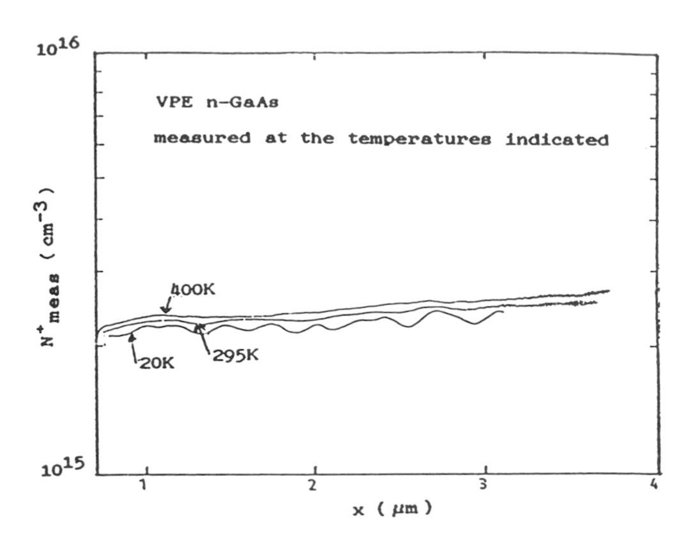

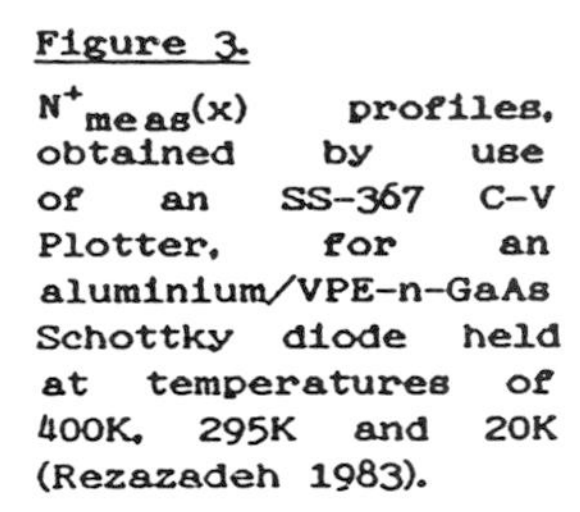
Figure 3. $N^+_{meas}(x)$ profiles, obtained by use of an SS-367 C-V Plotter, for an aluminium/VPE-n-GaAs Schottky diode held at temperatures of 400K, 295K and 20K (Rezazadeh 1983).

of electrical breakdown in the Schottky–semiconductor junction at an applied reverse bias of 15V, despite this being smaller than the theoretical breakdown voltage for a doping concentration of 2–3 x 10^{15} cm^{-3} in n–GaAs. The C–V profiling method has been very much applied, as in Figure 3, to study the the quality and doping profiles of layers of semiconductors, especially of silicon and gallium arsenide, grown under different experimental conditions by various techniques.

The effect of the measurement temperature in the data of Figure 3, related to the presence of deep–level electron traps, will be considered in Section 3.4.

3.3.2 Effects of Gradients in Doping Concentration

As has been explained in Section 2.2, the differential capacitance of a reverse–biased p^+–n or Schottky junction is a measure of the charge that moves into the external circuit when the reverse bias voltage is increased or decreased by a small amount. Therefore (Kennedy et al 1968, Johnson and Panousis 1971, Wu et al. 1975, Blood and Orton 1978) the depth profile $N^+_{meas}(x)$ obtained from C–V data is not the profile $N_d(x)$ of the chemical donors, but is, instead, the depth profile n(x) of the majority–carrier, free electrons that results from the balance of electron diffusion and electric–field drift in the $N_d(x)$ depth distribution. If $N_d(x)$ is independent of x, then (in the absence of deep electron traps) n(x) is equal to $N_d(x)$. But if $N_d(x)$ has any sharp gradient with respect to depth x, then electron diffusion causes n(x) to change more slowly with x than $N_d(x)$, and in this circumstance the C–V profile $N^+_{meas}(x)$ is not the same as $N_d(x)$. The magnitude of this effect depends on the Debye length L_D given by the expression

$$L_D = (\epsilon_r \; \epsilon_0 \; k_B \; T \; / \; (e^2 \; N_d))^{0.5} \qquad (15),$$

in which k_B is Boltzmann's constant and T is the sample temperature. Near a spatially abrupt change in N_d occurring at x_{abr}, n(x) varies approximately as $\exp(-(x-x_{abr})^2/2L_D^2)^{0.5}$ (Blood 1986) as illustrated in Figure 4 for step and plateau profiles of $N_d(x)$.

Thus the exact shape of $N_d(x)$ cannot be determined to a spatial resolution of better than a few times the Debye length. The effect can be considered also as an inaccuracy in the depletion region model outlined in Section 2, the inaccuracy being

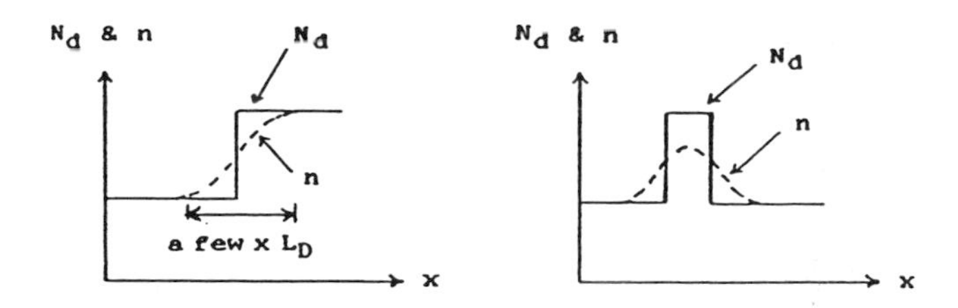

Figure 4. Spreading of the free electron concentration profile n(x) in the vicinity of a sharp change in donor-doping concentration N_d.

caused by the neglect of the spread of the free electrons into the depletion region due to the kinetic energy of the electrons. For silicon and gallium arsenide, which have dielectric constants of approximately 12 and 12.5 respectively, expression (15) gives Debye length values at 300K of approximately 0.4, 0.04, 0.004 microns for doping concentrations of $1x10^{14}$, $1x10^{16}$ and $1x10^{18}$ cm^{-3} respectively; we see therefore that for lightly doped semiconductors in which N_d changes over distances of a tenth of a micron or smaller the profile $N^+{}_{meas}(x)$ obtained from a C–V measurement by means of expressions (13) and (14) is significantly different from the $N_d(x)$ distribution. Expression (15) shows that some improvement in spatial resolution can be obtained by reducing the sample temperature, but the presence of deep level carrier traps then needs to be considered (Section 3.4). Kennedy and O'Brian 1969 have shown that a better approximation to the true $N_d(x)$ doping profile can be obtained from $N^+{}_{meas}(x)$ by adding or subtracting a term dependent on the curvature of $N^+{}_{meas}(x)$; Johnson and Panousis 1971 have discussed the applicability of that analysis.

3.3.3 Effects of Series Resistance and Leakage Current

As has been outlined in Section 3.2.2, the standard method for determining the differential capacitance of the reverse biased semiconductor junction is to measure the component of the junction current that is 90° out of phase with the small, applied, AC test voltage $V_c(t)$. Therefore a resistance in series with the junction capacitance (eg, the resistance of the ohmic contact or in the semiconductor itself of low doping) will produce error in the capacitance measurement due to a resistance-induced change in the phase angle (Wiley and Miller 1975). Furthermore, reverse-bias and leakage currents in the junction capacitance will lead to a voltage drop across the series resistance, and the voltage V_a across the junction may, in the worst circumtances, be significantly less than the externally applied voltage. These points have been given additional consideration by Blood and Orton 1978 and by Blood 1986. Analysis in the latter paper shows that even for the most highly doped semiconductor materials that can be studied by C–V profiling, any series resistance less than about 30 ohms will give no serious error in capacitance measurement at 1 MHz.

3.3.4 Electro-Chemical Profiling

A further development of the standard C–V profiling method is the use of a liquid electrolyte, instead of a metal layer, to form the Schottky barrier junction on the p–type or n–type semiconductor to be assessed, the important additional feature of this being that the electrolyte can also controllably etch away the semiconductor; this allows determination of the $N^+{}_{meas}(x)$ profile to larger depths from the initial surface than the maximum value of the depletion thickness x_d as limited by electrical breakdown. The principle of such an electro-chemical profiler was described in detail by Ambridge and Faktor 1975, and their design formed the basis of the instrument manufactured commercially by Syncryst Ltd as the "Post-Office Profile Plotter", and subsequently and currently by Bio–Rad Ltd as the PN4200 and PN4300 Profile Plotters. Mayes 1985 has given a very clear outline of the theory of and procedures for electro–chemical profiling using the PN4200/4300 Plotter, and Blood 1986 has discussed in detail the major principles and applications of the method.

It is found that the electrolyte (eg aqueous concentrated sodium hydroxide works well for GaAs) acts very much like the metal in a metal/semiconductor Schottky junction, and that a depletion layer is formed in the semiconductor, essentially as has been described in Section 2. Again, by measurement of the differential capacitance C_S and dC_S/dV_a as a function of reverse bias V_a the C–V profile $N^+_{meas}(x)$ can be found using expressions (13) and (14). The Bio–Rad PN4200 and PN4300 Plotters can use test frequencies of 1 to 25.5 kHZ and of 0.01–2.55 kHz to measure C_S and dC_S/dV_a respectively, and employ an analogue circuit similar to that of Baxandall 1971, as described in Section 3.2.2, to evaluate $N^+_{meas}(x_d)$. In these instruments, the sample is held in an electro–chemical cell (Figure 5) in which the electrolyte is in contact with a defined area of the front surface of the semiconductor, spring–loaded sharp pins are used to produce an essentially ohmic back contact, and the circuit connection to the electrolyte is made by a platinum electrode.

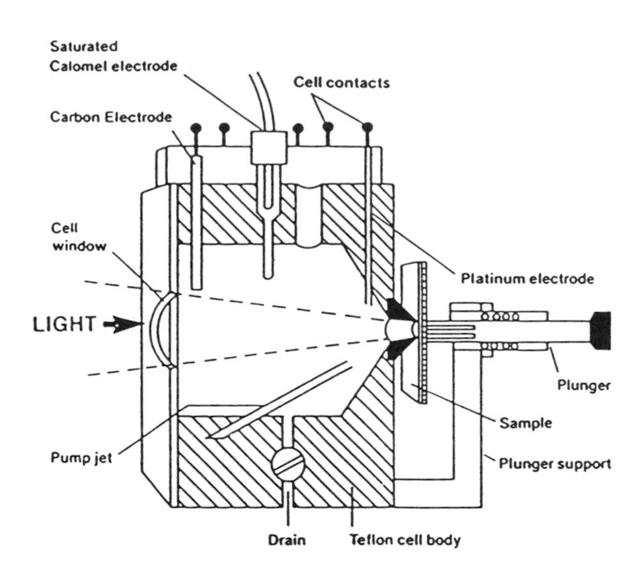

Figure 5. Cell for holding a semiconductor sample as used in the Bio–Rad PN4200/4300 Electro–Chemical C–V Profile Plotter. The platinum electrode is employed for the C(V) measurement, and the carbon and calomel electrodes are used for the electro–chemical etching and as a reference electrode respectively.

Electro–chemical etching of the semiconductor requires a sufficient hole concentration to be present in the semiconductor. Therefore in the dark n–type material is not etched, but etching can be achieved, with the junction under reverse bias, by illuminating the semiconductor surface, through the electrolyte, by light of photon energy greater than the band gap, the etch rate being proportional to the reverse current flowing; continuous determination of N^+_{meas} can be made upon n–type material while the etching continues. For p–material, the etching occurs even without illumination under forward bias (ie, positive bias on the semiconductor); thus for such material the measurement procedure involves alternate forward–biasing to accomplish the etching and reverse–biasing for determination of N^+_{meas}.

By these means the profile $N^{+}_{meas}(x)$ can be measured to depths of many microns from the initial semiconductor surface. Furthermore, since the etching, rather than the reverse bias V_a, is used to increase the depth x, V_a can be kept to a low value, such as 0.5 Volt, and this can enable semiconductor materials of high doping concentration, upto greater than 10^{18} cm^{-3}, to be assessed without the limitation of electrical breakdown in the depletion region.

3.3.5 Studies using Metal–Insulator–Semiconductor Structures

Since a depletion region can be produced in the semiconductor material of a metal–insulator–semiconductor structure by application of a bias potential difference of appropriate polarity across between the metal and the semiconductor, the principles described in Section 2 are again available in such samples for determination of a C–V profile, with the measured capacitance being that of the insulator capacitor in series with that of the semiconductor depletion region. MOS structures on silicon have indeed been much investigated by capacitance methods, information on the doping etc in the silicon being obtainable by capacitance measurements at high frequencies, and on oxide–silicon interface states and minority carrier lifetimes from measurements of capacitance at low frequencies (10–100Hz) or of capacitance–time transients. Sections in the book by Sze deal with this large topic, and commercial leaflets (eg, that from Bio–Rad by Pearce and Mayes 1987 and that from Keithley Instruments Division, 1987) give outlines and equipment details.

3.4 C–V Profiling in the Presence of Deep Traps

3.4.1 Deep Traps: Definitions and Natures

Deep traps in the present context are electron or hole traps which have electronic energy levels, in the semiconductor energy gap, that are well separated in energy from the respective band edges (ie, well below the conduction band edge for electron traps and well above the valence band edge for hole traps). Usually the definition is made somewhat more quantitative by saying that a deep level is one having an energy difference from the appropriate band edge that is larger than that of the shallow chemical–doping levels, ie more than about 0.05eV for silicon and than about 0.01eV for gallium arsenide). They are designated as acceptor traps if they are negative when they have trapped an electron (ie when the electronic level is filled with an electron) and electrically neutral when they have lost the electron (empty level); correspondingly, donor traps are those that are neutral when they have trapped an electron and positive when the electron has been emitted.

These deep traps arise from various kinds of impurity atom, defect or defect–impurity complex in the semiconductor lattice; they may be introduced unintentionally (or even intentionally) into the semiconductor material during growth of the bulk crystal or epitaxial layer, or during subsequent processing stages. In addition to decreasing carrier concentrations by carrier trapping, they affect mobility, minority–carrier lifetime, radiative efficieny in LED's and lasers, etc.

3.4.2 Band Bending and Trap Occupancy

By means of Figure 6 we begin to examine how the presence of deep traps can affect the interpretation of experimental $N^{+}_{meas}(x)$ profiles obtained from a C–V measurement.

We consider here the effects for an n–type semiconductor of doping donor

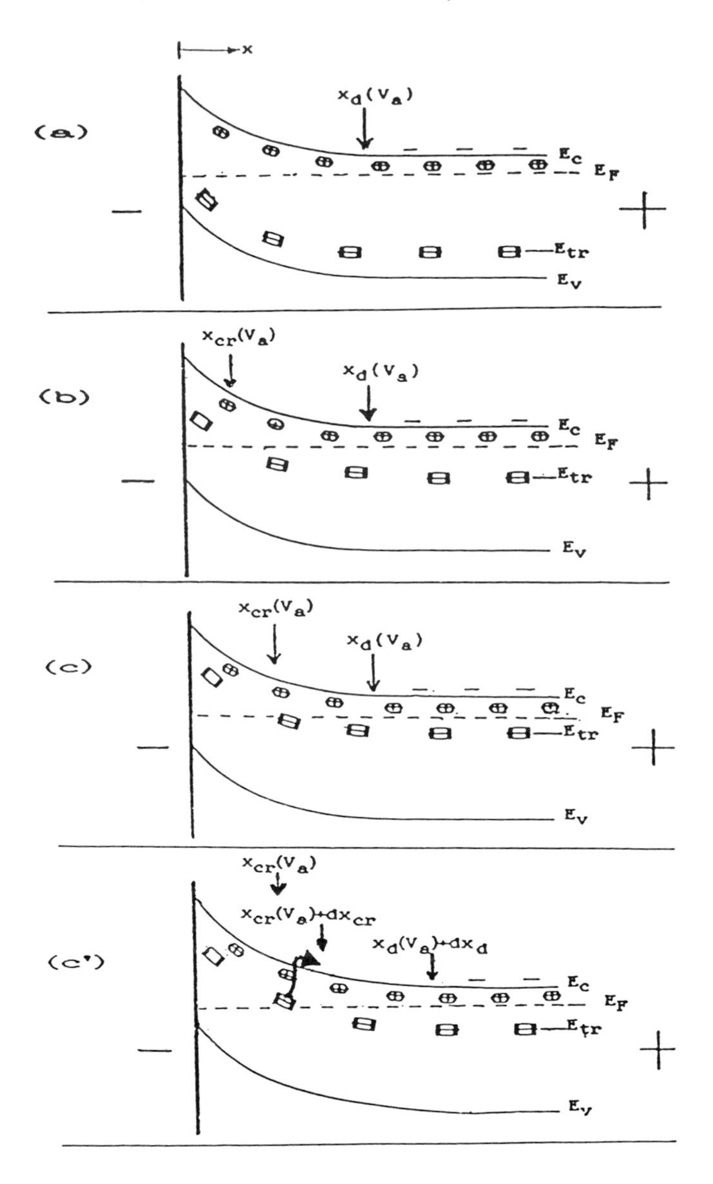

Figure 6. Electron energies as functions of distance x for materials containing acceptor-type electron-trapping levels E_{tr}. Between (c) and (c'), the applied reverse bias has been slightly increased, and traps in the spatial interval x_{cr} to $(x_{cr} + dx_{cr})$ then emit electrons to the conduction band.

concentration N_d, and we deal initially with traps that are acceptors (see Section 3.4.1). We envisage that the electronic energy diagrams of Figure 6 result from application of a reverse-bias V_a, and we shall consider the effect of an AC modulation voltage of amplitude $V_{mod}(t=0)$ (section 3.2.2) that is used to determine dC_s/dV_a. Figures 6a, 6b and 6c are drawn with inclusion of the energy levels of acceptor electron traps of three different energies respectively, each trap being of concentration N_{tr}, where $N_{tr} < N_d$.

In Figure 6a, the trap is very deep in the energy gap, such that all the individual trapping centres throughout the depletion region of width x_d are below the Fermi energy E_F, and so are all filled and negative. At depths x greater than $x_d(V_a)$, the semiconductor has to be electrically neutral at equilibrium, and the concentration n of free electrons in the conduction band in that neutral region must therefore be given by the expression

$$n(x) = N_d(x) - N_{tr}(x) \quad (16)$$

provided (section 3.3.2) that there is no significant spatial gradient in the N_d or N_{tr} concentrations.

When the modulation voltage increases the total voltage across the junction, the right-hand side of the energy diagram 6(a) moves down slightly, the value of x_d increases, and electrons of concentration n in the semiconductor move into the external circuit. Thus the dC_s/dV_a is proportional to n, and the C–V profile obtained is

$$N^+_{meas}(x_d) = n(x_d) = N_d(x_d) - N_{tr}(x_d) \quad (17).$$

That is to say, in this situation, the C–V method does not measure N_d but instead measures $N_d - N_{tr}$. This is exactly equivalent to what has been said in Section 3.3, that if some residual chemical acceptors are present at concentration N_a in addition to the required chemical donors at concentration N_d, the C–V profiling method gives $N^+_{meas}(x_d)$ equal to $N_d(x_d) - N_a(x_d)$.

Figure 6b shows a trap energy level that is closer to the conduction-band edge than for the trap of Figure 6a, such that the Figure 6b trap levels that are near to the p^+–n or metal–n interface (ie for x less than $x_{cr}(V_a)$, where $x_{cr}(V_a)$ is the point at which the trap energy crosses E_F for the applied reverse bias V_a) are now above the Fermi level E_F and at equilibrium do not contain trapped electrons. (The spatial region between $x_{cr}(V_a)$ and x_d is often called the transition region). However we assume for Figure 6b that the energy difference $E_c - E_{tr}$ is large enough that at the sample measurement temperature being used the rate of emission of an electron from the trap into the conduction band is very slow (Let us assume, for example, that the mean time needed to remove the electron from the trap is greater than a second.).

When the modulation voltage increases the total reverse bias, the right hand side of the energy diagram 6(b) is lowered, the point $x_{cr}(V_a)$ moves to the right by a small distance dx_{cr}, and filled traps within that distance interval dx_{cr} are now above E_F. However, if as is usually the situation, the modulation frequency is higher than a few Hz (and remembering that we have assumed a trap emission time of larger than a second), the filled traps in the interval dx_{cr} will not be able to emit their electrons to the conduction band before the AC modulation voltage has reached its peak and has begun to decrease, and therefore the only electrons that can move from the semiconductor into the external circuit are those already in the conduction band at x_d, ie those of concentration $n(x_d)$ given still by the expression (16). Therefore again the measured C–V profile $N^+_{meas}(x_d)$ is given by expression (17).

By means of Figure 6c we now examine the more complicated situation for which the electron trap level E_{tr} is closer to the conduction band edge E_c, such that if, for any reason, there is a filled such trap above the Fermi energy E_F, the thermal emptying of the filled level by emission of its electron into the conduction band occurs rapidly. This situation has been discussed in detail by Kimerling 1974, and the major relevant points are outlined here. As for Figure 6a and 6b, increase of the total applied reverse potential, due to the modulation voltage of amplitude $V_{mod}(t=0)$, results in increase of $x_{cr}(V_a)$ by dx_{cr} and of $x_d(V_a)$ by dx_d. But for the case of Figure 6c, the trap levels that are above E_F in the spatial interval dx_{cr} are able to emit their trapped electrons to the conduction band (Figure 6c') before the alternating modulation voltage moves back from its maximum value to zero in a quarter of its AC cycle; those emitted electrons, now in the conduction band, then drift rapidly (towards the right in Figure 6c') to the edge of the depletion region and thence to the external positive electrode. The very significant effect of this is that the total electron charge moving out of and into the depletion region due to the AC modulation voltage is the sum of the free electron charge $n(x_d).dx_d$ at x_d and of the electron charge $N_{tr}.dx_{cr}$ emitted from the traps at x_{cr}. The result is that the N^+_{meas} value determined by expression (13) is larger than that obtained for the situations of Figure 6a and 6b for the same trap concentrations:

ie, $$N^+_{meas}(x) > n(x) \qquad (18)$$

ie, $$N^+_{meas}(x) > (N_d(x) - N_{tr}(x)) \qquad (19),$$

with x equal to x_d, as in expression (14), if, as is usual, the frequency (eg 100kHz–1MHz) at which C_s is being measured is large compared to the emission rate of electrons from the acceptor traps (since then only electrons at x_d are influential upon C_s).

Thus for deep acceptor trap levels close to the conduction band edge (to be exact, levels for which the emission rate of trapped carriers to the band edge is greater than the frequency of the modulation voltage), the experimental depth profile obtained from the C–V measurement is neither the doping concentration N_d, nor the equilibrium carrier concentration (equal to N_d-N_{tr}).

Further detailed consideration (Kimerling 1974) gives the result that, in this situation, the profile $N^+_{meas}(x_d)$ obtained from the C–V data is given by the expression

$$N^+_{meas}(x_d) = N_d(x_d) - N_{tr}(x_d) + N_{tr}(x_{cr})\ (x_{cr}/x_d)\ (dx_{cr}/dx_d) \qquad (20).$$

If the doping concentrations and trap concentrations are each spatially uniform, then dx_{cr} is equal to dx_d, and expression (20) reduces to

$$N^+_{meas}(x_d) = N_d(x_d) - N_{tr}(x_d) + N_{tr}(x_{cr})\ (x_{cr}/x_d) \qquad (21).$$

For use of expression (21), the value of x_{cr} needs to be known, and this depends on the trap energy E_{tr} in accordance with the expression (spatially uniform concentrations again being assumed)

$$x_{cr} = x_d - \left[\frac{2\ \epsilon_r\ \epsilon_o\ (E_F - E_{tr})}{e^2\ (N_d - N_{tr})}\right]^{0.5} \qquad (22).$$

But, as is seen, evaluation of expression (22) for use in (21) requires knowledge of N_d-N_{tr} and E_F (which depends on N_d-N_{tr}), and there is the obvious problem that N_d-N_{tr} cannot be known until expression (21) is evaluated. It is clear that iterative analysis procedures need to be used to obtain full information from C–V data.

However, as described in Section 3.4.3 below, some very helpful simplifications result when one considers $N^+_{meas}(x)$ measured towards large x and at low temperatures.

3.4.3 $N^+_{meas}(x)$ Profiles and the Effect of Measurement Temperature

Expression (21) can be rearranged in the form

$$N^+_{meas}(x_d) = N_d(x_d) - N_{tr}(x_d)\,(x_d - x_{cr})/x_d \tag{23}$$

With change of applied reverse bias V_a, the smallest value of x_{cr} is zero, ie for small x_d, and $N^+_{meas}(x_d)$ then becomes equal to (N_d-N_{tr}); for larger V_a, expression (22) shows that (x_d-x_{cr}) is constant, and therefore $N^+_{meas}(x_d)$ approaches N_d at large x_d.

This is illustrated in Figure 7 by means of an evaluation of expression (23). The calculation is made for n–GaAs containing a donor doping concentration of 1.0×10^{15} cm^{-3} and a small concentration of acceptor traps of energy level E_c–0.25eV; x_d and x_{cr} have been calculated using formulae previously given. The evaluation is made for a measurement temperature of 293K, at which, as is required for the validity of (23), the electron emission rate from the trap would certainly be larger than the assumed modulation frequency of 1kHz. It is seen that for small depletion depths N^+_{meas} is equal to N_d-N_{tr}, but that N^+_{meas} closely approaches the value of N_d for x_d greater than about 5μm. The scale, shown on Figure (7), of applied reverse voltage V_a assumes a built–in potential V_{bi} of 0.7 Volt, and it is valuable to note the typical result that the depletion–region thickness for zero applied bias already takes N^+_{meas} well away from N_d-N_{tr} towards N_d. Thus application of a (small) forward bias across the junction is needed to enable the N_d-N_{tr} value of N^+_{meas} to be attained.

The data line marked '(eg, at 80K)' in Figure 7 assumes a low measurement temperature such that the electron thermal–emission rate from the trap at E_c–0.25eV would be much less than the modulation frequency; in consequence expression (17) is valid, and the value of N^+_{meas} obtained from the C–V data is equal to N_d-N_{tr} at all measured depths.

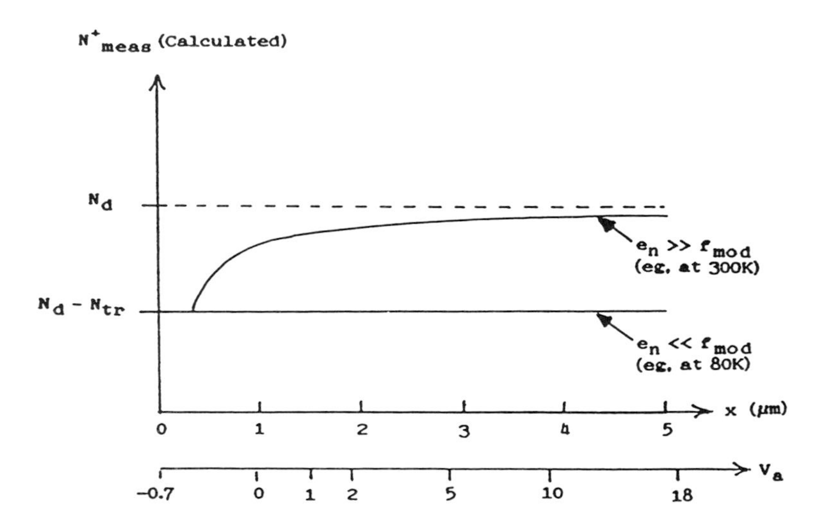

Figure 7. Calculated profiles $N^+_{meas}(x)$ for n–GaAs of doping concentration N_d of 1.0×10^{15} cm^{-3} and containing a small concentration N_{tr} of acceptor electron traps lying at E_c–0.25eV; at 300K, (E_c-E_F) is 0.153eV and (x_d-x_{cr}) is 0.366μm. A value of 0.7 Volt is assumed for the built–in potential V_{bi}.

Figure 7 thus demonstrates that C–V profiling of a semiconductor junction over a reasonable depth interval and over a range of temperatures can allow both chemical–doping concentrations and trap concentrations in the semiconductor to be determined. A major point is that for an n–type semiconductor sample containing a concentration N_d of chemical donors and a total concentration N_{tr} of acceptor traps having a variety of deep energy levels, the equilibrium, 293K, free electron concentration n, equal to $N_d - N_{tr}$, can be determined by the C–V method only by having the sample at low temperature; eg, for gallium arsenide, temperatures as low as 20K are needed for this purpose if defect traps as shallow as E_c–0.05eV are present.

The discussion above has considered acceptor traps in an n–type material. Analogous effects occur for donor traps, but of course these are neutral when occupied by an electron and postive when empty. The result (Kimerling 1974) is that the shapes of $N^+_{meas}(x)$ at high and low temperatures are the same as those of Figure 7 (provided, again, that the chemical doping and trap concentrations are each spatially uniform), but the value of $N^+_{meas}(x)$ reaches N_d for small x at both temperatures (instead of $N_d - N_{tr}$), remains at N_d at all x for measurement at sufficiently low temperature, and approaches $N_d + N_{tr}$ at large x for measurement at sufficiently high temperature.

As an example of the effect of grown–in defects in a semiconductor we refer back to Figure 3. The observed decreases of N^+_{meas} in the epitaxially–grown n–GaAs for the lower measurement temperatures shows, in accordance with Figure 7, the presence of some deep electron traps, their total concentration being of the order of 1×10^{14} cm^{-3}. The spatial oscillations apparent in the 20K data (and slightly visible in the 293K data) suggest the introduction rate of those traps had changed in an oscillatory way with time during the epitaxial growth process, perhaps due to variations of the substrate temperature.

A further set of experimental C–V profiling data, Figure 8, show the effect on N^+_{meas} of measurement at different temperatures when carrier traps of more than one energy level are present. The data are for a VPE n–GaAs layer before and after irradiation by 500 keV protons, which have a range of about 5μm in that semiconductor. The difference at any given spatial depth between the N^+_{meas} values measured at 295K before and after irradiation indicates that the proton irradiation had produced some rather deep traps whose electron emission rates even at 295K were slow compared to the 1kHz modulation frequency in the C–V measurement. Similarly, the difference between the

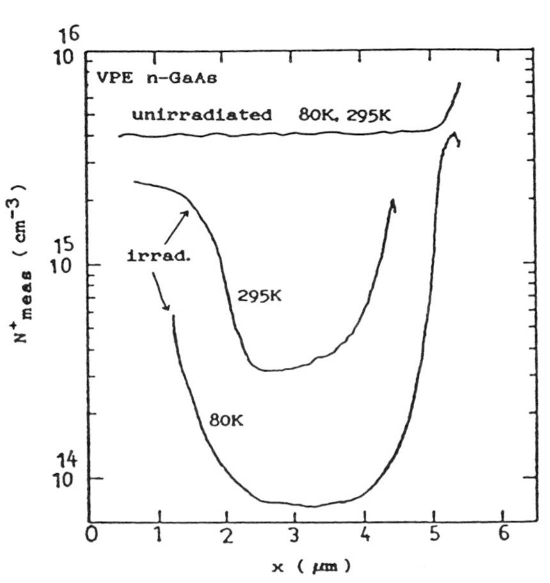

Figure 8.

$N^+_{meas}(x)$ profiles of n–GaAs measured at 295K and 80K, before and after irradiation at 295K by 500 keV protons to a dose of 6×10^{11} cm^{-2}; (Rezazadeh 1983).

N^+_{meas} profiles obtained at 295K and 80K on the irradiated material demonstrates the presence there of other, shallower, electron traps whose electron emission rates were high at 295K but low (less than 1kHz) at 80K.

Changes in N^+_{meas} profiles with temperature, such as those of Figures 3, 7, and 8, can be considered less quantitatively as what is called "thermally-stimulated capacitance". It is clear from the relationship between N^+_{meas} and the measured capacitance C_s that observed reductions of N^+_{meas} as the measurement temperature is decreased are associated with decreases of C_s at lower temperatures. Therefore experimental $C_s(T)$ data can often give a quick indication of the presence of deep carrier traps in the semiconductor material.

3.4.4 Photo-Capacitance Measurements

The method of studying carrier traps by measurements of photon-induced changes of capacitance of a reverse-biased junction should be mentioned at this stage. The principle is that if the junction is cooled to a low enough temperature (80K or less, depending on the energy levels of the traps present) in the dark under zero or low reverse bias, the traps stay occupied even when the reverse bias is applied or increased, since the traps cannot emit their trapped electrons or holes at the low temperature (assuming the absence of significant electric-field-assisted emission). If now, with the semiconductor maintained cold, the reverse-biased junction region is illuminated with photons, the photon energy being progressively increased from about 0.05eV or lower upto the band-gap energy of the semiconductor, the effect will be that the traps will be successively emptied, thus causing changes in the reverse-bias capacitance, as the photon energy is scanned through the appropriate energy difference E_c-E_{tr} or $E_{tr}-E_v$ (for electron and hole traps respectively). Thus information on the presence of traps at various energy levels can be obtained, but determination of the concentrations of the traps is difficult since the optical excitation cross-sections of the traps are often not known. Photo-capacitance studies have been made by Hughes 1972 on GaAs, and by Kukimoto et al 1973 and Hamilton 1974 on GaP.

Depth profiling of the concentrations of the individual trap can be accomplished by carrying out a series of photocapacitance measurements for different reverse biases or reverse bias intervals, but, for the reason stated in the previous paragraph, comparisons of the concentrations of different traps cannot usually be made.

4. DEEP LEVEL TRANSIENT SPECTROSCOPY (DLTS)

4.1 Majority-Carrier Capacitance-Mode DLTS

4.1.1 Capacitance Transients

As has been shown in Figures 3 and 8, as experimental examples, and in Figure 7 as a calculated example, the concentrations N^+_{meas} obtained from C-V measurements for different temperatures of a reverse-biased semiconductor junction can give quantitative information on the presence of deep carrier traps in the semiconductor material. But, as can be seen in Figure 3, trap concentrations less than a few per cent of the chemical doping concentration are difficult to observe by that method. What is needed is a technique that is much more sensitive to the presence of carrier traps; Deep Level Transient Spectroscopy (Lang 1974, Lang and Kimerling 1975, Miller et al 1977) is the powerful method that serves that purpose, its sensitivity being as good as 10^{-4} of the doping concentration under optimum conditions.

DLTS depends on the principle, applying also to the C-V profiling method, that thermally excited emptying of electrons or holes from traps in a semiconductor junction held in reverse bias leads to movement of those released charges into the external circuit. The number of electrons or holes released from the traps per second depends on the temperature, and decays exponentially with time as the traps become progressively empty; measurement of the decay time as a function of temperature allows the transition energy, E_c-E_{tr} in the case of an electron trap (Figure 6) or $E_{tr}-E_v$ in the case of a hole trap, to be determined. The release of the trapped charges can be observed experimentally as a thermally-induced transient current from the junction or as a transient change in the junction capacitance. The major particular and crucial further feature of the DLTS technique is that the filling and thermal emptying of the traps is arranged to be repetitive, and this permits very sensitive experimental analysis of the output current or capacitance transients by means of electronic averaging methods. It turns out for both practical and theoretical reasons that current-transient DLTS is generally less useful and informative than capacitance-transient DLTS, and it is therefore the latter that will be described in some detail here. This Section, 4.1, treats majority-carrier traps, and as in previous sections of this paper, acceptor electron traps in a one-sided (p^+-n or metal-n) n-type semiconductor will be considered as the example.

Figure 9 shows the states of the deep traps and the corresponding capacitances at various stages during one trap-filling and trap-emptying cycle of a DLTS experiment. The temperature of the semiconductor is assumed to be high enough that the time for release of electrons from the traps is fairly short (eg a release time-constant of 10ms). As we shall see a little later, Figure 9 is somewhat simplified, but it suffices to show the essential origin of the capacitance transient. In Figure 9a, a reverse bias V_{a1} is being applied across the junction (see also Figure 10(a)), and the depletion region width is x_{d1}. At equilibrium, the traps at $x > x_{d1}$ are filled because of the high density of free electrons in that neutral region, but those within the depletion region ($x < x_{d1}$) are empty and the capacitance C_1 of the junction is equal to $\epsilon_r\epsilon_0 A/x_{d1}$. At a time that we define as zero, the applied reverse bias is suddenly increased to V_{a2} (Figure 9b and 10(a)); the effect of this is that the depletion width increases immediately to x'_{d2}, with the consequence that the measured capacitance is taken down to C'_2. However the electrons on the traps in the spatial region between x_{d1} and x'_{d2} are now no longer at thermal equilibrium (because the free electrons have been removed from there), and thermally induced electron-emission from the traps to the conduction band begins to occur. Because this causes the space charge in the new part of the depletion region to increase from $e(N_d-N_{tr})$ towards eN_d, the capacitance progressively increases (in accordance with the discussions in Sections 2 and 3) to the value C_2 that is the equilibrium capacitance for the applied bias V_{a2} (Figure 9c).

The overall effect is that a capacitance transient $\Delta C(t)$ has been produced (Figure 9c), of time dependence that is characteristic of the release of the electrons from the traps, and we can write

$$\Delta C(t) = C_2 - C(t) = \Delta C(t{=}0)\exp(-t/\tau_e) \quad (24).$$

with $$\tau_e = (e_n)^{-1} \quad (25)$$

where e_n is the rate of electron emission from the trap, which, in accordance with standard semiconductor theory (principle of detailed balance), is given by

$$e_n = \sigma_n v_{th} N_c \exp(-(E_c-E_{tr})/k_BT) \quad (26),$$

in which σ_n is the electron-capture cross-section of the traps, v_{th} is the mean velocity of conduction band elecrons, and N_c is the density of electronic states in the conduction band.

It is clear from the previous paragraph that $\Delta C(t{=}0)$ is larger for greater trap concentrations. The exact relationship between $\Delta C(t{=}0)$ and N_{tr} depends on the magnitude of N_{tr} compared with N_d and also on whether the trap concentration is

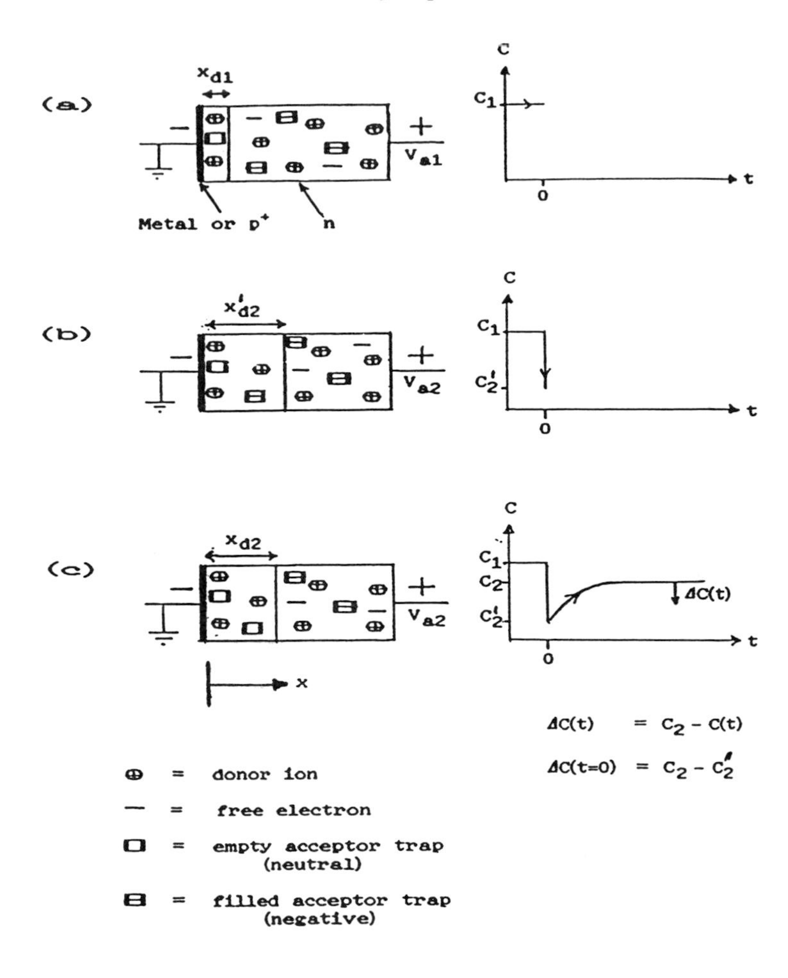

Figure 9. Variation with time of acceptor-electron-trap emptying and of the capacitance of a one-sided p^+-n or metal - n-semiconductor junction following increase of the applied reverse bias from V_{a1} to V_{a2}.

spatially uniform. The simplest case, that of a uniform trap concentration that is much smaller than N_d (eg 10^{-2} times N_d or less) can be treated in the following way. We return to expression (7) and write it for the equilibrium condition of Figure 11c under applied bias of V_{a2} as

$$C_2 = A\ (\epsilon_r\ \epsilon_0\ e\ /\ (2\ (V_{a2}+V_{bi}))\ )^{0.5}\ N_d^{0.5} \qquad (27).$$

Therefore

$$dC_2/dN_d = 0.5\ A\ (\epsilon_r\ \epsilon_0\ e\ /\ (2\ (V_{a2}+V_{bi}))\ )^{0.5}\ N_d^{-0.5} \qquad (28)$$

and thus

$$dC_2/C_2 = 0.5\ dN_d/N_d \qquad (29).$$

We can now consider dN_d as representing the trap concentration N_{tr} and dC_2 as being the total capacitance transient $\Delta C(t{=}0)$.

Therefore we reach the expression

$$\Delta C(t{=}0)/C_2 = 0.5\ N_{tr}/N_d \qquad (30),$$

and this is an important (but approximate) equation relating the capacitance transient of a DLTS experiment to the trap concentration. We return in Section 4.1.5 to a more quantitative examination of the relationship between $\Delta C(t{=}0)$ and N_{tr}.

4.1.2 Obtaining a DLTS Defect Spectrum

Figure 10(a) shows the sequence of applied biases employed during each cycle of a DLTS measurement, and, following from Figure 9 and expressions (24) to (26), Figure 10(b) shows the forms of the capacitance transients expected at three different temperatures; for the smaller applied bias V_{a1}, the capacitance is large and equal to C_1 (Figure 9(a)), and when the larger bias, V_{a2}, is applied, slow capacitance transients are produced at low temperatures and fast transients at high temperatures. If a capacitance meter such as the Boonton 72B (Section 3.2.2), which has a response time of 1ms, is used for measuring the capacitance, then the time scale shown on Figure 10 would typically extend upto about 100–200ms. For any particular trap, characterised by its values of σ_n and E_c-E_{tr}, expressions (24) to (26) define the time constant of the transient as a function of temperature.

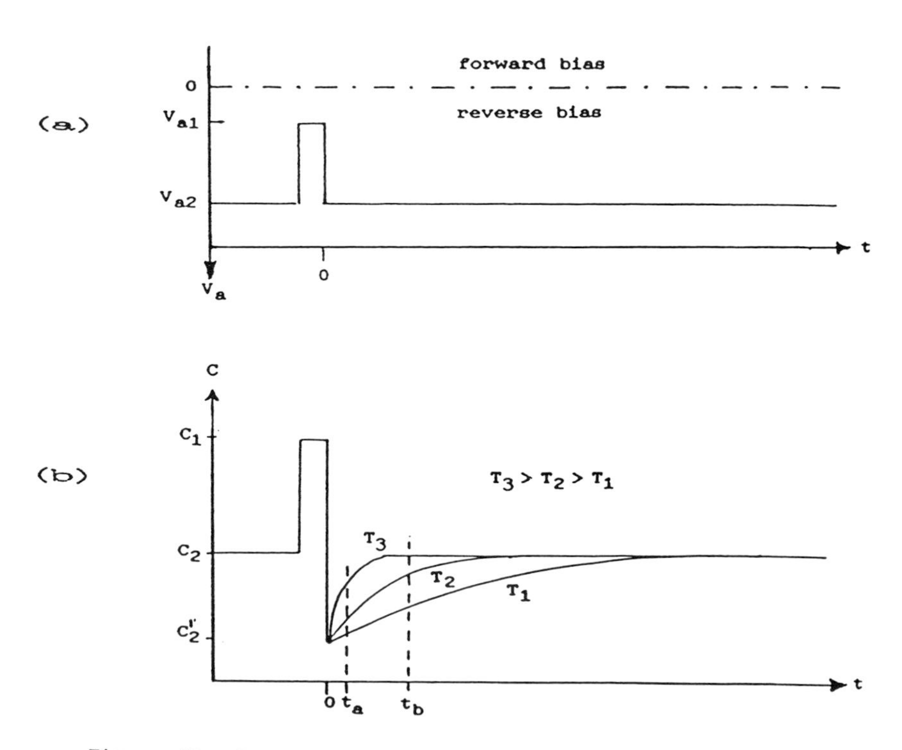

Figure 10. For each cycle of a DLTS measurement: (a), the applied bias, and (b), the capacitance transients that would occur at three temperatures where $T_3 > T_2 > T_1$. The times t_a and t_b are explained in Section 4.1.3.

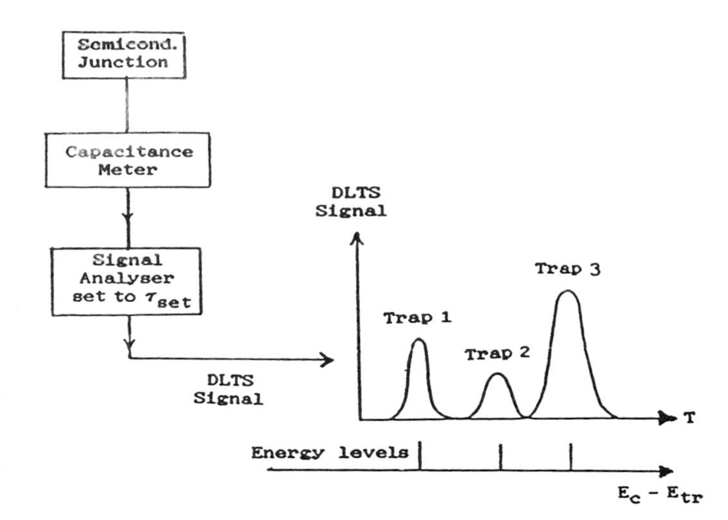

Figure 11.
Principle of the use of a Signal Analyser sensitive to a time-constant τ_{set} to obtain a DLTS trap spectrum.

In the DLTS method, a voltage proportional to $\Delta C(t)$ is fed into a circuit that I shall call the Signal Analyser (Figure 11), that is set to be sensitive especially to a chosen time-constant τ_{set} (or "rate window" τ_{set}^{-1}), and capacitance transients are produced from the semiconductor junction reverse-biased at V_{a2} (Figures 9 and 10) while the temperature of the junction is scanned over a range of temperatures, usually from low to high temperature. The result is that the signal analyser gives an output only near and at the temperature at which the trap emission time is equal to τ_{set} of the signal analyser. The signal analyser can be considered as a kind of band-pass electronic filter, but special in the sense that its internal circuit is designed to examine a signal decaying exponentially with time rather than a sinusoidal oscillation.

If there are several kinds of majority carrier trap present in the semiconductor material being assessed, the signal analyser gives an output at each temperature at which the emission time from a trap is equal to the signal analyser set time τ_{set}. Thus a DLTS spectrum (Figure 11) is produced in which each peak corresponds to a particular trap. Because of the strong temperature dependence of e_n upon E_c-E_{tr}, as shown in expression (26), because v_{th} and N_c have much weaker dependences on temperature, and because the cross sections σ_n for different traps are not often vastly different from each other, it is almost always the case that the DLTS-spectrum peaks at lower temperatures are those which have the smaller values of E_c-E_{tr} (or of $E_{tr}-E_v$ in investigations of hole traps in p-type semiconductors). Thus a DLTS temperature spectrum (Figure 11) can usually be interpreted as a spectrum of defect energy levels, with increasing trap emission from left (low temperature) to right on the spectrum.

Quantitatively, in terms of expression (26), putting e_n equal to τ_{set}^{-1} for each DLTS peak, the energy level E_{tr} of the trap is given by

$$E_c-E_{tr} = k_BT \log_e(\tau_{set}\sigma_n v_{th} N_c) \qquad (31).$$

4.1.3 DLTS Circuitry and Equipment

A block diagram of the electronic equipment arrangement needed for typical DLTS studies is shown in Figure 12. It is fairly straightforward to set up such a system, but the matters of accurate measurement of sample temperature and of the reduction of electronic noise need to be carefully considered.

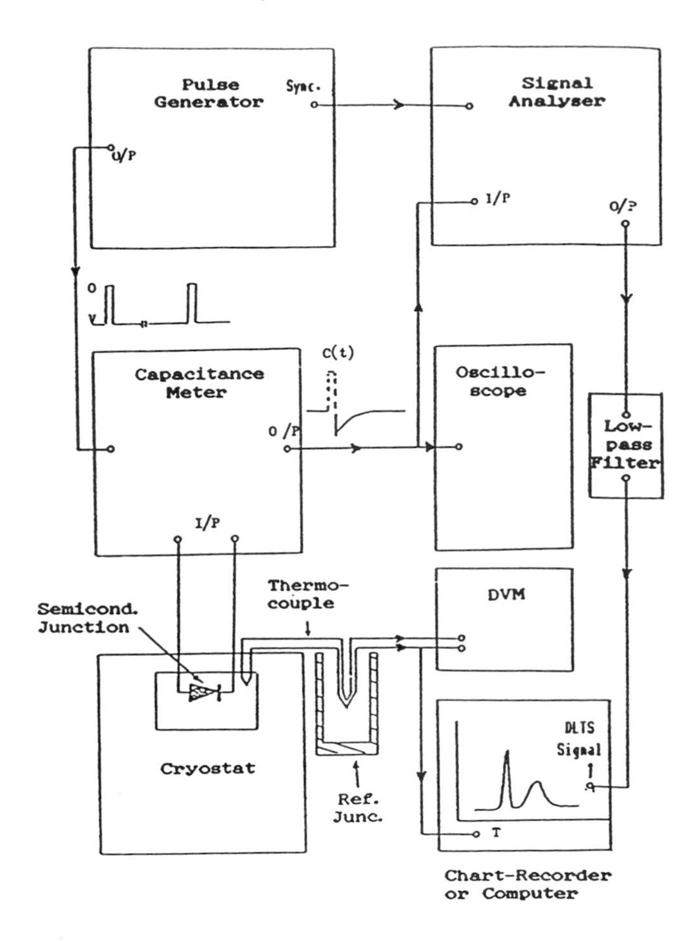

Figure 12.

Method of practical implementation of the DLTS principle given in Figure 11 (from Rezazadeh 1983).

The system needs to provide measurement of the sample temperature (an accuracy of 0.5K or better is required), and to allow its change from low temperature (80K or perhaps from about 20K) to room temperature or slightly higher in times of about 15 minutes to more than an hour; the rate of temperature rise has to be low enough to ensure that the sample held in the cryostat is at the same temperature as the thermo-couple or other device which is giving the temperature scale on the DLTS spectrum.

The pulse generator of Figure 12 produces the trap-filling and trap-emptying cycles, as described in section 4.1.1; the voltages produced by the pulse generator across the semiconductor junction and the corresponding capacitance transients are repetitive cycles of those that have been shown in Figure 10. The required durations of the filling pulses (ie the periods of reduced reverse bias, V_{a1}) depend on the carrier-trapping cross-sections of the traps being investigated; usually filling pulse durations of 100μs are sufficiently long, but for some known traps in semiconductors filling times of several ms are needed.

The operating principle of a typical ("box-car" type) DLTS signal analyser depends on the use of two externally settable times t_a and t_b (see Figure 10(b)) at which the incoming signal voltage $SV_C(t)$, proportional to the capacitance transient $\Delta C(t)$, is measured; the signal analyser circuit produces and outputs the difference $SV_C(t_b)-SV_C(t_a)$ of the voltages at the times t_b and t_a. It can be seen from Figure 10(b) that, both for very slow transients (semiconductor at low temperature, T_1) and for very fast transients (semiconductor at high temperature, T_3), that voltage difference is small, but that for transients (eg, for temperature of about T_2) having time constants τ_e about equal to t_b-t_a that voltage difference, and therefore the signal analyser output, will be large. Quantitative consideration (Lang 1974) gives the result

$$\tau_{set} = (t_b-t_a) \;/\; \log_e(t_b-t_a) \qquad (32)$$

for the effective time constant of such a signal analyser. A very useful circuit implementing this signal-analysing procedure has been given by Guldberg 1977.

Then (Figure 12), the voltages from the temperature transducer and the corresponding output voltages from the signal analyer are recorded by appropriate means (an X-Y chart recorder or, more likely at the present time, a computer-based data acquisition system) and the DLTS spectrum is obtained as the plot of the signal analyser output as a function of the temperature. Averaging of the signal analyser outputs over a number (eg, 10-100) of capacitance transient cycles is almost always useful so as to allow low amplitude transients (corresponding to small trap concentrations) to be analysed in the inevitable presence of electronic noise, and this averaging is a very important feature of the DLTS technique. The averaging can be accomplished either by means of a simple RC smoothing circuit between the signal analyser and the data aquisition unit (as shown in Figure 12) or by software-program averaging if a computer is used to store the data.

A well known commercial computer-based DLTS system is the DL4600 of Bio-Rad Ltd. This has a variety of features, including the use of three, rather than two, voltage-sampling times in the signal analyser. This allows two DLTS spectra (for two different values of τ_{set}) to be obtained in a single temperature scan, and this means that some information on the trap energy levels can be obtained in that one scan.

A crucial assumption in the whole principle for obtaining DLTS spectra is that the capacitance transients of the form shown in Figures 9(c) and 10(b) correspond accurately to exponentially decaying curves of the form (24)

$$\Delta C(t) = \Delta C(t{=}0)\ \exp(-t/\tau_e) \qquad (33),$$

and it is important to check the validity of this assumption for each trap under the given experimental conditions. For this purpose at the University of Sussex we use a Datalab DL910 Transient Recorder linked to a BBC Model B computer to measure in detail the capacitance of the semiconductor junction as a function of time (typically at 1000 sampling times in the interval -20ms to +100ms) at various temperatures for each carrier trap. The data are analysed by a computer program to provide, in accordance with the theoretical expression (33), a plot of $\log_e(\Delta C(t)/\Delta C(t{=}0))$ as a function of t; a near-straight-line plot demonstrates the validity of (33) for the trap being studied, and also provides an accurate value for the carrier-emission time τ_e at the particular temperature.

Before leaving this outline of DLTS circuitry, it is worth noting that for some kinds of semiconductor structure, conductance transients instead of capacitance transients can be sensitive to release of carriers from deep-level traps, and then DLTS spectra can be obtained using those conductance transients; for example, Maracas et al 1984 have by this means studied interface states in $GaAs/Al_xGa_{1-x}As$ layer structures. This DLTS method has been very much less used than the capacitance method, but can give valuable information on particular kinds of trap.

4.1.4 Examples of DLTS Spectra

Figures 13 for silicon and Figures 14 and 15 for gallium arsenide demonstrate the general nature of capacitance-mode DLTS spectra that can be obtained by the means described above.

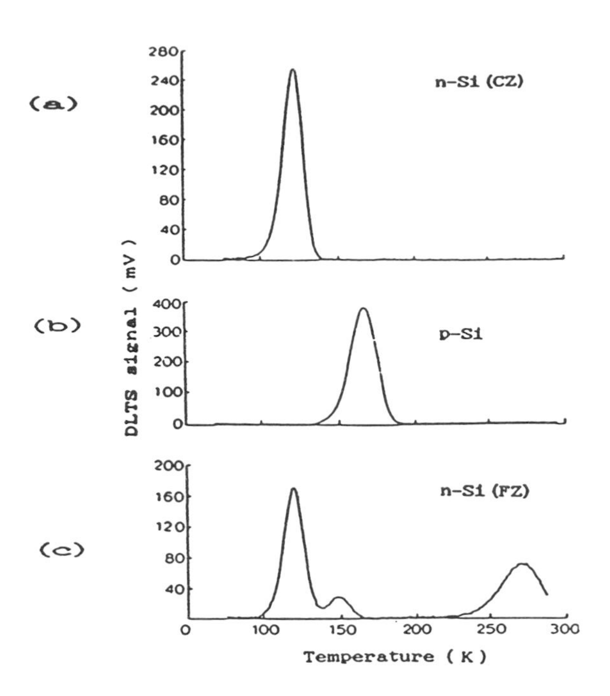

Figure 13. Majority-carrier DLTS spectra (τ_{set} = 14ms) for silicon into which platinum had been intentionally diffused: (a) and (c), Czochralski and Float-Zone n-Si, respectively; (b) p-Si. The main peak in each spectrum is identified with a carrier trap due to platinum. (Brotherton et al 1979).

In Figure 13 (Brotherton et al 1979) are shown majority-carrier DLTS spectra obtained for silicon in an investigation of carrier trapping at platinum atoms (these having been intentionally diffused into the silicon for the study). Figure 13a is a spectrum for n-Si, and so the peak corresponds to excitation of electrons from the platinum atom traps to the conduction band. Figure 13b is a spectrum observed for p-Si, and in this case, therefore, the DLTS peak is that for removal of holes from platinum atoms to the valence band. The results showed that platinum produces at least two energy levels in the silicon band-gap. Figure 13c is again a spectrum for n-Si, but here for float-zone silicon whereas Figure 13a was for pull-grown (Czochralski) silicon; in this case two extra peaks are seen, at about 150K and 270K, of which that at 150K was due to a processing-induced defect and the other to unintenional gold contamination.

A DLTS spectrum for n-GaAs grown by Vapour-Phase Epitaxy is given in Figure 14 (Rezazadeh and Palmer 1981) and this can be compared with the lower spectrum of Figure 15 (Nandhra 1986) which is for n-GaAs grown by Molecular Beam Epitaxy. It can be seen that the VPE and MBE growth methods produce different sets of electron traps in GaAs. The upper spectrum of Figure 15 is that obtained after proton irradiation of the MBE-grown n-GaAs, and a further set of electron-trapping, levels designated E1, E2 and E3, has now appeared.

DLTS measurements on semiconductors allow a wide range of defect and impurity carrier traps, originating in the growth or processing stages, to be investigated.

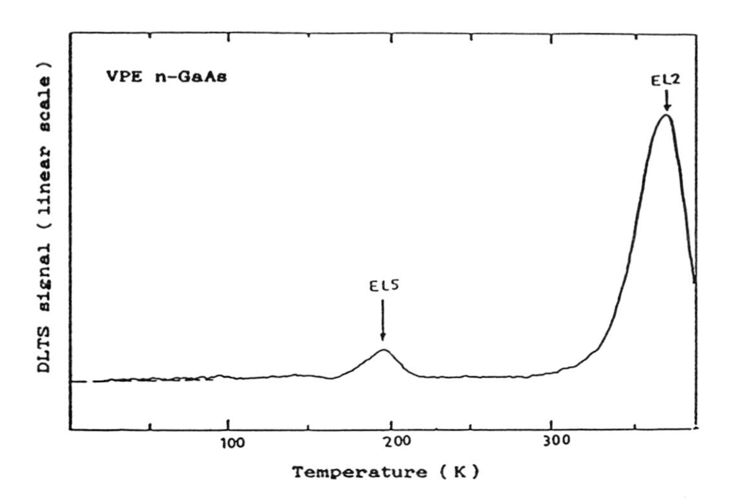

Figure 14. Majority-carrier DLTS spectrum ($\tau_{set} = 50ms$) for VPE n-GaAs, showing the presence of grown-in traps EL5 and EL2 (Rezazadeh and Palmer 1981).

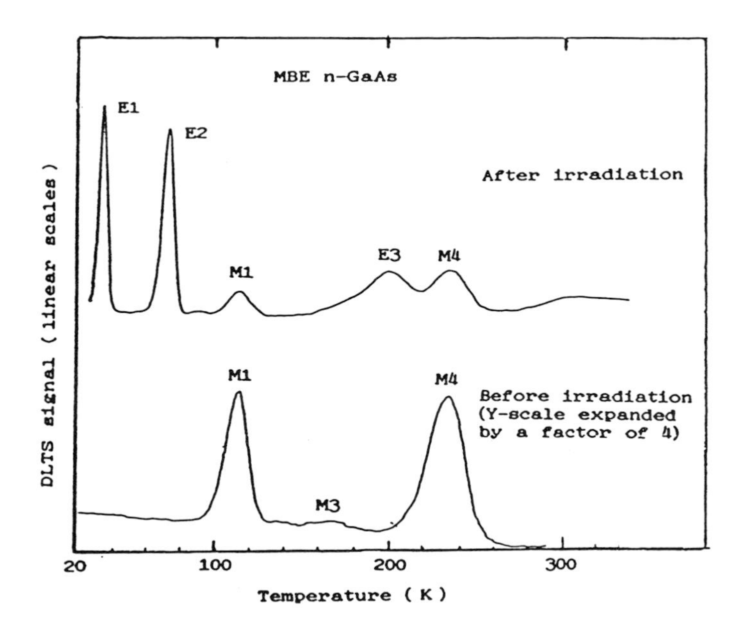

Figure 15. Majority-carrier DLTS spectra ($\tau_{set} = 17ms$) for MBE n-GaAs before and after irradiation with 1.0 MeV protons at 293K to a dose of 1.86×10^{12} $H^{+}cm^{-2}$. The M1, M3 and M4 levels are electron traps very commonly found in GaAs grown by Molecular Beam Epitaxy. The E1, E2, E3 and E4 levels are electron traps due to irradiation-induced defects. (Nandhra 1986).

4.1.5 Determination of Trap Energy Levels and of Carrier Capture Cross Sections

The method for experimental determination of the energy level of a carrier trap is based on expression (31), itself arising from (26). The principle is that the DLTS peak for any given trap will appear at different temperatures on the DLTS spectrum for different chosen values of the time constant τ_{set} of the DLTS signal analyser. Analysis of the change of peak temperature with τ_{set} allows E_c-E_{tr} and σ_n (for electron traps) or $E_{tr}-E_v$ and σ_p (for hole traps), to be measured.

An example of this method is given in Figure 16 (from Rezazadeh 1983) for the electron-trapping defect called E3 that is produced in GaAs by irradiation. The DLTS spectra shown are those obtained for five values of τ_{set} in the range 10 to 50ms; higher temperatures correspond to larger electron emission rates and therefore to shorter emission time constants.

From semiconductor theory, the free-electron thermal velocity v_{th} and the conduction-band effective density of states N_c, of expression (26) or (31), are proportional to $T^{1/2}$ and $T^{3/2}$ respectively. Therefore an Arrhenius-type plot of $\log_e(e_n/T^2)$ as a

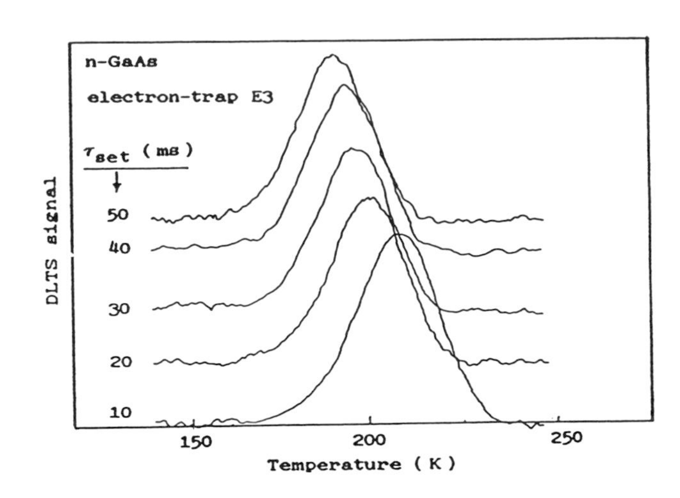

Figure 16. Experimental data showing how the temperature of the DLTS peak for the E3 electron trap in n-GaAs depends on the signal analyser time constant τ_{set} (Rezazadeh 1983).

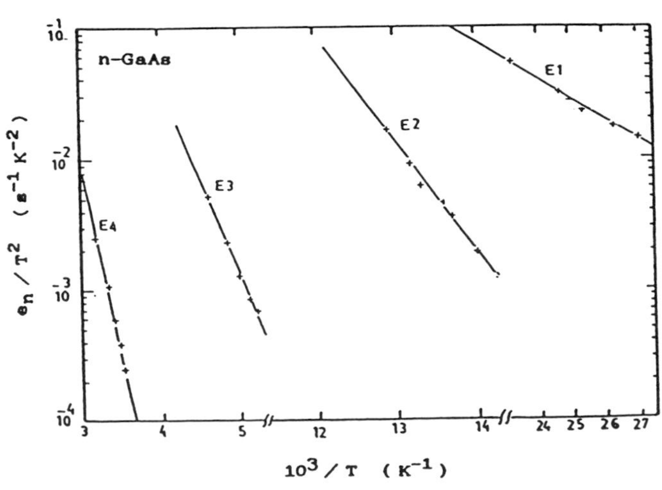

Figure 17. Arrhenius plots from DLTS data, like those of Figure 16, for the E1, E2, E3 and E4 electron traps in n-GaAs (Rezazadeh 1983).

function of T^{-1} is expected to be a straight line having a slope equal to $(E_c-E_{tr})/k_B$ and an intercept on the e_n/T^2 axis $(T^{-1}=0)$ proportional to σ_n. Arrhenius plots of this kind from the Figure 16 data for the E3 trap and from other analagous data for the E1, E2 and E4 traps also in irradiated n-GaAs are given in Figure 17. The good straight lines demonstrate the validity of the analysis. The values of E_c-E_{tr} and σ_n for the E3 trap were found to be 0.31±0.01eV and $(2.5\pm0.5)\times10^{-16}cm^2$ respectively (Rezazadeh 1983).

Under optimum conditions, values of E_c-E_{tr} or of $E_{tr}-E_v$ can be determined with imprecisions of about 1 to a few %.

4.1.6 Determination of Trap Concentrations

From expression (30) previously given, which arose from Figure 9, the approximate concentration N_{tr} of each particular carrier trap can be found from its capacitance transient $\Delta C(t=0)$, measured in a DLTS experiment, by means of the formula

$$N_{tr} = 2\,N_d \frac{\Delta C(t=0)}{C_2} \tag{34}$$

where C_2 is the junction capacitance at equilibrium at the reverse bias V_{a2} applied during the phase of carrier emission from the trap. The reason why expressions (30) and (34) are only approximate is now explained, and a formula for determining N_{tr} that is valid for many practical assessments of semiconductor materials is given.

Figure 18 shows in the usual way the energy band bending for an n-type semiconductor for two different applied reverse biases V_{a1} and V_{a2}, where V_{a1} is smaller than V_{a2}. These are to be considered, respectively, as the filling condition (V_{a1}) for an acceptor trap whose energy level E_{tr} is indicated, and the emission condition (V_{a2}) for that trap. The change between (a) and (b) of Figure 18 for DLTS measurement is similar to that between (c) and (c') of Figure 6 for C(V) profiling, except that Figure 18 involves a relatively much larger increase in applied reverse bias. As in Figure 6, the energy level of the trap indicated in Figure 18 crosses the Fermi level E_F at an x value called x_{cr}, and at equilibrium the traps are occupied by electrons for x values less than x_{cr}; the value of x_{cr} is x_{cr1} when V_{a1} is applied and is x_{cr2} when V_{a2} is applied.

Upon change of the applied bias from V_{a1} to V_{a2} the traps in the spatial interval $x_{cr2}-x_{cr1}$ become above E_F and these begin to emit their trapped electrons. This causes a capacitance transient as has been explained in Section 4.1.1 with reference to Figure 9; but Figure 18 shows that the correct spatial interval of electron emission from the traps is in fact $x_{cr2}-x_{cr1}$ rather than the interval $x_{d2}-x_{d1}$ that was assumed in Figure 9. The effect of this is that for a given trap concentration N_{tr} the true value of $\Delta C(t=0)$ is smaller than given by expression (30); in consequence, there has to be a correction factor greater than unity in the numerator of the RHS of expression (34).

The correct expression replacing (34) can be found from the analysis given by Lefevre and Schulz 1977, and is as follows

$$N_{tr} = 2\,N_d \frac{\Delta C(t=0)}{C_2} \left[\frac{x_{d2}^2}{x_{cr2}^2 - x_{cr1}^2} \right] \tag{35}$$

ie, a correction factor of $(x^2_{d2} / (x^2_{cr2}-x^2_{cr1}))$ is applied to the RHS of expression (34).

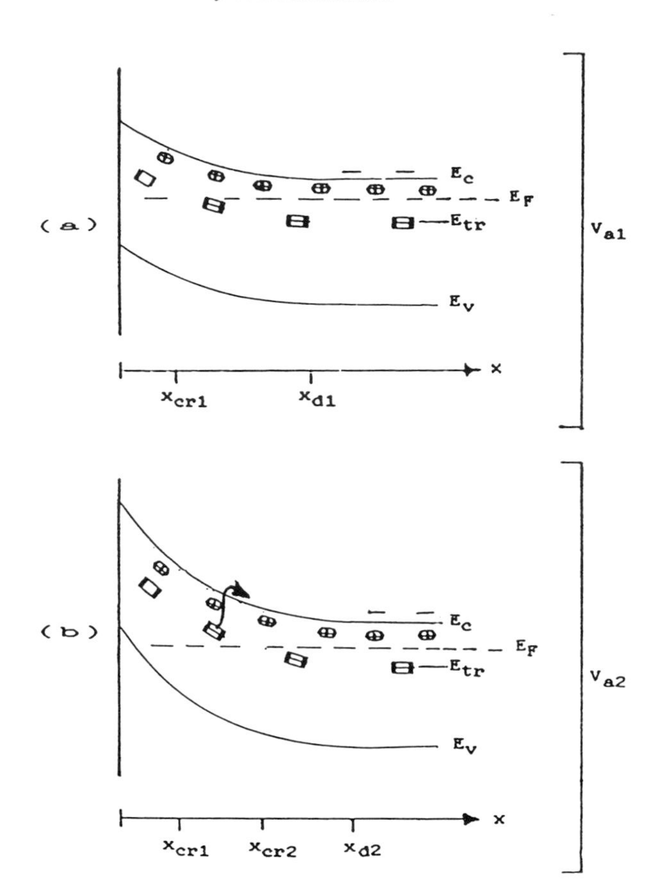

Figure 18.

The thermal emission of electrons from traps, when the applied bias is increased from V_{a1} to V_{a2}, that needs to be taken into account in a DLTS experiment in relating the observed transient capacitance amplitude to the trap concentration: the electrons released from traps in the interval x_{cr1} to x_{cr2} contribute to the observed transient.

For use of (35), x_{d2} can be determined directly from the measured capacitance $C_s(V_{a2})$ via (7), and the two values of x_{cr} are to be found from (22) (valid for uniform doping and trap concentrations). But as has been discussed in Section 3.4.2, evaluation of expression (22) requires knowledge of the trap concentration N_{tr}, which it is of course the aim of the DLTS measurement to determine. Therefore, as in use of expression (21), an iterative procedure is required for full evaluation of (35), and this definitely makes the whole analysis rather complicated, especially if more than one deep trap is present.

A helpful practical procedure is to choose the experimental conditions so as to ensure that the correction factor is as close as possible to unity and is therefore almost independent of x_{cr1} and x_{cr2}; this permits effective use of an estimated value of N_{tr} on the RHS of (22). A correction factor value near to unity can be achieved by making x_{d1} as small as conveniently possible (eg by having V_{a1} equal to zero) and by making x_{d2} large (eg, by setting V_{a2} as close as possible, consistent with stable behaviour, to the reverse-bias breakdown voltage of the junction). For example, for the trap E3 at E_c–0.3eV in a sample of n–GaAs having a majority carrier concentration of 1.0×10^{15} cm^{-3} at the temperature of 200K at which the trap can give a DLTS peak in a typical measurement, the value of the correction factor is 3.90 if the applied reverse biases are 1.0 and 2.0 Volts for trap filling and emptying respectively, but is only 1.37 if the biases used are zero and 10 Volts.

Finally concerning determination of trap concentrations, it should be emphasized that because the derivation of expression (30) involved a mathematical differentiation, even the expression (35) is valid only for the condition that the trap concentration is small compared with the doping concentration. If that condition is not fulfilled, an expression involving the squares of the relevant capacitances has to be used (Lefevre and Schulz 1977). Then, also, the use of a constant-capacitance DLTS method (Johnson et al 1979) can be valuable. In this potentially useful, but sometimes operationally difficult, variation of the standard DLTS system, the input to the signal analyser is arranged to be the reverse voltage needed to keep the capacitance constant as the traps emit their electrons or holes; the constancy of the capacitance ensures that the spatial region in which the traps are being examined does not change during the carrier-emission process.

4.2 Minority-Carrier Capacitance Mode DLTS (MCTS)

4.2.1 Introduction

Section 4.1 has described the principles and practice of majority-carrier DLTS. In that DLTS method, by keeping the semiconductor junction continuously under reverse bias but repetitively changing the magnitude of the reverse bias potential, electron-capturing traps in n-type material and hole-capturing traps in p-type material can be cyclically filled and emptied, and their energy levels, carrier cross sections and concentrations can be measured. However, since for many applications of semiconductors the presence or absence of minority carrier traps is crucial to the performance of the semiconductor device, it is important to know how Deep Level Transient Spectroscopy can be applied also to the study of such traps.

The DLTS technique for minority carrier traps is very similar in principle and practical procedure to that for majority traps, the difference being that the trap-filling stage has to be such that minority traps are filled. This minority carrier filling can be achieved either electrically (Lang 1974, Lang and Kimerling 1975, Auret and Nel 1987), or optically (Brunwin et al 1979, Hamilton et al 1979). The acronym MCTS, meaning Minority Carrier Transient Spectroscopy, is used to designate either of these DLTS methods.

4.2.2 MCTS using Electrical-Pulse Filling

The electrical-filling MCTS method normally uses a p^+-n semiconductor junction for investigating hole traps in n-type material or an n^+-p junction for studies of electron traps in p-type material. The description here will be given in terms of hole traps in an n-type semiconductor.

For such MCTS measurements, Figure 19, the experimental arrangement is as has been shown in Figures 10a, 11 and 12 for majority carrier DLTS, except that during the filling-pulse stage the junction is taken somewhat into forward bias. The effect is that forward-bias current flows at that time, producing an injection of holes from the p^+ part of the junction into the n-type material where capture of the holes can occur at any hole traps present.

At the end of the hole-trap filling stage, the junction is returned to the reverse bias V_{a2} (Figure 19a). But because of the captured holes on traps in the n-material the positive space charge per unit volume in the n-material is greater than it would be if there were no holes captured on traps, this being true whether the traps are (in electron terms) donors or acceptors. Consequently the capacitance of the junction is larger than it would be in the absence of the trapped holes. But, since now the concentration of free holes is very low, the traps will, if the temperature is sufficiently high, immediately begin to emit their trapped holes to the valence band, and the junction capacitance decreases as shown in Figure 19b.

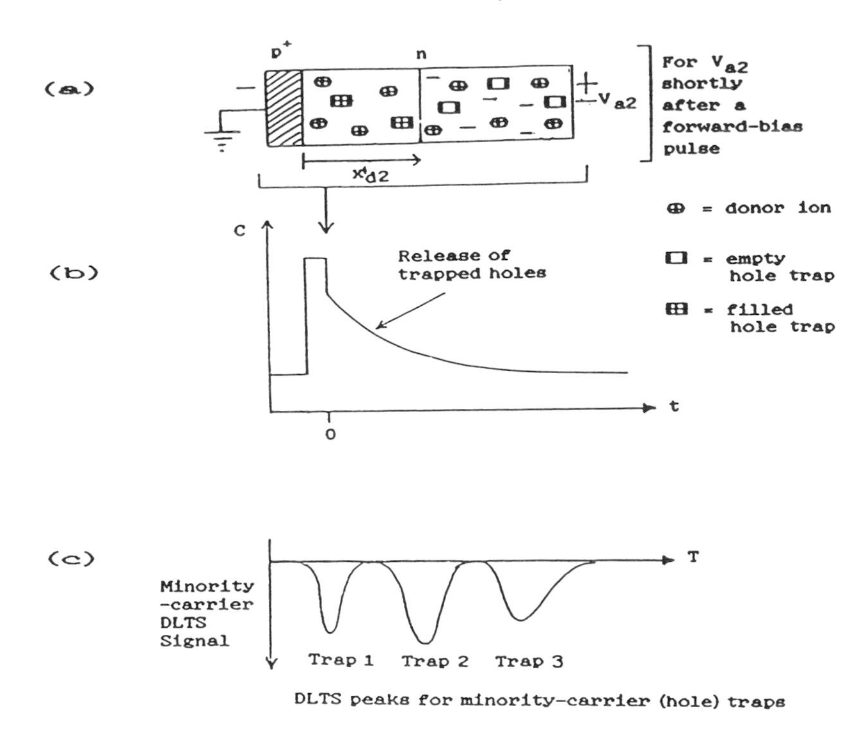

Figure 19. Minority-carrier DLTS (MCTS):
(a) Holes trapped in the reverse-bias depletion region x_{d2} after a forward-bias hole-injection pulse.
(b) Transient capacitance due to release of the trapped holes.
(c) Minority-carrier spectrum if the signal analyser produces a negative output for a transient that decreases with time.

The time dependence of the capacitance during the emission of the holes from the traps is expected to be exponential in form, and the usual type of signal analyser, involving sampling times t_a and t_b (Figure 10b), can be used provided that its circuitry can deal with an input signal that decreases as a function of time. Usually this is so, and it is convenient if the signal analyser then produces an output voltage that is of polarity opposite to that produced by the rising capacitance due to a majority-carrier trap (Figure 10b). In this situation the output of the signal analyser as a function of the temperature of the semiconductor junction, ie the MCTS spectrum, is as shown in Figure 19c, in which each (negative - going) peak corresponds to an individual minority-carrier trapping level.

If, in addition to the minority carrier traps, the semiconductor material contains some majority-carrier traps, the latter also become occupied during the filling phase of the DLTS cycle, and the capacitance transient will be a decreasing function of time (emission of minority carriers) at some temperatures and a rising function of time (emission of majority carriers) at (usually) other temperatures. If the signal analyser responds with different output polarities to the two kinds of transient, the DLTS spectrum can then show both negative and positive peaks due to minority-carrier and majority-carrier traps respectively.

As has been mentioned above, electrical-pulsing MCTS is usually carried out using p^+-n or n^+-p semiconductor junctions. It has been pointed out, however, by Auret and

Nel 1987 that forward biasing of metal–semiconductor Schottky junctions can inject significant concentrations of minority carriers into the semiconductor, for capture by minority–carrier traps, provided that the Schottky barrier is high and that large forward currents are used. They have effectively applied this method, using gold/silicon Schottky structures, to the study of hole traps in n–silicon.

4.2.3 MCTS using Optical Pulsing

For Minority–Carrier Transient Spectroscopy using optically induced filling of the minority–carrier traps, the semiconductor junction, usually a Schottky junction, is maintained at a constant reverse bias, and the sample is repetitively illuminated, usually through the Schottky metal layer, by light of photon energy just larger than the band–gap of the semiconductor. Such light produces free holes and free electrons by electron excitation across the band–gap, and these can be captured by hole traps and electron traps in the depletion region. In the usual way, as the sample temperature is raised, the carriers can be released from the traps and the capacitance transients produced are employed to provide the DLTS spectrum of the various traps.

An important matter in optical–filling MCTS is that, since the photons have energies larger than the semiconductor band–gap, the absorption coefficient for them can be rather large and consequently it may be that only a part of the depletion region will be effectively illuminated. However by careful choice of the photon wavelength and of the depletion region thickness, it can sometimes be arranged that the light can illuminate not only the depletion region but also some thickness of semiconductor beyond the depletion region; minority carriers created in that further material may then diffuse back to the edge of the depletion region and be moved well into that region by the electric field therein. Also, the duration of the optical pulse must be sufficiently large that enough minority carriers are created during the pulse to fill the minority carrier traps; the duration required depends of course on the intensity of the light being used, but often durations of at least several ms have to be used. It is useful to monitor the capacitance of the junction as a function of the duration of the optical pulse so as to know when saturation of the trap filling occurs. Under suitable conditions, optical–pulsing MCTS can be an effective means of investigating the presence, concentrations and electronic properties of minority–carrier traps in semiconductors, including of irradiation–induced traps (Siyanbola and Palmer 1988).

4.3 Capacitance–Mode Optical DLTS

A method of Deep Level Transient Spectroscopy that is different from but related to those described in the previous sections is that of Optical DLTS (Mitonneau et al 1977). In this method, as in optical–pulse MCTS, the junction is maintained under steady reverse bias and repetitively illuminated by optical pulses, but the photon energy used is now smaller than the semiconductor band–gap. The effect is that band–to–band transitions do not occur, but instead electron traps in the gap can be filled by optical excitation of electrons from the valence band, and hole traps can be filled by excitation of holes from the conduction band. At the end of the optical pulse, thermal removal of the trapped carriers can take place at appropriate temperatures, and a DLTS spectrum of the traps can be obtained in the usual way.

The Optical–DLTS method has an advantage compared to optical–pulse MCTS that, since band–to–band photon absorption does not occur, there is not the potential problem of a short penetration distance of the incident light; indeed it can even be possible to illuminate the depletion region by sending the light through the substrate that is supporting the semiconductor layer being investigated. However, the Optical–DLTS method has the disadvantage that absolute concentrations of the traps are difficult to measure because of lack of knowledge of the optical–filling cross–sections. Furthermore the optical filling cross–sections of some of the traps may be so low that those traps cannot be detected at all. But if the concentrations of the traps are known from use of other DLTS methods, then Optical–DLTS can be employed to obtain information about the magnitudes of the optical–filling cross–sections (Loualiche et al 1984).

5. STUDIES OF SEMICONDUCTOR HETERO–STRUCTURES

Semiconductor hetero–structure layers are of increasing fundamental and device importance. They can be investigated by the same capacitance–voltage and DLTS methods as described above for homo–structure materials, and this is a developing field of research investigation. Space limitations here allow only a very brief outline of this topic.

It is clear from expression (7), for a p^+–n or Schottky junction on a uniformly doped n–type semiconductor, that in a plot of $1/C_s^2$ as a function of applied reverse bias V_a, the intercept voltage, for $1/C_s^2$ equal to zero, is $-V_{bi}$; thus the value of V_{bi} can be experimentally determined. The same formula can be applied in the case of a p^+–n or p–n^+ hetero–structure interface (eg, Ge/GaAs or GaAs/$Al_xGa_{1-x}As$ etc), and again the built–in potential difference across the junction can be found experimentally. From the doping concentrations on the two sides of such a junction the Fermi energies with respect to the band edges can be calculated, and then since at equilibrium the Fermi energies are coincident in energy across the junction such experiments allow determination of the discontinuities ("band offsets") of the conduction band and the valence band edges at the interface between the two hetero–structure materials. The method can even be applied to an isotype n^+–n or p^+–p heterostructure interface if the $C_s(V_a)$ measurements are made at sample temperatures (eg perhaps 80K) that are low enough to reduce sufficiently the excitation of carriers over the potential barriers. Forrest 1987 has comprehensively reviewed the principles and practical considerations of such C–V measurements on semiconductor hetero–structures.

Transient capacitances and DLTS data can also be obtained for the thermal release of electrons and holes from the potential wells (the so–called 'giant traps') due to the semiconductor layers of lower band–gap in hetero–structure materials. The activation energies for carrier release can be measured in the usual way (Section 4.1.5 above), but there can be strong effects on the carrier-emission rates due to the electric field arising from the reverse–bias being applied to the semiconductor junction. For thin quantum wells, the effect of layer thickness on the carrier binding energies can be observed. Lang et al 1986 have reported DLTS results and thermal activation energies for the release of holes from a $Ga_{0.47}In_{0.53}As$ quantum–well layer in an InP p^+–n junction, and Lang 1987 has reviewed some principles and results of such studies. DLTS investigations of lattice–defect and impurity carrier–traps in hetero–structure materials can also be made.

6. SUMMARY

Measurement of the small–signal capacitance of a reverse–biased p^+–n or n^+–p semiconductor junction or metal–semiconductor Schottky junction as a function of the reverse bias allows determination for the n–semiconductor or p–semiconductor of a spatial concentration distribution $N_{meas}(x)$, where x is the distance from the junction interface. In the absence of deep–energy–level carrier traps and if the chemical–doping concentration does not change rapidly with distance, $N_{meas}(x)$ is equal to that doping concentration distribution and therefore equal also to the carrier–concentration distribution n(x). This facility to determine the spatial distribution of the doping and carrier concentration by a single measurement is the very valuable feature of the C(V) method, and contrasts with a measurement of current-transport properties parallel to the surface which gives only an average value of the carrier concentration in the semiconductor layer being assessed. In the presence of deep-level carrier traps, the value of N_{meas} depends also on the concentrations of those traps and on the position of their energy levels in the semiconductor band gap. Measurements of N_{meas} as a function of temperature and detailed consideration of its dependence on x can allow the values of the doping and trap concentrations to be separately determined.

The technique of Deep Level Transient Spectroscopy (DLTS) employs experimental analysis, as a function of semiconductor-junction temperature, of the transient capacitances that result from sudden changes in the reverse-bias applied to the junction. DLTS is a sensitive and versatile method for determination of the concentrations, energy levels and carrier-trapping cross-sections of majority-carrier traps having electronic energy levels that are deeper from the respective energy-band edges than the chemical-doping levels. However, the magnitude of the observed capacitance transient for any individual trap depends on the trap energy level, and so considerable care has to be used in assessing the trap concentrations from the measured transient capacitances or from the DLTS spectra. Under optimum conditions traps present at concentrations of 10^{-4} of the doping concentrtion can be detected and studied. DLTS studies of minority-carriers can be made by use of either electrical-injection-pulse or optical-pulse production of minority carriers; however the concentration sensitivity is poorer than that for majority-carrier DLTS.

These capacitance and transient-capacitance methods can be used also to study and characterise hetero-structure semiconductor materials.

Equipment units for assembling of capacitance-voltage profiling and DLTS systems and complete commercial systems of various kinds for these purposes are available.

ACKNOWLEDGEMENTS

The author wishes to thank Dr P Blood of the Philips Research Laboratories, Redhill, UK, for many valuable discussions concerning these characterisation techniques.

REFERENCES

Ambridge T and Faktor M M, 1975, Inst. Phys. Conf. Ser. 24 320-330
Auret F D and Nel M, 1987, J. Appl.Phys. 61 2546-2549
Bar-Lev A, "Semiconductors and Electronic Devices" (Prentice/Hall)
Baxandall P J, Colliver D J and Fray A F, 1971, J. Phys. E 4 213-221
Bio-Rad Microscience Ltd, Maylands Avenue, Hemel Hempstead, Herts, UK (formerly Polaron Equipment Ltd)
Binet M, 1975, Electronics Lett. 11 580
Bleicher M and Lange E, 1973, Solid State Electronics 16 375-380
Blood P, 1986, Semicond. Sci. Technol. 1 7-27.
Blood P and Orton J W, 1978, Rep. Prog. Phys. 41 157-257
Boonton Electronics Corporation, Randolph, NJ, USA (now represented in the UK by Aspen Electronics Ltd, Ruislip, Middlesex, UK)
Brotherton S D, Bradley P and Bicknell J, 1979, J. Appl. Phys. 50 3396-3403
Brunwin R, Hamilton B, Jordan P and Peaker A R, 1979, Electronics Letters 15 349-350
Carroll J E "Physical Models for Semiconductor Devices" (Arnold)
Datalab: Data Laboratories Ltd, Mitcham, Surrey, UK
Forrest S R, 1987, Chap 8 (pages 311-375) of "Heterojunction Band Discontinuities" ed. by F Capasso and G Margaritondo (North Holland)
Guldberg J, 1977, J. Phys. E 10 1016-1018
Hamilton B, 1974, Inst. of Phys. Conf. Ser. 22 218-225
Hamilton B, Peaker A R and Wight D R, 1979, J. Appl. Phys. 50 6373-6385

Hughes F D, 1972, Acta Electronica 15 43-53
Hewlett-Packard, Palo-Alto, CA, USA
Irvine A C and Palmer D W, 1988, Physics Division, University of Sussex, UK
JAC Electronics Ltd, Camberley, Surrey, UK.
Johnson N M, Bartelink D J, Gold R B and Gibbons J F, 1979,
J. Appl. Phys. 50 4828-4833
Johnson W C and Panousis P T, 1971, IEEE Trans. Electron Devices ED-18 965-973
Keithley Instruments Division, 1987, Notes on use of the 590 and 595 C-V systems
Keithley Instruments Inc, Cleveland, Ohio, USA
Kennedy D P, Murley P C and Kleinfelder W, 1968, IBM J. of Res. Devel. 12 399-409
Kennedy D P and O'Brian R R, 1969, IBM J. of Res. Devel. 13 212-214
Kimerling L C, 1974, J. Appl. Phys. 45 1839-1845
Kukimoto H, Henry C H and Merritt F R, 1973, Phys. Rev. B7 2486-2499
Lang D V, 1974, J. Appl. Phys. 45 3023-3032
Lang D V, 1987, Chap 9 (pages 377-396) of "Heterojunction Band Discontinuities"
ed. by F Capasso and G Margaritondo (North Holland)
Lang D V and Kimerling L C, 1975, Inst. Phys. Conf. Ser. 23 581-588
Lang D V, Sergent A M, Panish M B and Temkin H, 1986, Appl. Phys. Lett. 49 812
Lefevre H and Schulz M, 1977, Appl. Phys. 12 45-53
Loualiche S, Nouailhat A, Guillot G and Lannoo M, 1984, Phys. Rev. B30 5822
Maracas G N, Laidig W D and WIttman H R, 1984, J. Vac. Sci. Technol. B2 599-603
Materials Development Corporation, Reseda, California, USA
Mayes I C, 1985, Bio-Rad Semiconductor Notes, Note 201
McKelvey J P "Solid State and Semiconductor Physics" (Harper and Row)
Miller G L, Lang D V and Kimerling L C, 1977, Ann. Rev. Mater. Sci. for 1977, 377-448
Mitonneau A, Martin G M and Mircea A, 1977, Inst. Phys. Conf. Ser. 33a 73-83
Nandhra P S, 1986, D.Phil Thesis, University of Sussex, UK
Oxford Instruments Ltd, Oxford, UK.
Pearce N O and Mayes I C, 1987, Bio-Rad Semiconductor Notes, Note 402.
Rezazadeh A A and Palmer D W, 1981, Inst. Phys. Conf. Ser. 59 317-322
Rezazadeh A A, 1983, D.Phil Thesis, University of Sussex
Rhoderick E H, "Metal-Semiconductor Contacts" (Clarendon Press)
Rideout V L, 1978, Thin Solid Films 48 261-291
Severin P J and Poodt G J, 1972, J. Electrochem. Soc. 119 1384
Shandon Southern Instruments Ltd, Camberley, Surrey, UK
Siyanbola W O and Palmer D W, 1988, Physics Division, University of Sussex, UK
Syncryst Ltd, East Molesey, Surrey, UK
Sze S M "Physics of Semiconductor Devices" (Wiley)
Wiley J D and Miller G L, 1975, IEEE Trans. Electron Devices ED-22 265-272
Wu C P, Douglas E C and Mueller C W, 1975,
IEEE Trans. Electron Devices ED-22 319-329

High Resolution Electron Microscopy of Semiconductors

J L Hutchison

Department of Metallurgy and Science of Materials, University of Oxford, Parks Road, Oxford OX1 3PH, U.K.

ABSTRACT: This paper presents a basic outline of imaging processes involved in high resolution electron microscopy. The concept of a contrast transfer function is introduced and the effects of lens aberrations and instabilities on image quality illustrated. The application of high resolution electron microscopy is reviewed firstly with reference to perfect crystals, and then with examples drawn from studies of dislocation cores, precipitates, and various types of epitaxial interfaces in semiconductors.

1. INTRODUCTION

Transmission electron microscopy is now widely used in materials research and, in their high resolution configuration, modern instruments are capable of resolution in the 0.15 - 0.2 nm range. The aim of this contribution is to provide an introduction to the basic concepts which lie behind high resolution imaging, and then to give a broad overview of some of its applications in the study of semiconductors. We hope thus to show that the technique is capable of producing a wealth of information on the microstructures of these important materials, at close to the atomic level.

2. HOW A HIGH RESULTION IMAGE IS FORMED IN THE ELECTRON MICROSCOPE.

Seeing image detail in a TEM depends upon there being amplitude differences in the electron waves which reach the fluorescent screen (or photographic film). There are two distinct methods of creating this contrast:

a) Insertion of a small aperture in the back focal plane of the objective lens (this plane is where the diffracted beams exiting from a crystalline specimen are focussed into a diffraction pattern). As a first approximation, only the primary beam is allowed to contribute to the image. Regions of the specimen which strongly scatter electrons outside this aperture will thus appear dark in the image, whereas weakly scattering parts will appear with light contrast. In the case of crystalline specimens this, together with various dark-field imaging techniques, forms the basis of diffraction contrast, the extensive theory for which may be found in a number of standard text-books, see, for example, Hirsch et al. (1977). Because of the small size of the objective aperture, the resolution is limited to some 1 - 1.5 nm at best and in order to achieve higher resolution it is necessary to increase the aperture size, thereby allowing more than one beam to contribute to the image.

b) For image detail smaller than about 1 nm phase contrast becomes the dominant imaging mechanism and it may be understood qualitatively in the following way: if we regard the electron beam incident upon the specimen as a plane wave, electrons passing regions of positive potential, i.e. atomic nuclei, are accelerated, the wavelength λ is reduced and the phase is advanced by an amount proportional to the potential. The exit wave has a distribution of relative phase values determined by the variations in potential field traversed by different parts of the wave. For a thin crystal, amplitude changes due to inelastic scattering will be small and we may regard the the specimen as a phase object; if phase changes are also small (i.e. the electrons undergo only single scattering through the crystal) the specimen corresponds to a weak phase object - a useful approximation which is rarely achieved in practice.

In order to actually "see" contrast from a weak phase object it is necessary to introduce some additional phase shift so as to bring the phase relationships of selected diffracted beams exactly into or out of phase with respect to the primary beam. The recognition that the spherical aberration of the objective lens, and also its defocus, both impose phase shifts on diffracted beams and may be used to optimise phase contrast provided the key to this, and the classic analysis by Scherzer (1949) produced the following formula:

$$\chi \text{ (phase shift)} = \frac{\pi}{2} C_s \lambda^3 u^4 + \pi \Delta f u^2 . \lambda$$

where C_s is the coefficient of spherical aberration, u the spatial frequency and Δf, the objective lens defocus. Note that this function contains some fixed terms: λ and Cs are determined by the microscope itself; the skill of the operator lies in controlling the defocus accurately. A plot of sin χ against u (Fig. 1) illustrates graphically the way in which, for a given defocus value, the contrast, which is shown by sin χ, varies.

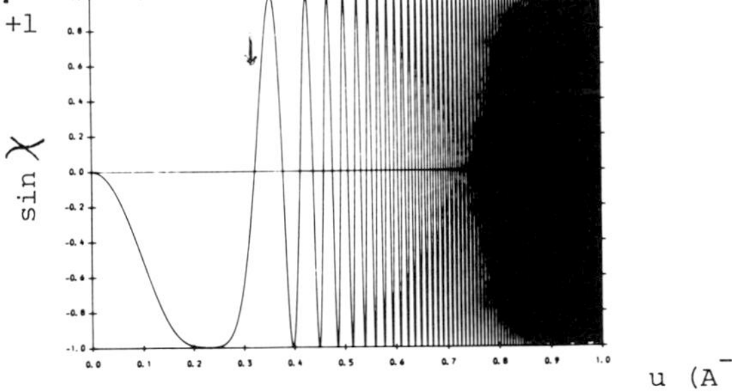

Figure 1. Contrast transfer function plot for a 100 kilovolt electron microscope, with Cs = 0.7 mm, and a defocus of -50nm.

Where sin χ is -1, the phase contrast is maximum and negative, i.e. for a weak phase object scattering centres appear dark. For sin χ =+1, the contrast is maximum and positive - scattering centres appear light. From Fig.1 it is clear that contrast varies both in intensity and in sign, as spatial frequency varies. The contrast also varies with defocus. In order to produce the best possible representation of an object, the function sin χ ought to be close to unity over as large a range of reciprocal space as possible. The particular defocus which optimises this, producing an extended, negative plateau in the contrast transfer function, the so-called "Scherzer defocus", is given by the expression

$$\Delta f = \sqrt{(C_s \lambda)}$$

corresponding to a transfer function similar to that shown in Fig. 1. Note the broad, negative plateau in the plot. The first cross-over point (arrowed) corresponds to zero contrast, and defines the resolution down to which there is no contrast reversal. This is the "Scherzer" or structure resolution and is defined by:

$$d_o = 0{\cdot}7 C_s^{\frac{1}{4}} \lambda^{\frac{3}{4}}$$

Note that this defocus and the corresponding resolution are determined only by C_s and λ, i.e. they are essentially fixed for a particular microscope.

The only ways of improving structure (or point) resolution are to reduce C_s by designing more sophisticated objective lens polepieces, or to reduce λ by increasing the accelerating voltage. Both trends are seen in the newest instruments, with C_s being reduced to as little as 0.4 mm (at 200 kilovolts), and 400 kV instruments now in routine operation. Typical resolutions as defined above are now in the 0.15 - 0.2 nm range, with the prospect of 0.1 nm still some way ahead.

3. PRACTICAL LIMITATIONS

The above description makes a number of assumptions, some of which are not strictly valid. Firstly, it was assumed that the specimen would be sufficiently thin as to represent a weak phase object. This unfortunately applies to foils only a few nm in thickness, and is experimentally difficult to achieve. For most specimens, dynamical diffraction (multiple scattering) effects must be taken into account, with the result that an intuitive, one-to-one interpretation of image contrast in terms of projected structure is rarely feasible: computer simulation of high resolution images as back-up is now regarded as an essential part of image interpretation.

Secondly, the contrast transfer function derivation assumed that the illumination was fully coherent, and also failed to take account of chromatic aberration terms, electrical and mechanical instabilities. The combined effect of these is to superimpose "damping envelopes" on the CTF Fig. 2 shows the effect of adding finite divergence to the illumination, and of adding a chromatic spread of focus which includes energy spread, chromatic aberration, electrical instabilities, etc. The resolution at which this attenuation effect reduces sin χ to a detectable (nominally 14 %) limit maybe taken as another indicator of resolution: the so-called "information limit". All information down to this limit may in principle be retrieved, although it may require extensive image processing. As a guide, information limits around 0.13 - 0.12 nm have been demonstrated on a few state-of-the art HREMs. In this context it is important to note that to obtain the optimum peformance from a modern high resolution instrument, particular attention must be paid to operating conditions in an effort to minimise the damping effects on the CTF; beam current, ilumination conditions, and also mechanical stablitiies all contribute, and in many cases the vibration levels of the microscope room itself (and also acoustic noise levels) may all conspire against high resolution. It is perhaps significant that many of the highest resolution images are recorded late at night...!

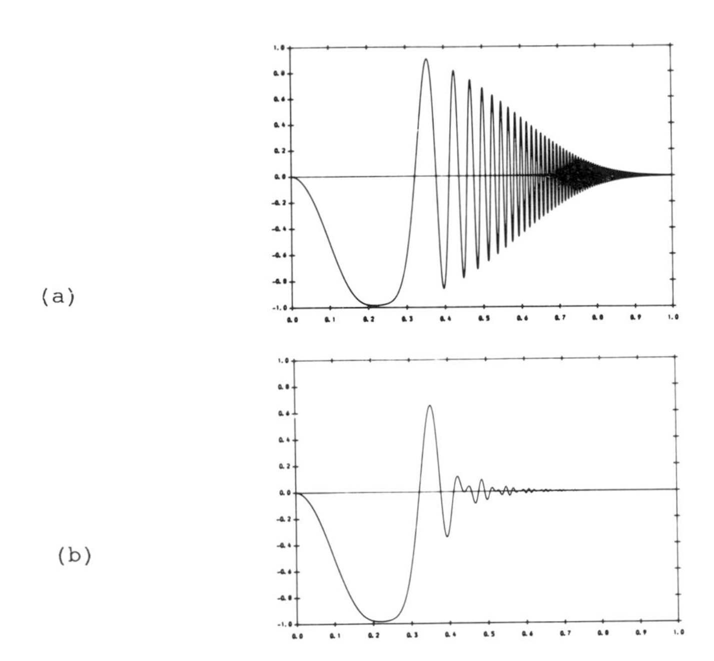

Fig. 2 Contrast transfer function for 100 kV, with C_s = 0.7 mm, including (a) chromatic spread of focus of 3 nm, and (b) also an incident beam divergence of 1 mrad.

Note: It is perhaps appropriate here to include a cautionary note about claims which may appear in electron microscope manufacturers' brochures about "guaranteed resolution"; "point" resolution and "information limit" are clearly two completely different criteria, and to add further difficulty, "lattice" or "line resolution", often quoted by manufacturers, may arise from cross-interference of two diffracted beams across the optic axis to generate "half-spacings" which may be very finely spaced, down to less than 0.1 nm ! Although providing a good indication of an instrument's electrical and mechanical stability, it must be pointed out that these fringe patterns do not give useful information on the quality of the electron optical imaging system.

4. APPLICATIONS OF HREM TO SEMICONDUCTORS

Because of the need to have structural features in the specimen aligned along the incident beam direction it is necessary to set the specimen into a major zone-axis orientation, <110> and <100> being the most widely used for semiconductors. There is an obvious need to have the specimen as uniformly thin as possible, and free from damaged or altered material created during the thinning process.Considerable effort is now being put into the development of suitable thinning techniques, both for bulk materials and for actual device samples; see for example papers by Cullis et al (1985) and Chew and Cullis (1987) for a description of new approaches to ion-thinning of compound semiconductors, and Anderson et al (1989) for an up-to-date review of specimen preparation from device materials.

4.1 Perfect crystals

Before describing studies of real, defective materials, it is useful to consider how far "structure imaging" of semiconductors has now reached, and to appreciate some of the difficulties encountered.

In the <110> projection of the diamond f.c.c lattice of silicon and germanium, and in the sphalerite lattice of most compound semiconductors, the {111} and {200} lattice planes are the most widely spaced, and thus the most accessible to the electron microscopist. (It is also useful that many of the defects systems in these materials lie on the {111} planes and may thus be imaged edge-on).

However the challenge for the microscopist seeking "atomic resolution" in these materials is in the fact that in the <110> projection the single atomic columns appear in projection with a separation of $a_o/4$ which for Si is 0.136 nm, well below the structure resolution of the best available conventional microscopes. The artefacts which arise in so-called atomic images of Si (Izui etal. 1977) have been explored by computer simulation Hutchison et al (1981), Hutchison and Waddington (1988) and it has been shown that particular, non-standard combinations of defocus and specimen thickness are required for images which reveal contrast features corresponding to correct atomic positions (Hutchison et al,1986) The challenge of identifying the polarity in III/V and II/VI compound semiconductors by HREM and computer simulation is now being pursued in a number of centres, with promising results, see Wright et al (1988) and Glaisher et al (1989).

Surfaces of CdTe crystals have been imaged in profile by Smith et al (1989), who found evidence for reconstructions on the {111} and {100} surfaces; it was possible to deduce {111} surface polarity from the image contrast, a feature employed subsequently by Hutchison et al (1989) to assign polarity to lattice defects in the same material.

For <100> oriented specimens the dominant planes are {200} and {220}, the latter being around 0.2 nm. Here the possibility arises for distinguishing cation and anion sub-lattices in compound semiconductors, since they occupy separate columns in this projection (Ourmazd et al, 1988). For this to be unambiguous specimen thickness and defocus must both be known accurately.

4.2. Dislocation core structures

A number of dislocation core structures have been elucidated by high resolution imaging, attention being first focussed on dislocations in deformed silicon and germanium, examples being 30° partial dislocation cores in Si (Anstis et al, 1981, Bourret et al, 1981) where it was shown by careful matching with computer simulatons that the "glide" model rather than the "vacancy shuffle" model was the more likely configuration. More recent analyses have included undissociated 60° dislocations in Si, (Hutchison et al, 1983), while both dissociated and undissociated dislocations in GaAs have been studied by Tanaka and Jouffrey (1984) and Gerthsen et al (1988). Recently, Lu and Smith (1989) have analysed in detail various dissociated dislocations which occur as a result of Ar^+ ion- thinning damage in CdTe.
Undissociated 60° dislocations are sometimes found as growth defects in chemically thinned CdTe, such as the example shown in Fig. 3 (Hutchison et al, 1989)

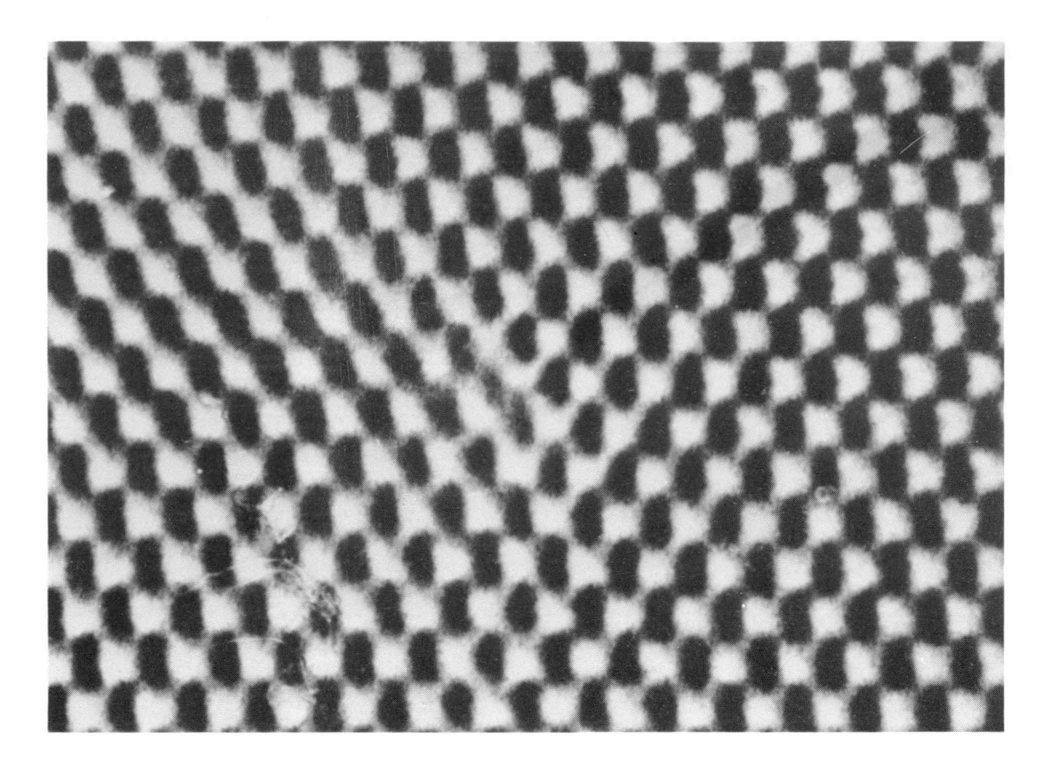

Fig. 3. Undissociated 60° growth dislocation in chemically thinned CdTe crystal; <110> - oriented foil.

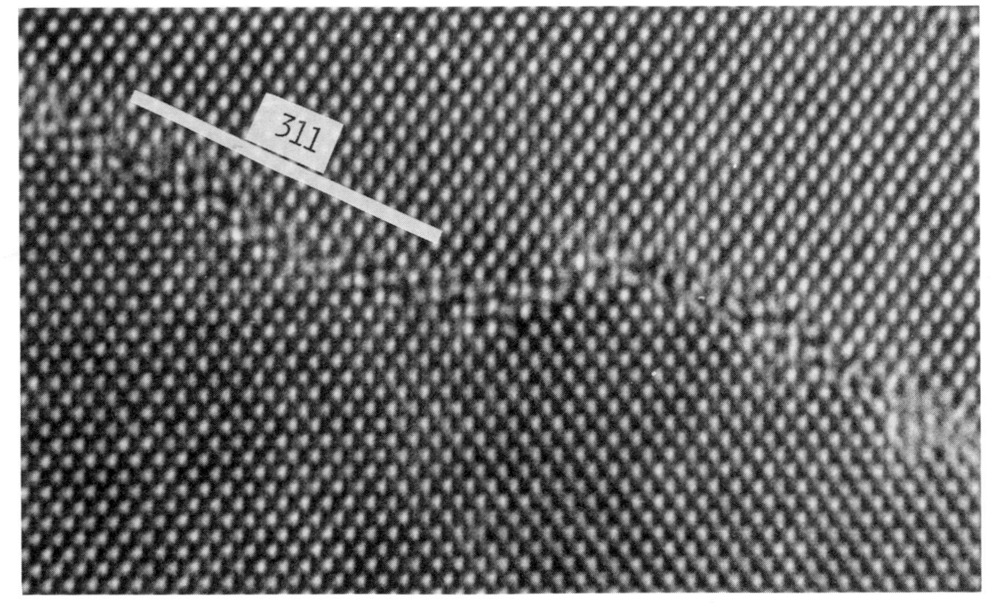

Fig. 4. Extended {311} - defect in damaged layer in O^+-implanted silicon.

4.3. Precipitation in Cz-silicon

Thermal annealing of Cz-silicon at a range of temperatures and for various times leads to the formation of a number of distinct defect types which have been characterised by HREM. They have been described by various groups, for example Bourret et al (1984) andBergholz et al (1985).
A recent paper by Bergholz et al (1989) is noteworthy in combining results of an HREM study with those from small-angle neutron scattering and also infra-red spectroscopy; a remarkable consistency was found in the results obtained from very diverse techniques.

Perhaps the most controversial defects in this area are the so-called "ribbon-like" or "rod-like" defects, occurring usually on {311} planes, and generally regarded as precipitates of coesite, a high pressure form of SiO_2 (Bourret et al, 1984, Bergholz et al, 1985). They also occur in a damaged layer in ion-implanted implanted silicon , see for example Lambert and Dobson (1981). An example of particularly clear defects is shown in Fig. 4. More recently, Bourret (1987) has produced new evidence that these defects may instead be a hexagonal form of silicon, although the matter is not yet fully resolved, Bergholz and Hutchison (1988); Carpenter and Bergholz (unpublished work) claim to have found EELS signals for oxygen in {311} defects in annealed Cz-Si.

4.4 Interfaces in semiconductor systems

This is an important area, and also a challenging one for the high resolution microscopist. Systems of technological interest include metallic layers on semiconductors, semiconductors on insulators, quantum well structures and oxide layers, etc. There are particular difficulties in preparing optimum cross-sections for HREM, and the paper by Anderson et al.(1989) presents a good review of recent practical developments. There are also difficulties in obtaining useful images, and interpreting the contrast; some of these are described by Hutchison (1987)

In the simplest case of heteroepitaxy, that involving lattice-matched components, the problem for the microscopist is to create contrast differences between the two layers. In the case of GaAs/Al,GaAs multi-layers, differences in scattering factors for Ga and Al-containing structures (Petroff et al, 1978) can be effectively exploited in <100> oriented cross-sections, where the 200-type fringes are composition-sensitive (de Jong et al. 1986). In <110> cross-sections, conditions are less straightforward and extensive simulations are often required to ascertain the particular combinations of foil thickness and defocus which must be used to give distinct contrast in the two components. This was successfully done for the CdTe/(100)InSb system where the two materials had the same average atomic number, as well as being almost exactly lattice-matched (Hutchison et al. (1986).
In the case of strained layer superlattice systems, HREM provides unique information on the type and distribution of misfit dislocations, an example of which is shown in Fig. 5. This GaAs/InP interface contains both undissociated 60° and Lomer dislocations, the latter being remarkably sharp. The 60° cores appear slightly less well defined, possibly as a result of impurity decoration. A much larger mismatch (14%) in the CdTe/(100)GaAs interface is accommodated by a series of well defined 60° and Lomer dislocations, appropriately spaced about every 7 lattice planes apart as described by Cullis et al (1985). Clearly, information on such closely spaced dislocations would be inaccessible to other techniques.
Silicide/silicon epitaxy has been the subject of numerous investigations

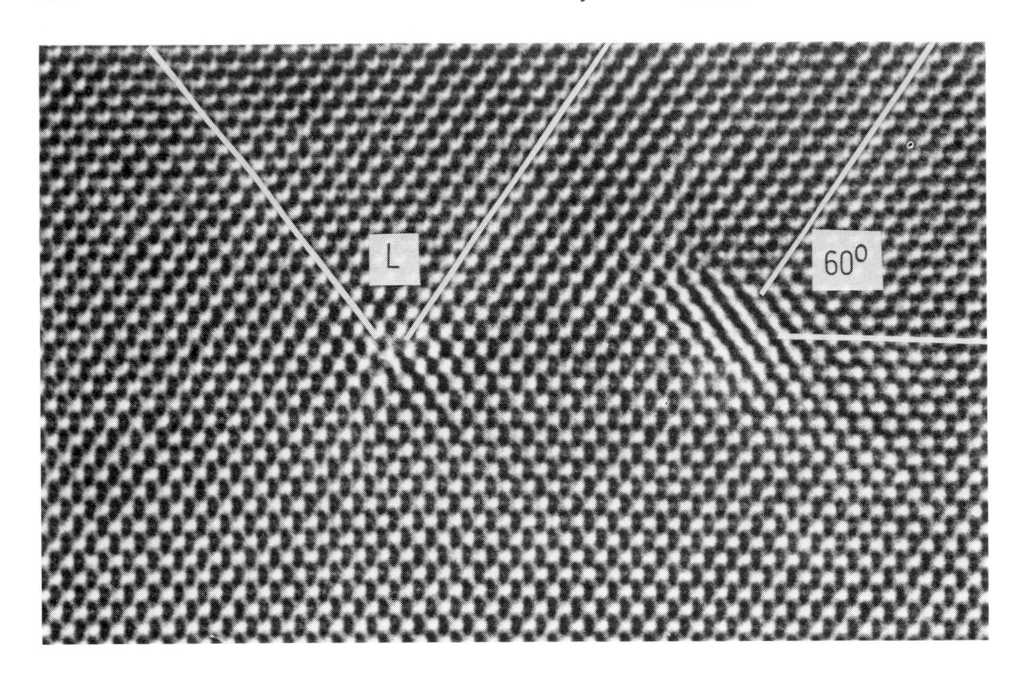

Fig. 5. Lomer (L) and 60° dislocations lying along a misfit boundary between $Ga_{0.47}In_{0.53}As$/InP layers.

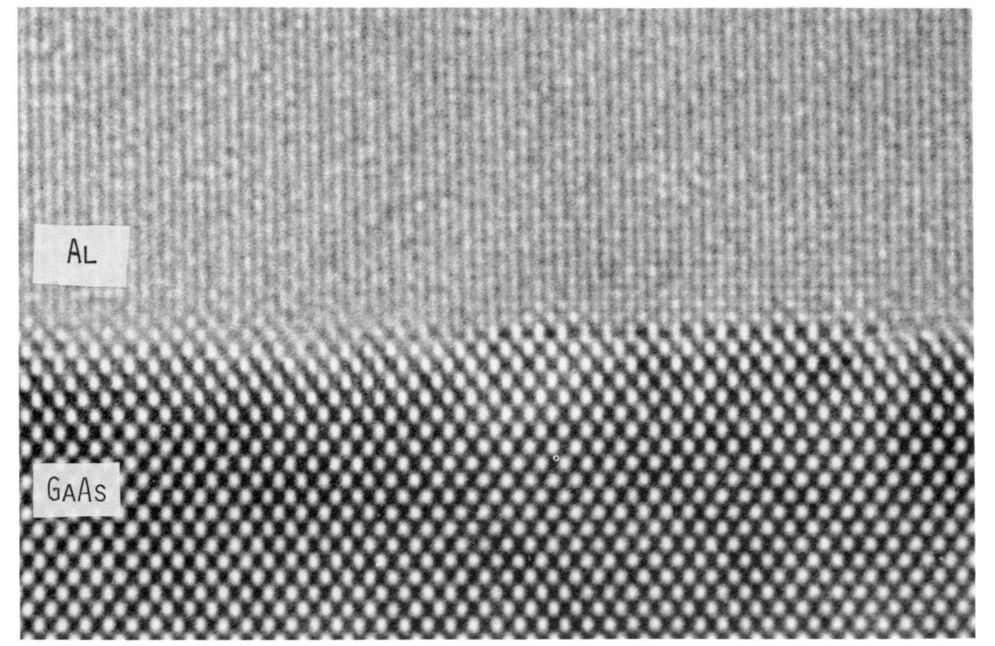

Fig. 6. Epitaxial interface of Al/(001)GaAs, with the relationship Al[001]//GaAs[001], with Al[100]//GaAs[110].

which have yielded atomic models for interfaces, such as $NiSi_2$/(111)Si and $NiSi_2$/(100)Si, reported by Cherns et al(1982, 1984). More recently, the PtSi/(111)Si interface has been analysed by Kawarada et al (1986)

An intriguing example of a metal/semiconductor interface is found in the (111)Al/(111)Si system; here this misfit is 30% and HREM has been employed by LeGoues et al. (1986) to produce plausibe atomic models for the two different interfacial structures found, although the authors used tilted illumination for this study.

Recently Kendrick et al (1989) have undertaken a study of the Al/(100)GaAs interface, using a specially modified program to simulate realistic models for the boundary. Preliminary results indicate that two related interfacial structures may be present in different areas. Fig. 6 reveals the remarkable sharpness of this interface, in which a lattice mismatch of some 29 $\%_{GaAs}$ is accommodated by a 45° rotation of the Al layer to bring 200_{Al} planes into register with 220 planes of the GaAs substrate.

As epitaxial layers become ever smaller and devices more complex, so the necessity to excercise full control over the growth processes becomes more urgent. With this there comes the need for an atomic-scale probe with which the perfection of these layers can be assessed, hopefully with some feed-back into the growth process. HREM is emerging as an indispensible tool in this area, particularly when used in conjunction with sensitive microanalytical techniques (Long, 1989)

5. WHAT FOR THE FUTURE ?

Microscopes have now become available which have resolutions in the 0.15-0.2 nm range. They have earned a place in microelectronics materials research and are producing information which is both unique and relevant. Further improvements in resolution, by whatever criterion it is measured, will be relatively small, but we can expect to see more combined use of HREM and other analytical techniques, particularly using small probes, essential for the fine multilayer structures now being produced. This will take place along two main routes: the continued development of illumination systems in conventional TEMs to produce small, high current density probes whilst retaining high spatial resolution capability (this may involve a complex arrangement of lenses that requires computer control to be properly useable) and the increasing use of field-emission-gun scanning transmission instruments (STEMs) which combine very sensitive analytical capabilities with good spatial resolution (Long, 1989)

6. CONCLUSION.

This review has of necessity been selective rather than comprehensive; it would be impossible to give adequate coverage to a rapidly growing field. However, we have tried to show some of the difficulties in producing high resolution images, and in interpreting their contrast. From the examples discussed it will be clear that HREM is playing an increasingly important part in inceasing our understanding of the microstructures which are important in today's - and tomorrow's - microelectronic materials.

ACKNOWLEDGEMENTS

I would like to thank my colleagues at Oxford for their many contributions to this work, and Kim Christensen and Robert Mallard in particular for supplying Figs. 4 and 5.

REFERENCES

Anderson R M, Klepeist S, Benedict J, Vandygrift WG and Orndoff M 1989 Inst.Phys.Conf.Ser. No. **100** 491
Anstis G R, Hirsch P B, Humphreys C J, Hutchison J L and Ourmazd A 1981 Inst.Phys Conf.Ser. No. **60** 15
Bergholz w, Binns M J, Booker G R, Hutchison J L, Kinder S H, Messoloras S, Newman R C, Stewart R J and Wilkes J G 1989 Phil. Mag **B59** 499
Bergholz W and Hutchison J L 1988 Proc.46th Ann.Mtg of EMSA ed. G W Bailey (San Fransisco Press Inc.) p.478
Bergholz W, Hutchison J L and Pirouz P 1985 J. Appl. Phys. **58** 3419
Bergholz w, Hutchison J L and Pirouz P 1986 J. Microscopy **141** 143
Bourret A 1987 Inst.Phys.Conf.Ser. No. **87** 39
Bourret A, Desseaux J and D'Anterroches C 1981 Inst.Phys.Conf.Ser.No.**60** 9
Bourret A, Thibault-Desseaux J and Seidmann B N 1984 J. Appl. Phys. **55** 825
Cherns D, Anstis G R, Hutchison J L and Spence J D H 1982 Phil. Mag. **A46** 849
Cherns D, Hetherington C J D and Humphreys C J 1984 Phil. Mag. **A49** 165
Chew N G and Cullis A G 1987 Ultramicroscopy **23** 175
Cullis A G, Chew N G and Hutchison J L 1985 Ultramicroscopy **17** 203
Cullis A G, Chew, N G, Hutchison J L, Irvine S J C and Geiss J 1985 Inst. Phys. Conf.Ser. No.**76** 29
Gerthsen D, Ponce F A and Anderson G B 1989 Phil. Mag. **A59** 1045
Glaisher R W, Spargo A E C and Smith D J 1989 Ultramicroscopy **27** 131
Hirsch P B, Howie A, Nicholson R B, Pashley D W and Whelan M J 1977 "Electron Microscopy of Thin Crystals" 2nd edition (New York, Krieger)
Hutchison J L 1987 Inst.Phys.Conf.Ser. No. **87** 1
Hutchison J L, Anstis G R, Humphreys C J and Ourmazd A 1981 Inst. Phys.Conf.Ser. No. **61** 351
Hutchison J L, Anstis G R and Pirouz P 1983 Inst.Phys.Conf.Ser. No. **67** 21
Hutchison J L, Honda T. and Boyes E D 1986 JEOL News **24E** 9
Hutchison J L, Lyster M and Booker G R 1989 Inst Phys.Conf.Ser. No. **100** 29
Hutchison J L, and Waddington W G 1988 Ultramicroscopy **25** 93
Hutchison J L, Waddington W G, Cullis A G and Chew N G 1986 J. Microscopy **142** 153
Izui K, Furuno S, Nishida T and Otsu H 1978-79 Chem. Scripta **14** 99
de Jong A F, Coene W and Bender H 1987 Inst.Phys.Conf.Ser. No. **87** 9
Kawarada H, Ishida M, Nakamishi J and Ohdomari I 1986 Phil. Mag **A54** 729
Kendrick A J, Hutchison J L and Cherns D 1989 Inst.Phys.Conf.Ser. No. **98** Vol. 1 p383
LeGoues K F, Krakow W and Ho P S 1986 Phil. Mag. **A53** 833
Long N J 1989 Inst.Phys.Conf.Ser. No. **100** 59
Lu Ping and Smith D J 1989 Phil. Mag. A (in press)
Ourmazd A, Rentschler J R and Taylor D W 1986 Phys. Rev. Letters **24** 3073
Petroff P M, Gossard A C, Weimann W and Savage A 1978 J. Crystal Growth **44** 5
Scherzer O 1949 J. Appl. Phys. **20** 20
Smith D J, Glaisher R W and LuPing 1989 Phil. Mag. Letters **59** 69
Tanaka M and Jouffrey B 1984 Phil. Mag. **A50** 733
Wright A C, Ng T L and Williams J O 1988 Phil. Mag. Letters **57** 107

Subject Index